Mechatro

Electronic control systems in mechanical engineering

W. Bolton

Addison Wesley Longman Limited
Edinburgh Gate
Harlow
Essex
CM20 2JE, England
and Associated Companies throughout the world.

First published 1995
Second impression 1996

British Library Cataloguing in Publication Data
A catalogue entry for this title is available from the British Library

ISBN 0–582–25634–8

Produced through Longman Malaysia, PP

Contents

Preface

The integration of electronic engineering, electrical engineering, computer technology and control engineering with mechanical engineering is increasingly forming a crucial part in the design, manufacture and maintenance of a wide range of engineering products and processes. A consequence of this is thus the need for engineers and technicians to adopt an interdisciplinary and integrated approach to engineering. The term *mechatronics* is used to describe this integrated subject. A consequence of this approach is that engineers and technicans need skills and knowledge that are not confined to a single subject area. They need to be capable of operating and communicating across a range of engineering disciplines and linking with those having more specialised skills. This book is an attempt to provide a basic background to mechatronics and provide links through to more specialised skills.

The book covers the Business and Technology Education Council (BTEC) Mechatronics A and B units (1413G and 1414G), these units being core units of their HNC/HND courses and designed to fit alongside more specialist units such as those for design, manufacture and maintenance determined by the application area of the course.

W. Bolton

1 Mechatronics

1.1 What is mechatronics?

Consider the modern auto-focus, auto-exposure camera. To use the camera all you need to do is point it at the subject and press the button to take the picture. The camera automatically adjusts the focus so that the subject is in focus and automatically adjusts the aperture and shutter speed so that the correct exposure is given. Consider a truck smart suspension. Such a suspension adjusts to uneven loading to maintain a level platform, adjusts to cornering, moving across rough ground, etc. to maintain a smooth ride. Consider an automated production line. Such a line may involve a number of production processes which are all automatically carried out in the correct sequence and in the correct way. The automatic camera, the truck suspension, and the automatic production line are examples of a marriage between electronic control systems and mechanical engineering. The term *mechatronics* is used for this integration of electronics, control engineering and mechanical engineering.

In the design now of cars, robots, machine tools, washing machines, cameras, and very many other machines, such an integrated and interdisciplinary approach to engineering design is increasingly being adopted. The integration across the traditional boundaries of mechanical engineering, electronics and control engineering has to occur at the earliest stages of the design process if cheaper, more reliable, more flexible systems are to be developed.

Mechatronics brings together areas of technology involving sensors and measurement systems, drive and actuation systems, analysis of the behaviour of systems, control systems, and microprocessor systems. That essentially is a summary of this book. This chapter is an introduction to the topic, developing some of the basic concepts in order to give a framework for the rest of the book in which the details will be developed.

1.2 Systems

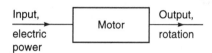

Fig. 1.1 An example of a system

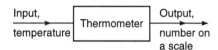

Fig. 1.2 An example of a measurement system

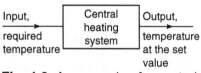

Fig. 1.3 An example of a control system

1.3 Measurement systems

Mechatronics involves, what are termed, systems. A *system* can be thought of as a black box which has an input and an output. It is a black box because we are not concerned with what goes on inside the box but only the relationship between the output and the input. Thus, for example, a motor may be thought of as a system which has as its input electric power and as output the rotation of a shaft. Figure 1.1 shows a representation of such a system.

A *measurement system* can be thought of as a black box which is used for making measurements. It has as its input the quantity being measured and its output the value of that quantity. For example, a temperature measurement system, i.e. a thermometer, has an input of temperature and an output of a number on a scale. Figure 1.2 shows a representation of such a system.

A *control system* can be thought of as a black box which is used to control its output to some particular value or particular sequence of values. For example, a central heating control system has as its input the temperature required in the house and as its output the house at that temperature, i.e. you dial up the required temperature on the thermostat or controller and the heating furnace adjusts itself to produce that temperature. Figure 1.3 shows a representation of such a system.

Measurement systems can, in general, be considered to be made up of three elements (as illustrated in figure 1.4):

1 A *sensor* which responds to the quantity being measured by giving as its output a signal which is related to the quantity. For example, a thermocouple is a temperature sensor. The input to the sensor is a temperature and the output is an e.m.f. which is related to the temperature value. A Bourdon pressure gauge has a curled tube which straightens out to some extent when the pressure inside it is increased.

2 A *signal conditioner* which takes the signal from the sensor and converts it into a condition which is suitable for either display, or, in the case of a control system, for use to exercise control. Thus, for example, the output from a thermocouple is a rather small e.m.f. and might be fed through an amplifier to obtain a bigger signal. The amplifier is the signal conditioner. A Bourdon pressure gauge has the curled tube output magnified by gearing to give a larger output.

3 A *display system* where the output from the signal conditioner is displayed. This might, for example, be a pointer moving across a scale or a digital readout.

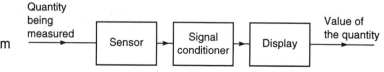

Fig. 1.4 A measurement system and its constituent elements

Consider, for example, an audio cassette player. This has an input from magnetic tape and an output of sound from a loudspeaker. The inputs to the system are patterns of magnetism on a strip of magnetic tape. The sensor converts these into electrical signals. These signals are small and so the signal conditioner transforms them into bigger signals which are capable of operating loudspeakers.

Sensors are discussed in chapter 2 and signal conditioners in chapter 3. Measurement systems involving all elements are discussed in chapter 4. For further details of measurement systems, readers are referred to texts more specifically concerned with measurement, e.g. *Jones' Instrument Technology*, edited by B.E. Noltingk, volume 4, *Instrumentation Systems* (Butterworth 1987) or *Newnes Instrumentation and Measurement* by W. Bolton (Newnes 1991).

1.4 Control systems

Your body temperature, unless you are ill, remains almost constant regardless of whether you are in a cold or hot environment. To maintain this constancy your body has a temperature control system. If your temperature begins to increase above the normal you sweat, if it decreases you shiver. Both these are mechanism which are used to restore the body temperature back to its normal value. The control system is maintaining constancy of temperature. One way to control the temperature of a centrally heated house is for a human to stand near the furnace on/off switch with a thermometer and switch the furnace on or off according to the thermometer reading. That is a crude form of feedback control using a human as a control element. The term *feedback* is used because signals are fed back from the output in order to modify the input. The more usual feedback control system has a thermostat which automatically switches the furnace on or off. This control system is maintaining constancy of temperature.

If you go to pick up a pencil from a bench there is a need for you to use a control system to ensure that your hand actually ends up at the pencil. This is done by you observing the position of your hand relative to the pencil and making adjustments in its position as it moves towards the pencil. This control system is controlling the positioning and movement of your hand.

Feedback control systems are widespread, not only in nature and the home but also in industry. There are many industrial process and machines where control, whether by humans or automatically, is required. For example, there is process control where such things as temperature, liquid level, fluid flow, pressure, etc. are maintained constant. Thus in a chemical process there may be a need to maintain the level of a liquid in a tank to a particular level or to a particular temperature. There are also control systems which involve consistently and accurately positioning a moving part or maintaining a constant speed. This might be, for example, a motor designed to run at a constant speed or perhaps a machining operation in which the position, speed and operation of a tool is automatically controlled.

1.4.1 Open- and closed-loop systems

There are two basic forms of control system, one being called *open loop* and the other *closed loop*. The difference between these can be illustrated by a simple example. Consider an electric fire which has a selection switch which allows a 1 kW or a 2 kW heating element to be selected. If a person used the fire to heat a room, he or she might just switch on the 1 kW element if the room is not required to be at too high a temperature. The room will heat up and reach a temperature which is only determined by the fact the 1 kW element was switched on and not the 2 kW element. If there are changes in the conditions, perhaps someone opening a window, there is no way the heat output is adjusted to compensate. This is an example of open-loop control in that there is no information fed back to the element to adjust it and maintain a constant temperature. The heating system with the electric fire could be made a closed loop system if the person has a thermometer and switches the 1 kW and 2 kW elements on or off, according to the difference between the actual temperature and the required temperature, to maintain the temperature of the room constant. In this situation there is feedback, the input to the system being adjusted according to whether its output is the required temperature. This means that the input to the switch depends on the deviation of the actual temperature from the required temperature, the difference between them determined by a comparison element – the person in this case. Figure 1.5 illustrates these two types of systems.

To illustrate further the differences between open- and closed-loop systems, consider a motor. With an open-loop system

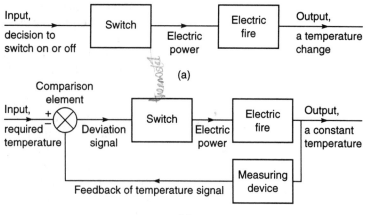

Fig. 1.5 Heating a room: (a) an open-loop system, (b) a closed-loop system

the speed of rotation of the shaft might be determined solely by the initial setting of a knob which affects the voltage applied to the motor. Any changes in the supply voltage, the characteristics of the motor as a result of temperature changes, or the shaft load will change the shaft speed but not be compensated for. There is no feedback loop. With a closed-loop system, however, the initial setting of the control knob will be for a particular shaft speed and this will be maintained by feedback, regardless of any changes in supply voltage, motor characteristics or load. In an open-loop control system the output from the system has no effect on the input signal. In a closed-loop control system the output does have an effect on the input signal, modifying it to maintain an output signal at the required value.

Open-loop systems have the advantage of being relatively simple and consequently low cost with generally good reliability. However, they are often inaccurate since there is no correction for error. Closed-loop systems have the advantage of being relatively accurate in matching the actual to the required values. They are, however, more complex and so more costly with a greater chance of breakdown as a consequence of the greater number of components.

1.4.2 Basic elements of a closed-loop system

Figure 1.6 shows the general form of a basic closed-loop system. It consists of the following elements:

1   *Comparison element*   The comparison element compares the required or reference value of the variable condition being controlled with the measured value of what is being achieved and produces an error signal. It can be regarded as adding the reference signal, which is positive, to the measured value signal, which is negative in this case:

Error signal = reference value signal − measured value signal

The symbol used, in general, for an element at which signals are summed is a segmented circle, inputs going into segments. The inputs are all added, hence the feedback input is marked as

Fig. 1.6 The elements of a closed-loop control system

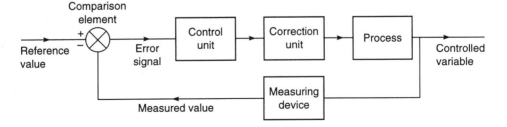

negative and the reference signal positive so that the sum gives the difference between the signals. A *feedback loop* is a means whereby a signal related to the actual condition being achieved is fed back to modify the input signal to a process. The feedback is said to be *negative feedback* when the signal which is fed back subtracts from the input value. It is negative feedback that is required to control a system. *Positive feedback* occurs when the signal fed back adds to the input signal.

2   *Control element* The control element decides what action to take when it receives an error signal. It may be, for example, a signal to operate a switch or open a valve. The control plan being used by the element may be just to supply a signal which switches on or off when there is an error, as in a room thermostat, or perhaps a signal which proportionally opens or closes a valve according to the size of the error. Control plans may be *hard-wired systems* in which the control plan is permanently fixed by the way the elements are connected together or *programmable systems* where the control plan is stored within a memory unit and may be altered by reprogramming it. Controllers are discussed in chapter 11.

3   *Correction element* The correction element produces a change in the process to correct or change the controlled condition. Thus it might be a switch which switches on a heater and so increases the temperature of the process or a valve which opens and allows more liquid to enter the process. The term *actuator* is used for the element of a correction unit that provides the power to carry out the control action. Correction units are discussed in chapters 5 and 6.

4   *Process element* The process is what is being controlled. It could be a room in a house with its temperature being controlled or a tank of water with its level being controlled.

5   *Measurement element* The measurement element produces a signal related to the variable condition of the process that is being controlled. It might be, for example, a switch which is switched on when a particular position is reached or a thermocouple which gives an e.m.f. related to the temperature.

With the closed loop system illustrated in figure 1.5 for a person controlling the temperature of a room, the various elements are:

Controlled variable	–	the room temperature
Reference value	–	the required room temperature
Comparison element	–	the person comparing the measured value with the required value of temperature

Error signal	–	the difference between the measured and required temperatures
Control unit	–	the person
Correction unit	–	the switch on the fire
Process	–	the heating by the fire
Measuring device	–	a thermometer

An automatic control system for the control of the room temperature could involve a temperature sensor, after suitable signal conditioning, feeding an electrical signal to the input of a computer where it is compared with the set value and an error signal generated. This is then acted on by the computer to give at its output a signal, which, after suitable signal conditioning, might be used to control a heater and hence the room temperature. Such a system can readily be programmed to give different temperatures at different times of the day.

Figure 1.7 shows an example of a simple control system used to maintain a constant water level in a tank. The reference value is the initial setting of the lever arm arrangement so that it just cuts off the water supply at the required level. When water is drawn from the tank the float moves downwards with the water level. This causes the lever arrangement to rotate and so allow water to enter the tank. This flow continues until the ball has risen to such a height that it has moved the lever arrangement to cut off the water supply. It is a closed-loop control system with the elements being:

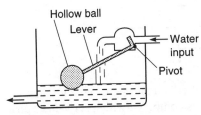

Fig. 1.7 The automatic control of water level

Controlled variable	–	water level in tank
Reference value	–	initial setting of the float and lever position
Comparison element	–	the lever
Error signal	–	the difference between the actual and initial settings of the lever positions
Control unit	–	the pivoted lever
Correction unit	–	the flap opening or closing the water supply
Process	–	the water level in the tank
Measuring device	–	the floating ball and lever

The above is an example of a closed-loop control system involving just mechanical elements. We could, however, have controlled the liquid level by means of an electronic control system. We thus might have had a level sensor supplying an

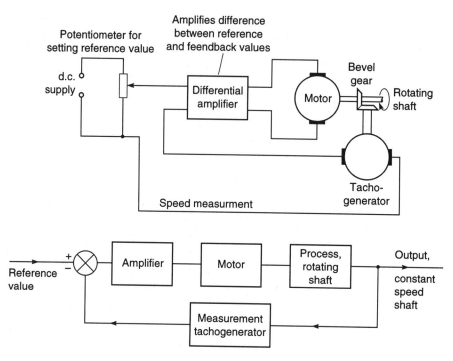

Fig. 1.8 Shaft speed control

electrical signal which is used, after suitable signal conditioning, as an input to a computer where it is compared with a set value signal and the difference between them, the error signal, then used to give an appropriate response from the computer output. This is then, after suitable signal conditioning, used to control the movement of an actuator in a flow control valve and so determine the amount of water fed into the tank.

Figure 1.8 shows a simple automatic control system for the speed of rotation of a shaft. A potentiometer is used to set the reference value, i.e. what voltage is supplied to the differential amplifier as the reference value for the required speed of rotation. The differential amplifier is used to both compare and amplify the difference between the reference and feedback values, i.e. it amplifies the error signal. The amplified error signal is then fed to a motor which in turn adjusts the speed of the rotating shaft. The speed of the rotating shaft is measured using a tachogenerator, connected to the rotating shaft by means of a pair of bevel gears. The signal from the tachogenerator is then fed back to the differential amplifier.

1.4.3 Sequential controllers

There are many situations where control is exercised by items being switched on or off at particular preset times or values to control processes and involve a step sequence of operations. After step 1 is complete then step 2 starts. When step 2 is complete then step 3 starts, etc. The term *sequential control* is used when control is such that actions are strictly ordered in a time sequence. This could be by sets of relays. Such mechanical switches are now more likely to have been replaced by microprocessors, such devices behaving like switches which are user programmable.

As an illustration of sequential control, consider the domestic washing machine. The machine has to carry out a number of operations in the correct sequence. These may involve a program consisting of a pre-wash cycle when the clothes in the drum are given a wash in cold water, followed by a main wash cycle when they are washed in hot water, then a rinse cycle when the clothes are rinsed with cold water a number of times, followed by spinning to remove water from the clothes. Each of these operations involves a number of steps, e.g. a pre-wash cycle involves opening a valve to fill the machine drum to the required level, closing the valve, switching on the drum motor to rotate the drum for a specific time, and operating the pump to empty the water from the drum. The system operating sequence is called a *program* and there will be a number of programs which can be selected, the program depending on the type of clothes being washed in the machine. The sequence of instructions in each program is predefined and built into the controller used.

Figure 1.9 shows the basic washing machine system and gives a rough idea of its constituent elements. The system that has generally been used for the washing machine controller involves a

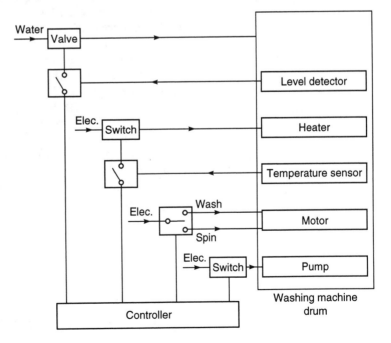

Fig. 1.9 Washing machine system

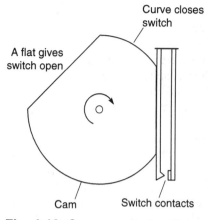

Curve closes switch

A flat gives switch open

Cam   Switch contacts

Fig. 1.10 Cam-operated switch

set of cam-operated switches, i.e. mechanical switches. Figure 1.10 shows the basic principle of one such switch. When the machine is switched on, a small electric motor slowly turns the controller cams so that each in turn operates electrical switches and so switches on circuits in the correct sequence. The contour of a cam determines the time at which it operates a switch. Thus the contours of the cams are the means by which the program is specified and stored in the machine. The sequence of instructions and the instructions used in a particular washing program are determined by the set of cams chosen.

For the pre-wash cycle an electrically operated valve is opened when a current is supplied and switched off when it ceases. This valve allows cold water into the drum for a period of time determined by the profile of the cam used to operate its switch. However, since the requirement is a specific level of water in the washing machine drum, there needs to be another mechanism which will stop the water going into the tank, during the permitted time, when it reaches the required level. In series with the cam-operated switch is a water level switch. This is a sensor that gives a signal when the water level has reached the preset level and switches off the current to the valve.

For the main wash cycle, the cam has a profile such that it starts in operation when the pre-wash cycle is completed. It switches a current into a circuit to open a valve to allow cold water into the drum. This is in series with a water level switch so that the water shuts off when the required level is reached. The cams then supply a current to activate a switch which applies a larger current to an electric heater to heat the water. A temperature sensor is used to switch off the current when the water temperature reaches the preset value. The cams then switch on the drum motor to rotate the drum. This will continue for the time determined by the cam profile before switching off. Then a cam switches on the current to a discharge pump to empty the water from the drum.

The rinse part of the operation is now switched as a sequence of signals to open valves which allow cold water into the machine, switch it off, operate the motor to rotate the drum, operate a pump to empty the water from the drum, and repeat this sequence a number of times.

The final part of the operation is when a cam switches on just the motor, at a higher speed than for the rinsing, to spin the clothes.

1.4.4 Microprocessor-based controllers

Microprocessors are now rapidly replacing the mechanical cam-operated controllers and being used in general to carry out control functions. They have the great advantage that a greater variety of

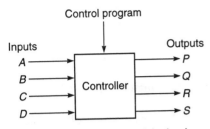

Inputs

A

B Controller

C

D

Control program

Outputs

P

Q

R

S

Fig. 1.11 Programmable logic controller

1.5 Digital control and logic gates

programs become feasible. The term *programmable logic controller* is used for a microprocessor-based controller which uses programmable memory to store instructions and to implement functions such as logic, sequence, timing counting and arithmetic to control events. Figure 1.11 shows the control action of a programmable logic controller, the inputs being signals from, say, switches being closed and the program used to determine how the controller should respond to the inputs and the output it should then give. Chapters 15 and 16 discuss microprocessors and their use as controllers.

Analogue control is when the control is continuous with input signals from sensors and output signals to actuators being continuously variable. The actuators themselves may be continuously variable, e.g. a valve, or simple on/off switches. Many control systems, however, involve *digital* signals where there are only two possible signal levels. These two levels may represent levels of on or off, open or closed, yes or no, true or false, +5 V or 0 V, etc. These two levels can be represented using the *binary* number system with the on, open, yes, true, +5 V levels being represented by 1 and the off, closed, no, false, 0 V levels by 0. This is the notation of *Boolean algebra*. With *digital control* the control is discontinuous. Thus, for example, we might have the water input to the domestic washing machine switched on if we have both the door to the machine closed and a particular time in the operating cycle has been reached. There are two input signals which are either yes or no signals and an output signal which is a yes or no signal. The controller is here programmed to only give a yes output if both the input signals are yes, i.e. if input A and input B are both 1 then there is an output of 1. Such an operation is said to be controlled by a *logic gate*, in this example an AND gate. There are many machines and processes which are controlled in this way.

1.5.1 Logic gates

The relationships between inputs to a logic gate and the outputs can be tabulated in a form known as a *truth table*. This specifies the relationships between the inputs and outputs. Thus for an AND gate with inputs A and B and a single output Q, we will have a 1 output when, and only when, $A = 1$ and $B = 1$. All other combinations of A and B will generate a 0 output. We can thus write the truth table as:

Inputs		Output
A	B	Q
0	0	0
0	1	0
1	0	0
1	1	1

Fig. 1.12 AND gate representation

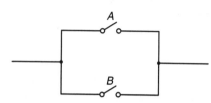

Fig. 1.13 OR gate representation

An example of an AND gate is an interlock control system for a machine tool such that if the safety guard is in place and gives a 1 signal and the power is on, giving a 1 signal, then there can be an output, a 1 signal, and the machine operates. We can visualise the AND gate as an electrical circuit involving two switches in series (figure 1.12). Only when switch A and switch B are closed is there a current.

An OR gate with inputs A and B gives an output of a 1 when A or B is 1. We can visualise such a gate as an electrical circuit involving two switches in parallel (figure 1.13). When switch A or B is closed then there is a current. The following is the truth table:

Inputs		Output
A	B	Q
0	0	0
0	1	1
1	0	1
1	1	1

A NOT gate has just one input and one output, giving a 1 output when the input is 0 and a 0 output when the input is 1. The NOT gate gives an output which is the inversion of the input and is called an *inverter*. The following is the truth table:

Input	Output
A	Q
0	1
1	0

The NAND gate can be considered as a combination of an AND gate followed by a NOT gate. Thus when input A is 1 and input B is 1 there is an output of 0, all other inputs giving an output of 1. It is just the AND gate truth table with the outputs inverted. An alternative way of considering the gate is as an AND gate with a NOT gate applied to invert both the inputs before they reach the AND gate. The following is the truth table:

Inputs		Output
A	B	Q
0	0	1
0	1	1
1	0	1
1	1	0

The NOR gate can be considered as a combination of an OR gate followed by a NOT gate. Thus when input A or input B is 1 there is an output of 0. It is just the OR gate with the outputs inverted. An alternative way of considering the gate is as an OR gate with a NOT gate applied to invert both the inputs before they reach the OR gate. The following is the truth table:

Inputs		Output
A	B	Q
0	0	1
0	1	0
1	0	0
1	1	0

The EXCLUSIVE-OR gate (XOR) can be considered to be an OR gate with a NOT gates applied to one of the inputs to invert it before the inputs reach the OR gate. Alternatively it can be considered as an AND gate with a NOT gate applied to one of the inputs to invert it before the inputs reach the AND gate. The following is the truth table:

Inputs		Output
A	B	Q
0	0	0
0	1	1
1	0	1
1	1	0

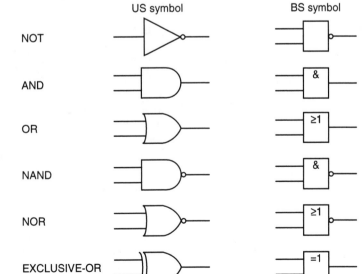

Fig. 1.14 Gate symbols

The symbols used to represent the functions of the above logic gates in logic diagrams are shown in figure 1.14. The addition of NOT to a gate is indicated by the O symbol. Two sets of symbols are shown, one being that specified by British Standards and the other by the United States, the latter being the ones most widely used.

1.5.2 Combining gates

It might seem that to make logic systems we require a range of gates. However, as the following shows, we can make up all the gates from just one. Consider the combination of three NOR gates shown in figure 1.15. The truth table, with the intermediate and final outputs, is as follows:

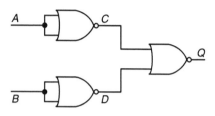

Fig. 1.15 Three NOR gates

A	B	C	D	Q
0	0	1	1	0
0	1	1	0	0
1	0	0	1	0
1	1	0	0	1

The result is the same as an AND gate. If we followed this assembly of gates by a NOT gate then we would obtain a truth table the same as an NAND gate.

A combination of three NAND gates is shown in figure 1.16. The truth table, with the intermediate and final outputs, is as follows:

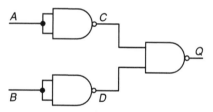

Fig. 1.16 Three NAND gates

A	B	C	D	Q
0	0	1	1	0
0	1	1	0	1
1	0	0	1	1
1	1	0	0	1

The result is the same as an OR gate. If we followed this assembly of gates by a NOT gate then we would obtain a truth table the same as a NOR gate.

The above two illustrations of gate combinations show how one type of gate, a NOR or a NAND, can be used to substitute for other gates, provided we use more than one gate. Gates can also be combined to make complex gating circuits and sequential circuits.

The number of gates which are available on a single chip varies from just a few in small-scale integrated circuits, e.g. four two-input NOR gates, to hundreds or more in microprocessors.

1.6 The mechatronics approach

The domestic washing machine referred to earlier in this chapter used cam-operated switches in order to control the washing cycle. Such mechanical switches are being replaced by microprocessors. This can be considered an example of a mechatronics approach in that a mechanical system has become integrated with electronic controls. As a consequence, a bulky mechanical system is replaced by a much more compact microprocessor system which is readily adjustable to give a greater variety of programs.

Mechatronics involves the bringing together of a number of technologies: mechanical engineering, electronic engineering, electrical engineering, computer technology, and control engineering. This can be considered to be the application of computer-based digital control techniques, through electronic and electric interfaces, to mechanical engineering problems. Mechatronics provides an opportunity to take a new look at problems, with mechanical engineers not just seeing a problem in terms of mechanical principles but having to see it in terms of a range of technologies. The electronics, etc., should not be seen as a bolt-on item to existing mechanical hardware. There needs to be a complete rethink of the requirements in terms of what an item is required to do.

There are many applications of mechatronics in the mass-produced products used in the home. Microprocessor-based controllers are to be found in domestic washing machines, dish washers, microwave ovens, cameras, camcorders, watches, hi-fi and video recorder systems, central heating thermostat controls, sewing machines, etc. They are to be found in cars in the active suspension, antiskid brakes, engine control, speedometer display, transmission, etc. A larger scale application of mechatronics is a flexible manufacturing engineering system (FMS) involving computer-controlled machines, robots, automatic material conveying and overall supervisory control.

Problems

1 Identify the sensor, signal conditioner and display elements in the measurement systems of (a) a mercury-in-glass thermometer, (b) a Bourdon pressure gauge.

2 Explain the difference between open- and closed-loop control.

3 Identify the various elements that might be present in a control system involving a thermostatically controlled electric heater.

4 The automatic control system for the temperature of a bath of liquid consists of a reference voltage fed into a differential amplifier. This is connected to a relay which then switches on or off the electrical power to a heater in the liquid. Negative feedback is provided by a measurement system which feeds a voltage into the differential amplifier. Sketch a block diagram of the system and explain how the error signal is produced.

5 Explain the function of a progammable logic controller.

6 Explain what is meant by sequential control and illustrate your answer by an example.

7 State steps that might be present in the sequential control of a dishwasher.

8 Explain what logic gates might be used to control the following situations:
   (a) The issue of tickets at an automatic ticket machine at a railway station.
   (b) A safety lock system for the operation of a machine tool.

9 Compare and contrast the traditional design of a watch with that of the mechatronics-designed product involving a microprocessor.

10 Compare and contrast the control system for the domestic central heating system involving a bimetallic thermostat and that involving a microprocessor.

2 Sensors and transducers

2.1 Sensors and transducers

The term *sensor* is used for an element which produces a signal relating to the quantity being measured. Thus in the case of, say, an electrical resistance temperature element, the quantity being measured is temperature and the sensor transforms and input of temperature into a change in resistance. The term *transducer* is often used in place of the term sensor. Transducers are defined as elements that when subject to some physical change experience a related change. Thus sensors are transducers. However, a measurement system may use transducers, in addition to the sensor, in other parts of the system to convert signals in one form to another form.

This chapter is about transducers and in particular those used as sensors. The terminology used to specify the performance characteristics of transducers is defined and the type of transducers commonly used in engineering is discussed.

2.2 Performance terminology

The following are some of the more common terms used to define the performance of transducers, and often measurement systems as a whole.

Range The range of a transducer is the limits between which the input can vary. Thus, for example, a load cell for the measurement of forces might have a range of 0 to 50 kN:

Error Error is the difference between the result of the measurement and the true value of the quantity being measured.

Error = measured value − true value

Thus if a measurement system gives a temperature reading of 25°C when the actual temperature is 24°C, then the error is +1°C. If the actual temperature had been 26°C then the error would have been −1°C. A sensor might give a resistance change of 10.2 Ω when the true change should have been 10.5 Ω. The error is −0.3 Ω. Errors can arise in a number of ways and common forms of error are listed in the next section.

Accuracy Accuracy is the extent to which the value indicated by a measurement system might be wrong. It is thus the summation of all the possible errors that are likely to occur, as well as the accuracy to which the transducer has been calibrated. A temperature-measuring instrument might, for example, be specified as having an accuracy of ±2°C. This would mean that the reading given by the instrument can be expected to lie within + or −2°C of the true value. Accuracy is often expressed as a percentage of the full range output or full-scale deflection. The percentage of full scale deflection term results from when the outputs of measuring systems was displayed almost exclusively on a circular or linear scale. A sensor might, for example, be specified as having an accuracy of ±5% of full range output. Thus if the range of the sensor was, say, 0 to 200°C, then the reading given can be expected to be within + or −10°C of the true reading.

Sensitivity The sensitivity is the relationship indicating how much output you get per unit input, i.e. ouput/input. For example, a resistance thermometer may have a sensitivity of 0.5 Ω/°C. This term is also frequently used to indicate the sensitivity to inputs other than that being measured, i.e. environmental changes. Thus there can be the sensitivity of the transducer to temperature changes in the environment or perhaps fluctuations in the mains voltage supply. A transducer for the measurement of pressure might be quoted as having a temperature sensitivity of ±0.1% of the reading per °C change in temperature.

Hysteresis error Transducers can give different outputs from the same value of quantity being measured according to whether that value has been reached by a continuously increasing change or a continuously decreasing change. This effect is called hysteresis. Figure 2.1 shows the type of transducer output which can occur. The hysteresis error is the maximum difference in output for increasing and decreasing values.

Non-linearity error For many transducers a linear relationship between the input and output is assumed over the working range, i.e. a graph of output plotted against input is assumed to give a straight line. Few transducers, however, have a truly linear relationship and thus errors occur as a result of the assumption of linearity. Various methods are used for the numerical expression of the non-linearity error. The differences occur in the determing the straight line relationship against which the error is specified. One method is to draw the straight line joining the output values at the

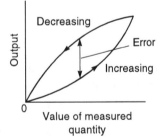

Fig. 2.1 Hysteresis

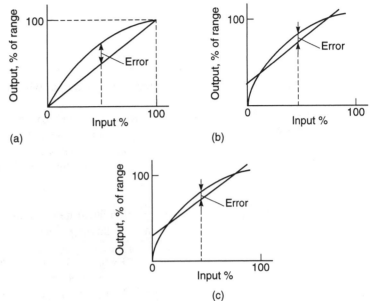

Fig. 2.2 Non-linearity error using: (a) end-range values, (b) best straight line for all values, (c) best straight line through zero point

end points of the range; another is to find the straight line by using the method of least squares to determine the best fit line when all data values are considered equally likely to be in error; another is to find the straight line by using the method of least squares to determine the best fit line which passes through the zero point. Figure 2.2 illustrates these three methods and how they can affect the non-linearity error quoted. The error is generally quoted as a percentage of the full range output. For example, a transducer for the measurement of pressure might be quoted as having a non-linearity error of ±0.5% of the full range.

Repeatability The repeatability of a transducer is its ability to give the same output for repeated applications of the same input value. This is without the transducer being disconnected from the input or any change in the environment in which the test is carried out. The error resulting from the same output not being given with repeated applications is usually expressed as a percentage of the full range output. A more realistic measure of errors that might be expected in practice is given by the reproducibility. A transducer for the measurement of angular velocity typically might be quoted as having a repeatability of ±0.01% of the full range at a particular angular velocity.

Reproducibility The reproducibility of a transducer is its ability to give the same output when used to measure a constant input and is measured on a number of occasions. Between the measurements the transducer is disconnected and reinstalled. The error resulting from the same output not being given is usually expressed as a percentage of the full range output.

Stability The stability of a transducer is its ability to give the same output when used to measure a constant input over a period of time. The term *drift* is often used to describe the change in output that occurs over time. The drift may be expressed as a percentage of the full range output. The term *zero drift* is used for the changes that occur in output when there is zero input.

Dead band The dead band or dead space of a transducer is the range of input values for which there is no output. For example, bearing friction in a flow meter using a rotor might mean that there is no output until the input has reached a particular velocity threshold.

To illustrate the above, consider the significance of the terms in the following specification of a strain gauge pressure transducer:

Ranges: 70 to 1000 kPa, 2000 to 70 000 kPa
Supply voltage: 10 V d.c. or a.c. r.m.s.
Full range output: 40 mV
Non-linearity and hysteresis: ±0.5% full range output
Temperature range: −54°C to +120°C when operating
Thermal zero shift: 0.030% full range output/°C
Thermal sensitivity: 0.030% full range output/°C

The range indicates that the transducer can be used to measure pressures between 70 and 1000 kPa or 2000 and 70 000 kPa. It requires a supply of 10 V d.c. or a.c. r.m.s. for its operation and will give an output of 40 mV when the pressure on the lower range is 1000 kPa and on the upper range 70 000 kPa. Non-linearity and hysteresis will lead to errors of ±0.5% of 1000, i.e. ±5 kPa on the lower range and ±0.5% of 70 000, i.e. ±350 kPa on the upper range. The transducer can be used between the temperatures of −54 and +120°C. When the temperature changes by 1°C the output of the transducer for zero input will change by 0.030% of 1000 = 0.3 kPa on the lower range and 0.030% of 70 000 = 21 kPa on the upper range. When the temperature changes by 1°C, the sensitivity of the transducer will change by 0.3 kPa on the lower range and 21 kPa on the upper range. This means that readings will change by these amounts when such a temperature change occurs.

2.2.1 Static and dynamic characteristics

The *static characteristics* are the values given when steady-state conditions occur, i.e. the values given when the transducer has settled down after having received some input. The terminology defined above refers to such a state. The *dynamic characteristics* refer to the behaviour between the time that the input value changes and the time that the value given by the transducer settles down to the steady-state value. Dynamic characteristics are stated in terms of the response of the transducer to inputs in particular forms. For example, this might be a step input when the input is suddenly changed from 0 to a constant value, or a ramp input when the input is changed at a steady rate, or a sinusoidal input of a specified frequency. Thus we might find the following terms (see chapter 8 for a more detailed discussion of dynamic systems):

Response time This is the time which elapses after a constant input, a step input, is applied to the transducer up to the point at which the transducer gives an output corresponding to some specified percentage, e.g. 95%, of the value of the input (figure 2.3). For example, if a mercury-in-glass thermometer is put into a hot liquid there can be quite an appreciable time lapse, perhaps as much as 100 s or more, before the thermometer indicates 95% of the actual temperature of the liquid.

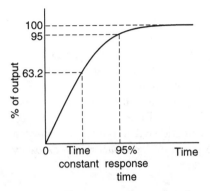

Fig. 2.3 Response to a step input

Time constant This is the 63.2% response time. A thermocouple in air might have a time constant of perhaps 40 to 100 s.

Rise time This is the time taken for the output to rise to some specified percentage of the steady-state output. Often the rise time refers to the time taken for the output to rise from 10% of the steady-state value to 90 or 95% of the steady-state value.

Settling time This is the time taken for the output to settle to within some percentage, e.g. 2%, of the steady state value.

To illustrate the above, consider the following data which indicates how an instrument reading changed with time, being obtained from a thermometer plunged into a liquid at time $t = 0$. The 95% response time is required.

Time (s)	0	30	60	90	120	150	180
Temp. (°C)	20	28	34	39	43	46	49

Time (s)	210	240	270	300	330	360
Temp. (°C)	51	53	54	55	55	55

Figure 2.4 shows the graph of how the temperature indicated by the thermometer varies with time. The steady-state value is 55°C and so, since 95% of 55 is 52.25°C, the 95% response time is about 228 s.

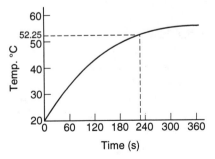

Fig. 2.4 Example

2.3 Examples of sensors

The following examples of transducers are grouped according to what they are being used to measure. The measurements considered are those frequently encountered in mechanical engineering, namely: displacement, velocity, force, pressure, fluid flow, liquid level, temperature, and proximity.

For a more comprehensive coverage of transducers the reader is referred to more specialist texts, e.g. the series *Jones' Instrument Technology* edited by B.E. Noltingk (Butterworth 1985), in particular volume 1, *Mechanical Measurements* and volume 2, *Measurement of Temperature and Chemical Composition*, and the concise *Newnes Instrumentation and Measurement Pocket Book* by W. Bolton (Newnes 1991). Examples of the types of specifications that might be expected from electrical transducers are given in *Transducer Handbook* by H.B. Boyle (Newnes 1992).

2.3.1 Displacement

Displacement, and position, measurements are in two groups: those concerned with linear displacements and those with angular displacements. Linear displacement sensors might be used to monitor the thickness or other dimensions of sheet materials, the separation of rollers, the position or presence of a part, the size of a part, etc. Angular displacement methods might be used to monitor the angular displacement of shafts.

Displacement sensors can be contacting by spring loading or mechanical connection with the item being monitored, or non-contacting. For example, for those linear displacement methods involving contact, there is usually a sensing shaft which is in direct contact with the object being monitored. The displacement of this shaft is then monitored by a sensor. The movement of the shaft may be used to cause changes in electrical voltage, resistance, capacitance, or mutual inductance. For angular displacement methods involving mechanical connection the rotation of a shaft might directly drive, through gears, the rotation of the transducer element. The following are examples of such displacement sensors.

Potentiometer

A *potentiometer* consists of a resistance element with a sliding contact which can be moved over the length of the element. Such elements can be used for linear or rotary displacements. The rotary potentiometer consists of a circular wire-wound track or a film of conductive plastic over which a rotatable sliding contact can be rotated (figure 2.5). The track may be a single turn or helical. With a constant input voltage V_s, between terminals 1 and 3, the output voltage V_o between terminals 2 and 3 is a fraction of the input voltage, the fraction depending on the ratio of the resistance R_{23}

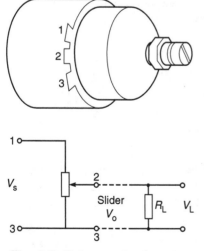

Fig. 2.5 Rotary potentiometer

between terminals 2 and 3 compared with the total resistance R_{13} between terminals 1 and 3, i.e. $V_o/V_s = R_{23}/R_{13}$. If the track has a constant resistance per unit length, i.e. per unit angle, then the output is proportional to the angle through which the slider has rotated. Hence an angular displacement can be converted into a potential difference.

The resolution of a wire track is limited by the diameter of the wire used and typically ranges from about 1.5 mm for a coarsely wound track to 0.5 mm for a finely wound one. Errors due to non-linearity of the track tend to range from less than 0.1% to about 1%. The track resistance tends to range from about 20 Ω to 200 kΩ. Conductive plastic has no resolution problems, errors due to non-linearity of track of the order of 0.05% and resistance values from about 500 Ω to 80 kΩ. The conductive plastic has a higher temperature coefficient of resistance than the wire and so temperature changes have a greater effect on accuracy.

An important effect to be considered with a potentiometer is the effect of a load R_L connected across the output. The potential difference across the load V_L is only directly proportional to V_o if the load resistance is infinite. For finite loads, however, the effect of the load is to transform what was a linear relationship between output voltage and angle into a non-linear relationship. The resistance R_L is in parallel with the fraction x of the potentiometer resistance R_p. This combined resistance is $R_L x R_p/(R_L + x R_p)$. The total resistance across the source voltage is thus

$$\text{Total resistance} = R_p(1 - x) + R_L x R_p/(R_L + x R_p)$$

Hence the fraction of the applied voltage dropped across the load resistance is

$$\frac{V_L}{V_s} = \frac{x R_L R_p/(R_L + x R_p)}{R_p(1 - x) + x R_L R_p/(R_L + x R_p)}$$

$$= \frac{x}{(R_p/R_L)x(1 - x) + 1}$$

If the load is of infinite resistance then we have $V_L = x V_s$. Thus the error introduced by the load having a finite resistance is

$$\text{Error} = x V_s - V_L = x V_s - \frac{x V_s}{(R_p/R_L)x(1 - x) + 1}$$

$$= V_s \frac{R_p}{R_L}(x^2 - x^3)$$

To illustrate the above, consider the non-linearity error with a potentiometer of resistance 500 Ω, when at a displacement of half its maximum slider travel, which results from there being a load of resistance 10 kΩ. The supply voltage is 4 V. Using the equation derived above

$$\text{Error} = 4 \times \frac{500}{10\,000}(0.5^2 - 0.5^3) = 0.025 \text{ V}$$

As a percentage of the full range reading, this is 0.625%.

Strain-gauged element

The electrical resistance strain gauge (figure 2.6) is a metal wire, metal foil strip, or a strip of semiconductor material which is wafer-like and can be stuck onto surfaces like a postage stamp. When subject to strain, its resistance R changes, the fractional change in resistance $\Delta R/R$ being proportional to the strain ε, i.e.

$$\frac{\Delta R}{R} = G\varepsilon$$

where G, the constant of proportionality, is termed the gauge factor. Since strain is the ratio (change in length/original length) then the resistance change of a strain gauge is a measurement of the change in length of the element to which the strain gauge is attached.

One form of displacement sensor has strain gauges attached to flexible elements in the form of cantilevers, rings or U-shapes. When the flexible element is bent or deformed as a result of forces being applied by a contact point being displaced, then the electrical resistance strain gauges mounted on the element are strained and so give a resistance change which can be monitored. The change in resistance is thus a measure of the displacement or deformation of the flexible element. Such arrangements are typically used for linear displacements of the order of 1 mm to 30 mm and have a non-linearity error of about ± 1% of full range.

To illustrate the above, consider an electrical resistance strain gauge with a resistance of 100 Ω and a gauge factor of 2.0. What

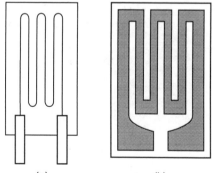

Fig. 2.6 Strain gauges:
(a) metal wire, (b) metal foil,
(c) semiconductor

(a)   (b)   (c)

is the change in resistance of the gauge when it is subject to a strain of 0.001? The fractional change in resistance is equal to the gauge factor multiplied by the strain, thus

Change in resistance = $2.0 \times 0.001 \times 100 = 0.2\ \Omega$

Capacitive element

The capacitance C of a parallel plate capacitor is given by

$$C = \frac{\varepsilon_r \varepsilon_0 A}{d}$$

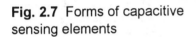

where ε_r is the relative permittivity of the dielectric between the plates, ε_0 a constant called the permittivity of free space, A the area of overlap between the two plates and d the plate separation. Capacitive sensors for the monitoring of linear displacements might thus take the forms shown in figure 2.7. In (a) one of the plates is moved by the displacement so that the plate separation changes; in (b) the displacement causes the area of overlap to change; in (c) the displacement causes the dielectric between the plates to change. The form shown in (d), often referred to as a *push–pull* displacement sensor, is particularly useful. It consists of two capacitors, between the movable central plate and the upper plate and between the central movable plate and the lower plate. The displacement moves the central plate between the two other plates. The result of, for example, the central plate moving up- wards is to decrease the plate separation of the upper capacitor and increase the separation of the lower capacitor. When the two capacitors are incorporated in an a.c. bridge, the bridge output voltage is proportional to the displacement. Such a sensor is typically used for monitoring displacements from a few millimetres to hundreds of millimetres. Non-linearity and hysteresis are about $\pm 0.01\%$ of full range.

Fig. 2.7 Forms of capacitive sensing elements

Differential transformers

The linear variable differential transformer, generally referred to by the abbreviation LVDT, consists of three coils symmetrically spaced along an insulated tube (figure 2.8). The central coil is the primary coil and the other two are identical secondary coils which are connected in series in such a way that their outputs oppose each other. A magnetic core is moved through the central tube as a result of the displacement being monitored.

When there is an alternating voltage input to the primary coil, alternating e.m.f.s are induced in the secondary coils. With the magnetic core central, the amount of magnetic material in each of the secondary coils is the same. Thus the e.m.f.s induced in each

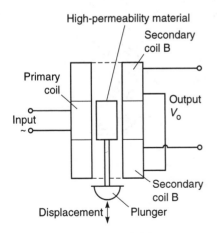

Fig. 2.8 Linear variable differential transformer

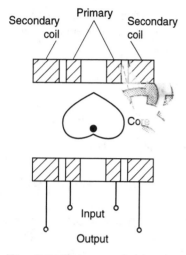

Fig. 2.9 Rotary variable differential transformer

coil is the same. Since they are so connected that their outputs oppose each other, the net result is zero output. However, when the core is displaced from the central position there is a greater amount of magnetic core in one coil than the other. The result is that a greater e.m.f. is induced in one coil than the other. There is then a net output from the two coils. Since a greater displacement means even more core in one coil than the other, the output, the difference between the two e.m.f.s increases the greater the displacement being monitored.

Typically, LVDTs have operating ranges from about ±2 mm to ±400 mm. Non-linearity errors are typically about ±0.25%. LVDTs are very widely used for monitoring displacements. The free end of the core may be spring loaded for contact with the surface being monitored, or threaded for mechanical connection.

A rotary variable differential transformer (RVDT) is available for the measurement of rotation (figure 2.9), operating on the same principle as the LVDT. The core is a cardoid-shaped piece of magnetic material and rotation causes more of it to pass into one secondary coil than the other. The range of operation is typically ±40° with the linearity error being about ±0.5% of the range.

Optical encoders

An encoder is a device that provides a digital output as a result of a linear or angular displacement. Position encoders can be grouped into two categories: incremental encoders and absolute encoders. Incremental encoders detect changes in rotation from some datum position while absolute encoders give the actual angular position.

Figure 2.10 shows the basic form of an *incremental encoder* for the measurement of angular displacement. A beam of light passes through slots in a disc and is detected by a suitable light sensor. When the disc is rotated, a pulsed output is produced by the sensor with the number of pulses being proportional to the

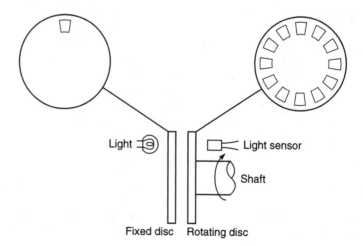

Fig. 2.10 Incremental encoder

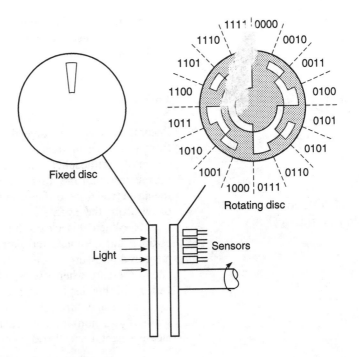

Fig. 2.11 Absolute encoder

angle through which the disc rotates. Thus the angular position of the disc, and hence the shaft rotating it, can be determined by the number of pulses produced since some datum position. The resolution is determined by the number of slots on the disc. Typically the number varies from 60 to over a thousand with multi-tracks having slightly offset slots in each track. With 60 slots occurring with 1 revolution then, since 1 revolution is a rotation of 360°, the resolution is 360/60 = 6°. The resolution thus varies from about 6° to 0.3° or better.

Figure 2.11 shows the basic form of an *absolute encoder* for the measurement of angular displacement. This gives an output in the form of a binary number of several digits, each such number representing a particular angular position. The rotating disc has four concentric circles of slots and four sensors to detect the light pulses. The slots are arranged in such a way that the sequential output from the sensors is a number in the binary code. A number of forms of binary code are used. Typical encoders tend to have up to 10 or 12 tracks. The number of bits in the binary number will be equal to the number of tracks. Thus with 10 tracks there will be 10 bits and so the number of positions that can be detected is 2^{10}, i.e. 1024, a resolution of 360/1024 = 0.35°.

To illustrate the above, consider an absolute encoder which has 7 tracks. What will be the number of unique positions specified by the code disc? Each track gives one of bits in the binary number. Thus we have 2^7 positions specified, i.e. 128.

The incremental encoder and the absolute encoder described

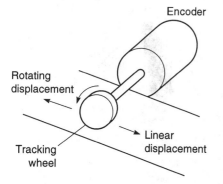

Fig. 2.12 Encoder used with a tracking wheel

above can be used for the monitoring of linear displacements if the linear displacement is converted to a rotary motion by means of a tracking wheel (figure 2.12).

Pneumatic sensor

Pneumatic sensors involve the use of compressed air. The basic form is shown in figure 2.13(a). Air at a constant pressure P_s above the atmospheric pressure flows through the orifice and escapes through the nozzle into the atmosphere. The pressure of the air between the orifice and the nozzle is measured. The escape of air from the nozzle is controlled by the displacement of the flapper. When this closes off the nozzle, i.e. $x = 0$, then no air escapes and the measured pressure equals P_s. As x increases, the measured pressure P decreases, becoming equal to the atmospheric pressure when x is very large. Figure 2.13(b) shows how P varies with the displacement x. The measured pressure can thus be used as a measure of displacement. The sensor has a high sensitivity, a non-linear characteristic, and a very small range of measurement, typically ±0.05 mm.

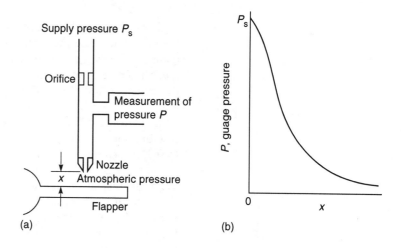

Fig. 2.13 Pneumatic sensor

(a)   (b)

2.3.2 Velocity

The following are examples of sensors that can be used to monitor linear and angular velocities.

Incremental encoder

The incremental encoder described in section 2.3.1 can be used for a measurement of angular velocity, the number of pulses produced per second being determined.

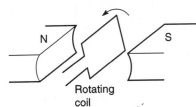

Fig. 2.14 Tachogenerator principle

Tachogenerator

This is used to measure angular velocity. It essentially is a small electric generator, consisting of a coil mounted in a magnetic field (figure 2.14). When the coil rotates an alternating e.m.f. is induced in the coil, the size of the maximum e.m.f. being a measure of the angular velocity. When used with a commutator a d.c. output can be obtained which is a measure of the angular velocity.

2.3.3 Force

A spring balance is an example of a force sensor in which a force, a weight, is applied to the scale pan and causes a displacement, i.e. the spring stretches. The displacement is then a measure of the force. Forces are commonly measured by the measurement of displacements, the following method illustrating this.

Strain gauge load cell

A very commonly used form of force-measuring transducer is based on the use of electrical resistance strain gauges to monitor the strain produced in some member when stretched, compressed or bent by the application of the force. The arrangement is generally referred to as a *load cell*. Figure 2.15 shows an example of such a cell. This is a cylindrical tube to which strain gauges have been attached. When forces are applied to the cylinder to compress it, then the strain gauges give a resistance change which is a measure of the strain and hence the applied forces. Typically such load cells are used for forces up to about 10 MN, the non-linearity error being about ±0.03% of full range, hysteresis error ±0.02% of full range and repeatability error ±0.02% of full range. Strain gauge load cells based on the bending of a strain-gauged metal element tend to be used for smaller forces, e.g. with ranges varying from 0 to 5 N up to 0 to 50 kN. Errors are typically a non-linearity error of about ±0.03% of full range, hysteresis error ±0.02% of full range and repeatability error ±0.02% of full range.

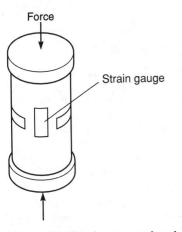

Fig. 2.15 Strain gauge load cell

2.3.4 Fluid pressure

Many of the devices used to monitor fluid pressure in industrial processes involve the monitoring of the elastic deformation of diaphragms, capsules, bellows and tubes (figure 2.16). Thus for a diaphragm (figure 2.16(a) and (b)), when there is a difference in pressure between the two sides then the centre of the diaphragm becomes displaced and this can be monitored by some form of displacement sensor, e.g. a strain gauge, as illustrated in figure 2.17. Corrugations in the diaphragm result in a greater sensitivity. Capsules (figure 2.16(c)) can be considered to be just two corrug-

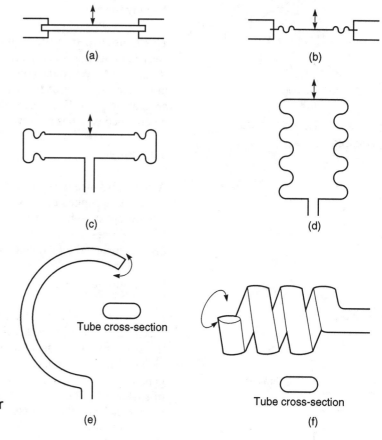

Fig. 2.16 Elastic transducers for pressure measurement

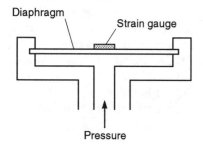

Fig. 2.17 Diaphragm pressure sensor

ated diaphragms combined and give even greater sensitivity. A stack of capsules is just a bellows (figure 2.16(c)) and even more sensitive. Figure 2.18 shows how a bellows can be combined with a LVDT to give a pressure sensor with an electrical output. Diaphragms, capsules and bellows are made from such materials as stainless steel, phosphor bronze, and nickel, with rubber and nylon also being used for some diaphragms. Pressures in the range of about 10^3 to 10^8 Pa can be monitored with such sensors.

A different form of deformation is obtained using a C-shaped tube (figure 2.16(e)), this being generally known as a *Bourdon tube*. This opens up to some extent when the pressure in the tube increases. A helical form of such tube (figure 2.16(f)) gives a greater sensitivity. The tubes are made from such materials as stainless steel and phosphor bronze and are used for pressures in the range 10^3 to 10^8 Pa.

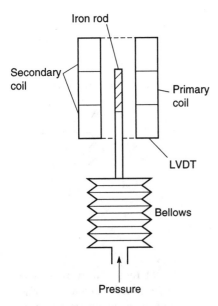

Fig. 2.18 Bellows pressure sensor

2.3.5 Liquid flow

The traditional methods of measuring the flow rate of liquids involves devices based on the measurement of the pressure drop occurring when the fluid flows through a constriction. Bernoulli's equation can then be used to determine the fluid velocity prior to the constriction. For a horizontal tube, where v_1 is the fluid velocity, P_1 the pressure and A_1 the cross-sectional of the tube prior to the constriction, v_2 the velocity, P_1 the pressure and A_2 the cross- sectional area at the constriction, then Bernoulli's equation gives

$$\frac{v_1^2}{2g} + \frac{P_1}{\rho g} = \frac{v_2^2}{2g} + \frac{P_2}{\rho g}$$

Since the mass of liquid passing per second through the tube prior to the constriction must equal that passing through the tube at the constriction, we have $A_1 v_1 \rho = A_2 v_2 \rho$. But the quantity Q of liquid passing through the tube per second is $A_1 v_1 = A_2 v_2$. Hence

$$Q = \frac{A_2}{\sqrt{1-(A_2/A_1)^2}} \sqrt{\frac{2(P_1-P_2)}{\rho}}$$

Thus the quantity of fluid flowing through the pipe per second is proportional to $\sqrt{}$(pressure difference). Measurements of the pressure difference can thus be used to give a measure of the rate of flow. There are many devices based on this principle, and the following example of the orifice plate is probably one of the most common.

Orifice plate

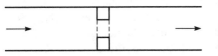

Fig. 2.19 Orifice plate

The orifice plate (figure 2.19) is simply a disc, with a central hole, which is placed in the tube through which the fluid is flowing. The pressure difference is measured between a point equal to the diameter of the tube upstream and a point equal to half the diameter downsteam. The orifice plate is simple, cheap, with no moving parts, and is widely used. It, however, does not work well with slurries. The accuracy is typically about ±1.5% of full range, it is non-linear, and does produce quite an appreciable pressure loss in the system to which it is connected.

Turbine meter

The turbine flowmeter (figure 2.20) consists of a multi-bladed rotor that is supported centrally in the pipe along which the flow occurs. The fluid flow results in rotation of the rotor, the angular

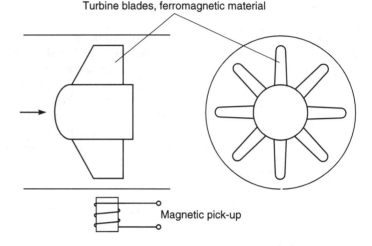

Fig. 2.20 Turbine meter

velocity being approximately proportional to the flow rate. The rate of revolution of the rotor can be determined using a magnetic pick-up which produces an induced e.m.f. pulse every time a rotor blade passes it as the blades are made from a magnetic material or have small magnets mounted at their tips. The pulses are counted and so the number of revolutions of the rotor can be determined. The meter is expensive with an accuracy of typically about ±0.3%.

2.3.6 Liquid level

The level of liquid in a vessel can be measured directly by monitoring the position of the liquid surface or indirectly by measuring some variable related to the height. Direct methods can involve floats; indirect methods include the monitoring of the weight of the vessel by, perhaps, load cells. The weight of the liquid is $Ah\rho g$, where A is the cross-sectional area of the vessel, h the height of liquid, ρ its density and g the acceleration due to gravity. Thus changes in the height of liquid give weight changes. More commonly, indirect methods involve the measurement of the pressure at some point in the liquid, the pressure due to a column of liquid of height h being $h\rho g$, where ρ is the liquid density.

Floats

A direct method of monitoring the level of liquid in a vessel is by monitoring the movement of a float. Figure 2.21 illustrates this with a simple float system. The displacement of the float causes a lever arm to rotate and so move a slider across a potentiometer. The result is an output of a voltage related to the height of liquid. Other forms of this involve the lever causing the core in a LVDT to become displaced, or stretch or compress a strain-gauged element.

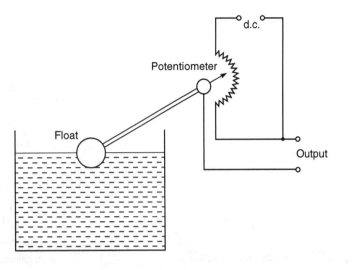

Fig. 2.21 Float system

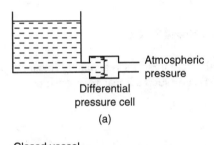

(a)

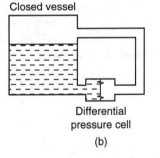

(b)

Fig. 2.22 Using a differential pressure sensor

Differential pressure

Figure 2.22 shows two forms of level measurement based on the measurement of differential pressure. In figure 2.22(a), the differential pressure cell determines the pressure difference between the liquid at the base of the vessel and atmospheric pressure, the vessel being open to atmospheric pressure. With a closed or open vessel the system illustrated in (b) can be used. The differential pressure cell monitors the difference in pressure between the base of the vessel and the air or gas above the surface of the liquid.

2.3.7 Temperature

Changes that are commonly used to monitor temperature are the expansion or contraction of solids, liquids or gases, the change in electrical resistance of conductors and semiconductors, thermo-electric e.m.f.s, etc. The following are some of the more useful methods that are relevant to temperature control systems.

Bimetallic strips

This device consists of two different metal strips bonded together. The metals have different coefficients of expansion and when the temperature changes the composite strip bends into a curved strip, with the higher coefficient metal on the outside of the curve. This deformation may be used as a temperature-controlled switch, as in the simple thermostat commonly used with domestic heating systems (figure 2.23).

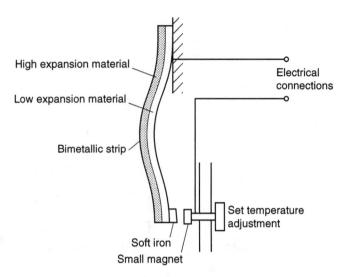

Fig. 2.23 Bimetallic thermostat

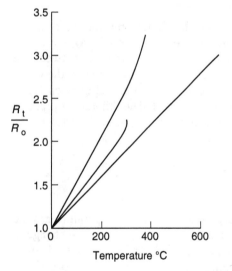

Fig. 2.24 Metals

Resistance temperature detectors (RTDs)

The resistance of most metals increases in a reasonably linear way with temperature (figure 2.24). For such a linear relationship

$$R_t = R_0(1 + \alpha t)$$

where R_t is the resistance at a temperature $t°C$, R_0 the resistance at $0°C$ and α a constant for the metal termed the temperature coefficient of resistance. Resistance temperature detectors (RTDs) are simple resistive elements in the form of coils of wire of such metals as platinum, nickel or nickel–copper alloys. Such detectors are highly stable and give reproducible responses over long periods of time.

Thermistors

Thermistors are small pieces of material made from mixtures of metal oxides, such as those of chromium, cobalt, iron, manganese and nickel. These oxides are semiconductors. The material is formed into various forms of element, such as beads, discs and rods (figure 2.25). The resistance of thermistors decreases in a very non-linear manner with an increase in temperature, as illustrated in figure 2.25. The change in resistance per degree change in temperature is considerably larger than that which occurs with metals. The resistance–temperature relationship for a thermistor can be described by an equation of the form

$$R_t = K \, e^{\beta/t}$$

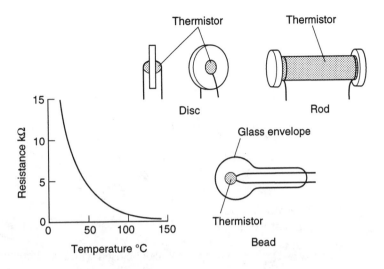

Fig. 2.25 Thermistors

where R_t is the resistance at temperature t, with K and β being constants.

Thermistors have many advantages when compared with other temperature sensors. They are rugged and can be very small, so enabling temperatures to be monitored at virtually a point. Because of their small size they respond very rapidly to changes in temperature. They give very large changes in resistance per degree change in temperature. Their main disadvantage is their non-linearity.

Thermocouples

If two different metals are joined together, a potential difference occurs across the junction. The potential difference depends on the metals used and the temperature of the junction. A thermocouple is a complete circuit involving two such junctions (figure 2.26). If both junctions are at the same temperature there is no net e.m.f. If, however, there is a difference in temperature between the two junctions, there is an e.m.f. The value of this e.m.f. E depends on the two metals concerned and the temperatures t of both junctions. Usually one junction is held at 0°C and then, to a reasonable extent, the following relationship holds:

$$E = at + bt^2$$

where a and b are constants for the metals concerned. Figure 2.27 shows how the e.m.f. varies with temperature for a number of commonly used pairs of metals. Standard tables are available for the metals usually used for thermocouples.

A thermocouple circuit can have other metals in the circuit and they will have no effect on the thermoelectric e.m.f. provided all their junctions are at the same temperature. A thermocouple

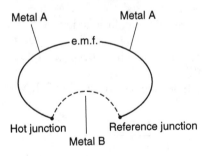

Fig. 2.26 A thermocouple

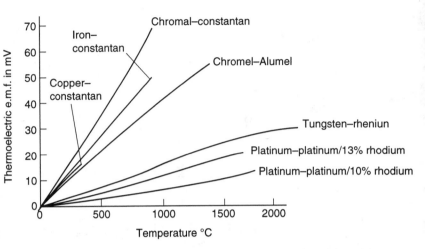

Fig. 2.27 Thermoelectric e.m.f. – temperature graphs

can be used with the reference junction at a temperature other than 0°C. The standard tables, however, assume a 0°C junction and hence a correction has to be applied before the tables can be used. The correction is applied using what is known as the *law of intermediate temperatures*, namely

$$E_{t,0} = E_{t,I} + E_{I,0}$$

The e.m.f. $E_{t,0}$ at temperature t when the cold junction is at 0°C equals the e.m.f. $E_{t,I}$ at the intermediate temperature I plus the e.m.f. $E_{I,0}$ at temperature I when the cold junction is at 0°C. To maintain one junction of a thermocouple at 0°C, i.e. have it immersed in a mixture of ice and water, is often not convenient. A compensation circuit can however be used to provide an e.m.f. which varies with the temperature of the cold junction in such a way that when it is added to the thermocouple e.m.f. it generates a combined e.m.f. which is the same as would have been generated if the cold junction had been at 0°C.

Commonly used thermocouples are shown in table 2.1, with

Table 2.1 Thermocouples

Type	Materials	Range °C	Sensitivity μV/°C
E	Chromel–constantan	0 to 980	63
J	Iron–constantan	−180 to 760	53
K	Chromel–alumel	−180 to 1260	41
R	Platinum–platinum/ rhodium 13%	0 to 1750	8
T	Copper–constantan	−180 to 370	43

the temperature ranges over which they are generally used and typical sensitivities. These commonly used thermocouples are given reference letters. For example, the iron–constantan thermocouple is called a type J thermocouple. The base-metal thermocouples, E, J, K and T, are relatively cheap but deteriorate with age. They have accuracies which are typically about ±1 to 3%. Noble-metal thermocouples, e.g. R, are more expensive but are more stable with longer life. They have accuracies of the order of ±1% or better. Thermocouples are generally mounted in a sheath to give them mechanical and chemical protection. The type of sheath used depends on the temperatures at which the thermocouple is to be used. In some cases the sheath is packed with a mineral which is a good conductor of heat and a good electrical insulator. The response time of an unsheathed thermocouple is very fast. With a sheath this may be increased to as much as a few seconds if a large sheath is used. In some instances a group of thermocouples are connected in series so that there are perhaps ten or more hot junctions sensing the temperature. The e.m.f. produced by each are added together. Such an arrangement is known as a thermopile.

To illustrate the above, consider a type E thermocouple which is to be used for the measurement of temperature with a cold junction at 20°C. What will be the thermoelectric e.m.f. at 200°C? The following is data from standard tables.

Temp. (°C)	0	20	200
e.m.f. (mV)	0	1.192	13.419

Using the law of intermediate temperatures

$$E_{200,0} = E_{200,20} + E_{20,0} = 13.419 - 1.192 = 12.227 \text{ mV}$$

Note that this is not the e.m.f. given by the tables for a temperature of 180°C with a cold junction at 0°C, namely 11.949 mV.

2.3.8 Proximity

Proximity devices are used to determine the proximity of one object relative to another, whether a particular location has been reached, or whether an item is present in a particular position. This might be whether a door or machine guard is closed, the position of a robot arm relative to a workpiece or the presence of a product on a conveyor belt and hence counting such objects. Proximity devices indicate the presence of objects by initiating a switching action.

Microswitch

This requires physical contact and a small operating force to close the contacts of a small switch. For example, in the case of determining the presence of an item on a conveyor belt, this might be actuated by the weight of the item on the belt depressing the belt and hence a spring-loaded platform under it, with the movement of this platform then closing the switch.

Magnetic reed switch

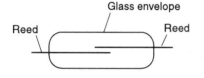

Fig. 2.28 Reed switch

Figure 2.28 shows the basic form of a reed switch. It consists of two magnetic switch contacts sealed in a glass tube. When a magnet is brought close to the switch, the magnetic reeds are attracted to each other and close the switch contacts. It is a non-contact proximity switch. Such a switch is very widely used for checking the closure of doors. It is also used with such devices as tachometers which involve the rotation of a toothed wheel past the reed switch. If one of the teeth has a magnet attached to it, then every time it passes the switch it will momentarily close the contacts and hence produce a current/voltage pulse in the associated electrical circuit.

Inductive proximity switch

This is a non-contact proximity switch. It consists of a coil wound round a core. When the end of this coil is close to a metal object its inductance changes. This change can be monitored by its effect on a resonant circuit. At some preset level, the change can be used to trigger a switch. It can only be used for the detection of the presence of metal objects and is best with ferrous metals.

Eddy current proximity switch

A coil is supplied with an alternating current and consequently produces an alternating magnetic field. If there is a metal object in close proximity to this alternating magnetic field, then eddy currents are induced in it. The presence of these eddy currents distorts the magnetic field responsible for their production. As a result, the impedance of the coil changes and so the amplitude of the alternating current. At some preset level, the change can be used to trigger a switch.

Photoelectric proximity switches

Photosensitive devices can be used to detect the presence of an opaque object by it breaking a beam of light, or infrared radiation, falling on such a device. Commonly used photosensitive devices are phototransistors, photodiodes and photoresistors. In the dark the phototransistor is in the cut-off region of its characteristic. When light strikes the base of the phototransistor, the transistor turns on. Photodiodes are connected into a circuit in reverse bias and so this circuit has a very high resistance. When light falls on the junction, the diode resistance drops and the current in the circuit rises appreciably. A common form of photoresistor is the cadmium–sulphide resistor. The resistance of this resistor depends on the intensity of the light falling on it. Such a resistor is widely used with automatic exposure cameras for the monitoring of the intensity of the light.

2.4 Selection of sensors

In selecting a sensor for a particular application there are a number of factors that need to be considered:

1   Identify the nature of the measurement required, e.g. the variable to be measured, its nominal value, the range of values, the accuracy required, the required speed of measurement, the reliability required, the environmental conditions under which the measurement is to be made.
2   Identify the nature of the output required from the sensor, this determining the signal conditioning requirements in order to give suitable output signals from the measurement.
3   Identify possible sensors, taking into account such factors as their range, accuracy, linearity, speed of response, reliability, maintainability, life, power supply requirements, ruggedness, availability, cost.

The selection of sensors cannot be taken in isolation from a consideration of the form of output that is required from the system after signal conditioning, and thus there has to be a suitable marriage between sensor and signal conditioner.

To illustrate the above, consider the selection of a sensor for the measurement of the level of a corrosive acid in a vessel. The level can vary from 0 to 2 m in a circular vessel which has a diameter of 1 m. The empty vessel has a weight of 100 kg. The minimum variation in level that is to be detected is 10 cm. The acid has a density of 1050 kg/m^3. The output from the sensor is to be electrical.

Because of the corrosive nature of the acid an indirect method of determining the level might be seen to be appropriate. Thus it is possibile to use a load cell, or load cells, to monitor the weight of the vessel. Such cells would give an electrical output. The weight of the liquid changes from 0 when empty to, when full, $1050 \times 2 \times \pi(1^2/4) \times 9.8 = 16.2$ kN. Adding this to the weight of the empty vessel gives a weight that varies from about 1 kN to 17 kN. The resolution required is for a change of level of 10 cm, i.e. a change in weight of $0.10 \times 1050 \times \pi(1^2/4) \times 9.8 = 0.8$ kN. If three load cells are used to support the tank then each will require a range of about 0 to 6 kN with a resolution of 0.27 kN. Manufacturers' catalogues can then be consulted to see if such load cells can be obtained.

Problems

1 Explain the significance of the following information given in the specification of transducers:
   (a) A piezoelectric accelerometer
      Non-linearity: ±0.5% of full range
   (b) A capacitive linear displacement transducer
      Non-linearity and hysteresis: ±0.01% of full range
   (c) A resistance strain gauge force measurement transducer
      Temperature sensitivity: ±1% of full range over normal environmental temperatures
   (d) A capacitance fluid pressure transducer
      Accuracy: ±1% of displayed reading
   (e) Thermocouple
      Sensitivity: nickel chromium–nickel aluminium:
      0.039 mV/°C when the cold junction is at 0°C
   (f) Gyroscope for angular velocity measurement
      Repeatability: ±0.01% of full range
   (g) Inductive displacement transducer
      Linearity: ±1% of rated load
   (h) Load cell
      Total error due to non-linearity, hysteresis and non-repeatability: ±0.1%

2 A copper–constantan thermocouple is to be used to measure temperatures between 0 and 200°C. The e.m.f. at 0°C is 0 mV, at 100°C it is 4.277 mV and at 200°C it is 9.286 mV. What will be the non-linearity error at 100°C as a percentage of the full range output if a linear relationship is assumed between e.m.f. and temperature over the full range?

3 A thermocouple element when taken from a liquid at 50°C and plunged into a liquid at 100°C at time $t = 0$ gave the following e.m.f. values. Determine the 95% response time.

Time (s)	0	20	40	60	80	100	120
e.m.f. (mV)	2.5	3.8	4.5	4.8	4.9	5.0	5.0

4 What is the non-linearity error, as a percentage of full range, produced when a 1 kW potentiometer has a load of 10 kW and is at one-third of its maximum displacement?

5 What will be the change in resistance of an electrical resistance strain gauge with a gauge factor of 2.1 and resistance 50 Ω if it is subject to a strain of 0.001?

6 You are offered a choice of an incremental shaft encoder or an absolute shaft encoder for the measurement of an angular

displacement. What is the principal difference between the results that can be obtained by these methods?

7 A shaft encoder is to be used with a 50 mm radius tracking wheel to monitor linear displacement. If the encoder produces 256 pulses per revolution, what will be the number of pulses produced by a linear displacement of 200 mm?

8 A rotary variable differential transformer has a specification which includes the following information:

Ranges: ±30°, linearity error ±0.5% full range
       ±60°, linearity error ±2.0% full range
Sensitivity: 1.1 (mV/V input)/degree
Impedance: Primary 750 Ω, Secondary 2000 Ω

What will be (a) the error in a reading of 40° due to non-linearity when the RDVT is used on the ±60° range, and (b) the output voltage change that occurs per degree if there is an input voltage of 3 V?

9 What are the advantages and disadvantages of the plastic film type of potentiometer when compared with the wire-wound potentiometer?

10 A pressure sensor consisting of a diaphragm with strain gauges bonded to its surface has the following information in its specification:

Ranges: 0 to 1400 kPa, 0 to 35 000 kPa
Non-linearity error: ±0.15% of full range
Hysteresis error: ±0.05% of full range

What is the total error due to non-linearity and hysteresis for a reading of 1000 kPa on the 0 to 1400 kPa range?

11 The water level in an open vessel is to be monitored by a differential pressure cell responding to the difference in pressure between that at the base of the vessel and the atmosphere. Determine the range of differential pressures the cell will have to respond to if the water level can vary between zero height above the cell measurement point and 2 m above it.

12 An iron–constantan thermocouple is to be used to measure temperatures between 0 and 400°C. What will be the non-linearity error as a percentage of the full-scale reading at 100°C if a linear relationship is assumed between e.m.f. and temperature?
E.m.f. at 100°C = 5.268 mV; e.m.f. at 400°C = 21.846 mV

13 A platinum resistance temperature detector has a resistance of 100.00 Ω at 0°C, 138.50 Ω at 100°C and 175.83 Ω at 200°C. What will be the non-linearity error in °C at 100°C if the detector is assumed to have a linear relationship between 0 and 200°C?

14 A strain gauge pressure sensor has the following specification. Will it be suitable for the measurement of pressure of the order

of 100 kPa to an accuracy of ±5 kPa in an enivronment where the temperature is reasonably constant at about 20°C?

Ranges: 2 to 70 MPa, 70 kPa to 1 MPa
Excitation: 10 V d.c. or a.c. (r.m.s.)
Full range output: 40 mV
Non-linearity and hysteresis errors: ±0.5 %
Temperature range: −54 to +120°C
Thermal shift zero: 0.030% full range output/°C
Thermal shift sensitivity: 0.030% full range output/°C

15 A float sensor for the determination of the level of water in a vessel has a cylindrical float of mass 2.0 kg, cross-sectional area 20 cm^2 and a length of 1.5 m. It floats vertically in the water and presses upwards against a beam attached to its upward end. What will be the minimum and maximum upthrust forces exerted by the float on the beam? Suggest a means by which the deformation of the beam under the action of the upthrust force ould be monitored.

16 Suggest a sensor that could be used as part of the control system for a furnace to monitor the rate at which the heating oil flows along a pipe. The output from the measurement system is to be an electrical signal which can be used to adjust the speed of the oil pump. The system must be capable of operating continuously and automatically, without adjustment, for long periods of time.

17 Suggest a sensor that could be used, as part of a control system, to determine the difference in levels between liquids in two containers. The output is to provide an electrical signal for the control system.

18 Suggest a sensor that could be used as part of a system to control the thickness of rolled sheet by monitoring its thickness as it emerges from rollers. The sheet metal is in continuous motion and the measurement needs to be made quickly to enable corrective action to be made quickly. The measurement system has to supply an electrical signal.

3 Signal conditioning

3.1 Signal conditioning

The output signal from the sensor of a measurement system has generally to be processed in some way to make it suitable for the next stage of the operation. The signal may be, for example, too small and have to be amplified, contain interference which has to be removed, be non-linear and require linearisation, be analogue and have to be made digital, be digital and have to be made analogue, be a resistance change and have to be made into a current change, be a voltage change and have to be made into a suitable size current change, etc. All these changes can be referred to as *signal conditioning*. For example, the output from a thermocouple is a small voltage, a few millivolts. A signal conditioning module might then be used to convert this into a suitable size current signal, provide noise rejection, linearisation, and cold junction compensation (i.e. allowing for the cold junction not being at 0°C).

3.1.1 Interfacing

Input and output devices are connected to a microprocessor system through *ports*. The term *interface* is used for the item that is used to make connections between devices and a port. Thus there could be inputs from sensors, switches, and keyboards and outputs to displays and actuators. The simplest interface could be just a piece of wire. However, the interface often contains signal conditioning and protection, the protection being to prevent damage to the microprocessor system. For example, inputs needed to be protected against excessive voltages or signals of the wrong polarity. Microprocessors require inputs which are digital, thus a conversion of analogue to digital signal is necessary if the output from a sensor is analogue. However, many sensors generate only a very small signal, perhaps a few millivolts. Such a signal is insufficient to be directly converted from analogue to digital without first being amplified. Signal conditioning might also be needed with digital signals to improve their quality. The interface may thus contain a number of elements. There is also the output from a microprocessor, perhaps to operate an actuator. A suitable interface is also required here. The actuator might require an analogue signal and so the digital output from the microprocessor needs converting to an analogue signal. There can also be a need for protection to stop any signal becoming inputted back through the output port to damage the microprocessor.

3.1.2 Signal-conditioning processes

The following are some of the processes that can occur in conditioning a signal:

1 *Protection* to prevent damage to the next element, e.g. a microprocessor, as a result of high current or voltage. Thus there can series current-limiting resistors, fuses to break if the current is too high, polarity protection and voltage limitation circuits (see section 3.3).
2 Getting the signal into the *right type of signal*. This can mean making the signal into a d.c. voltage or current. Thus, for example, the resistance change of a strain gauge has to be converted into a voltage change. This can be done by the use of a Wheatstone bridge (see section 3.5). It can mean making the signal digital or analogue (see section 3.6). It can also mean the right type of signal for transmission over a distance. Problems can occur with the transmission over a distance of low level d.c. signals or low level a.c. signals and modulation may thus be used. Modulation, and conversely, demodulation can thus be a function of signal conditioning units (see section 3.7).
3 Getting the *level* of the signal right. The signal from a thermocouple might be just a few millivolts. If the signal is to be fed into an analogue-to-digital converter for inputting to a microprocessor then it needs to be made much larger, volts rather than millivolts. Operational amplifiers are widely used for amplification (see section 3.2).
4 Eliminating or reducing *noise*. For example, filters might be used to eliminate mains noise from a signal (see section 3.4).
5 Signal *manipulation*, e.g. making it a linear function of some variable. The signals from some sensors, e.g. a flowmeter, are non-linear and thus a signal conditioner might be used so that the signal fed on to the next element is linear (see section 3.2.6).

The following sections outlines some of the elements that might be used in signal conditioning. Analogue-to-digital and digital-to-analogue systems, with their requirements, are discussed towards the end of the chapter.

3.2 The operational amplifier

The basis of many signal conditioning modules is the *operational amplifier*. The operational amplifier is a high gain d.c. amplifier, the gain typically being of the order of 100 000 or more, supplied as an integrated circuit on a silicon chip. It has two inputs, known as the inverting input (–) and the non-inverting input (+). The output depends on the connections made to these inputs. There are other inputs to the operational amplifier, namely a negative voltage supply, a positive voltage supply and two inputs termed offset null, these being to enable corrections to be made for the non-ideal behaviour of the amplifier (see section 3.7). Figure 3.1 shows the

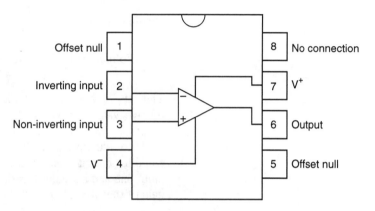

Fig. 3.1 Pin connections for a 741 operational amplifier

pin connections for a 741 type operational amplifier.

The following indicates the types of circuits that might be used with operational amplifiers when used as signal conditioners. For more details the reader is referred to more specialist texts, e.g. *Feedback Circuits and Op. Amps* by D.H. Horrocks (Chapman and Hall 1990) or *Analysis and Design of Analog Integrated Circuits* by P.R. Gray and R.G. Meyer (Wiley 1993).

3.2.1 Inverting amplifier

Figure 3.2 shows the connections made to the amplifier when used as an *inverting amplifier*. The input is taken to the inverting input through a resistor R_1 with the non-inverting input being connected to ground. A feedback path is provided from the output, via the resistor R_2 to the inverting input. The operational amplifier has a voltage gain of about 100 000 and the change in output voltage is limited to about ±10 V. The input voltage must then be between +0.0001 V and −0.0001 V. This is virtually zero and so point X is at virtually earth potential. For this reason it is called a *virtual earth*. The potential difference across R_1 is $(V_{in} - V_X)$. Hence, for an ideal operational amplifier with an infinite gain, and hence $V_X = 0$, the input potential V_{in} can be considered to be across R_1. Thus

$$V_{in} = I_1 R_1$$

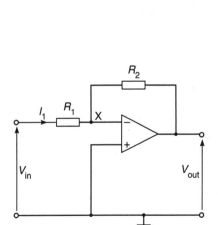

Fig. 3.2 Inverting amplifier

The operational amplifier has a very high impedance between its input terminals, for a 741 about 2 MΩ. Thus virtually no current flows through X into it. For an ideal operational amplifier the input impedance is taken to be infinite and so there is no current flow through X. Hence the current I_1 through R_1 must be the current through R_2. The potential difference across R_2 is $(V_X - V_{out})$ and thus, since V_X is zero for the ideal amplifier, the potential difference across R_2 is $-V_{out}$. Thus

$$-V_{out} = I_1 R_2$$

Dividing these two equations:

$$\text{Voltage gain of circuit} = \frac{V_{out}}{V_{in}} = -\frac{R_2}{R_1}$$

Thus the voltage gain of the circuit is determined solely by the relative values of R_2 and R_1. The negative sign indicates that the output is inverted, i.e. 180° out of phase, with respect to the input.

To illustrate the above, consider an inverting operational amplifier circuit which has a resistance of 1 MΩ in the inverting input line and a feedback resistance of 10 MΩ. What is the voltage gain of the circuit?

$$\text{Voltage gain of circuit} = \frac{V_{out}}{V_{in}} = -\frac{R_2}{R_1} = -\frac{10}{1} = -10$$

3.2.2 Non-inverting amplifier

Figure 3.3 shows the operational amplifier connected as a non-inverting amplifier. The output can be considered to be taken from across a potential divider circuit consisting of R_1 in series with R_2. The voltage V_X is then the fraction $R_1/(R_1 + R_2)$ of the output voltage.

$$V_X = \frac{R_1}{R_1 + R_2} V_{out}$$

Since there is virtually no current through the operational amplifier between the two inputs there can be virtually no potential difference between them. Thus, with the ideal operational amplifier, we must have $V_X = V_{in}$. Hence

$$\text{Voltage gain of circuit} = \frac{V_{out}}{V_{in}} = \frac{R_1 + R_2}{R_1} = 1 + \frac{R_2}{R_1}$$

A particular form of this amplifier is when the feedback loop is a short circuit, i.e. $R_2 = 0$. Then the voltage gain is 1. The input

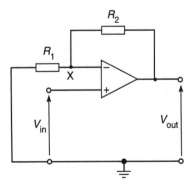

Fig. 3.3 Non-inverting amplifier

to the circuit is into a large resistance, the input resistance typically being 2 MΩ. The output resistance, i.e. the resistance between the output terminal and the ground line, is however much smaller, e.g. 75 Ω. Thus the resistance in the circuit that follows is a relatively small one and is less likely to load that circuit. Such an amplifier is referred to as a *voltage follower*.

3.2.3 Summing amplifier

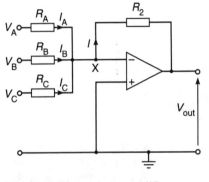

Fig. 3.4 Summing amplifier

Figure 3.4 shows the circuit of a summing amplifier. As with the inverting amplifier (section 3.2.1), X is a virtual earth. Thus the sum of the currents entering X must equal that leaving it. Hence

$$I = I_A + I_B + I_C$$

But $I_A = V_A/R_A$, $I_B = V_B/R_B$ and $I_C = V_C/R_C$. Also we must have the same current I passing through the feedback resistor. The potential difference across R_2 is $(V_X - V_{out})$. Hence, since V_X can be assumed to be zero, it is $-V_{out}$ and so $I = -V_{out}/R_2$. Thus

$$-\frac{V_{out}}{R_2} = \frac{V_A}{R_A} + \frac{V_B}{R_B} + \frac{V_C}{R_C}$$

The output is thus the scaled sum of the inputs, i.e.

$$V_{out} = -\left(\frac{R_2}{R_A}V_A + \frac{R_2}{R_B}V_B + \frac{R_2}{R_C}V_C\right)$$

If $R_A = R_B = R_C = R_1$ then

$$V_{out} = -\frac{R_1}{R_2}(V_A + V_B + V_C)$$

To illustrate the above, consider the design of a circuit that can be used to produce an output voltage which is the average the input voltages from three sensors. Assuming that an inverted output is acceptable, a circuit of the form shown in figure 3.4 can be used. Each of the three inputs must be scaled to 1/3 to give an output of the average. Thus a voltage gain of the circuit of 1/3 for each of the input signals is required. Hence, if the feedback resistance is 4 kΩ the resistors in each input arm will be 12 kΩ.

3.2.4 Integrating amplifier

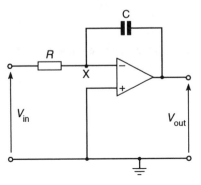

Fig. 3.5 Integrating amplifier

Consider an inverting operational amplifier circuit with the feedback being via a capacitor, as illustrated in figure 3.5. Current is the rate of movement of charge q and since, for a capacitor the charge $q = Cv$, where v is the voltage across it, then the current through the capacitor $i = dq/dt = C\, dv/dt$. The potential difference across C is $(v_X - v_{out})$ and since v_X is effectively zero, being the virtual earth, it is $-v_{out}$. Thus the current through the capacitor is $-C\, dv_{out}/dt$. But this is also the current through the input resistance R. Hence

$$\frac{v_{in}}{R} = -C\frac{dv_{out}}{dt}$$

Rearranging this gives

$$dv_{out} = -\left(\frac{1}{RC}\right)v_{in}\, dt$$

Integrating both sides gives

$$v_{out}(t_2) - v_{out}(t_1) = -\frac{1}{RC}\int_{t_1}^{t_2} v_{in}\, dt$$

$v_{out}(t_2)$ is the output voltage at time t_2 and $v_{out}(t_1)$ is the output voltage at time t_1. The output is proportional to the integral of the input voltage, i.e. the area under a graph of input voltage with time.

A differentiator circuit can be produced if the capacitor and resistor are interchanged in the circuit for the integrating amplifier.

3.2.5 Differential amplifier

A differential amplifier is one that amplifies the difference between two input voltages. Figure 3.6 shows the circuit. Since there is virtually no current through the high resistance in the operational amplifier between the two input terminals, there is no potential drop and thus both the inputs X will be at the same potential. The voltage V_2 is across resistors R_1 and R_2 in series. Thus the potential V_X at X is

$$\frac{V_X}{V_2} = \frac{R_2}{R_1 + R_2}$$

The current through the feedback resistance must be equal to that from V_1 through R_1. Hence

$$\frac{V_1 - V_X}{R_1} = \frac{V_X - V_{out}}{R_2}$$

This can be rearranged to give

$$\frac{V_{out}}{R_2} = V_X\left(\frac{1}{R_2} + \frac{1}{R_1}\right) - \frac{V_1}{R_1}$$

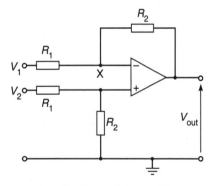

Fig. 3.6 Differential amplifier

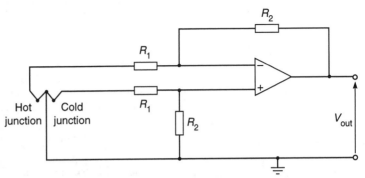

Fig. 3.7 Differential amplifier with a thermocouple

Hence substituting for V_X using the earlier equation,

$$V_{out} = \frac{R_2}{R_1}(V_2 - V_1)$$

The output is thus a measure of the difference between the two input voltages.

As an illustration of the use of such a circuit with a sensor, figure 3.7 shows it used with a thermocouple. The difference in voltage between the e.m.f.s of the two junctions of the thermocouple is being amplified. The values of R_1 and R_2 can, for example, be chosen to give a circuit with an output of 10 mV for a temperature difference between the thermocouple junctions of 10°C if such a temperature difference produces an e.m.f. difference between the junctions of 530 μV. For the circuit we have

$$V_{out} = \frac{R_2}{R_1}(V_2 - V_1)$$

$$10 \times 10^{-3} = \frac{R_2}{R_1} \times 530 \times 10^{-6}$$

Hence R_2/R_1 = 18.9. Thus if we take for R_1 a resistance of 10 kΩ then R_2 must be 189 kΩ.

3.2.6 Logarithmic amplifier

Some sensors have outputs which are non-linear. For example, the output from a thermocouple is not a perfectly linear function of the temperature difference between its junctions. A signal conditioner might then be used to linearise the output from such a sensor. This can be done using an operational amplifier circuit which is designed to have a non-linear relationship between its input and output so that when its input is non-linear the output is linear. This is achieved by a suitable choice of component for the feedback loop.

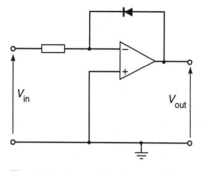

Fig. 3.8 Logarithmic amplifier

The logarithmic amplifier shown in figure 3.8 is an example of such a signal conditioner. The feedback loop contains a diode (or a transistor with a grounded base). The diode has a non-linear characteristic. It might be represented by $V = C \ln I$, where C is a constant. Then, since the current through the feedback loop is the same as the current through the input resistance and the potential difference across the diode is $-V_{out}$, we have

$$V_{out} = -C \ln (V_{in}/R) = K \ln V_{in}$$

where K is some constant. However, if the input V_{in} is provided by a sensor with an input t, where $V_{in} = A\ e^{at}$, with A and a being constants, then

$$V_{out} = K \ln V_{in} = K \ln (A\ e^{at}) = K \ln A + Kat$$

The result is a linear relationship between V_{out} and t.

3.2.7 Comparator

A comparator indicates which of two voltages is the larger. An operational amplifier used with no feedback or other components can be used as a comparator. One of the voltages is applied to the inverting input and the other to the non-inverting input (figure 3.9(a)). Figure 3.9(b) shows the relationship between the output voltage and the difference between the two input voltages. When the two inputs are equal there is, theoretically, no output. However, when the non-inverting input is greater than the inverting input by more than a small fraction of a volt then the output jumps to a steady positive saturation voltage of typically +10 V. When the inverting input is greater than the non-inverting input then the output jumps to a steady negative saturation voltage of typically −10 V. We can regard the circuit as being a logic circuit which gives a binary output of 0 when two voltages are equal and 1 when they differ.

Fig. 3.9 Comparator

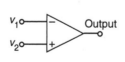

(a)

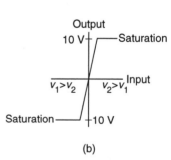

(b)

3.2.8 Amplifier errors

Operational amplifiers are not in the real world the perfect (ideal) element discussed in the previous sections of this chapter. A particularly significant problem is that of the *offset voltage*.

An operational amplifier is a high-gain amplifier which amplifies the difference between its two inputs. Thus if the two inputs are shorted we might expect to obtain no output. However, in practice this does not occur and quite a large output voltage might be detected. This effect is produced by imbalances in the internal circuitry in the operational amplifier. The output voltage can be made zero by applying a suitable voltage between the input terminals. This is known as the *offset voltage*. Many operational amplifiers are provided with arrangements for applying such an offset voltage via a potentiometer. With the 741 this is done by applying a voltage between pins 1 and 5 (see figure 3.1).

For more details of this, and other non-ideal characteristics, the reader is referred to the texts listed in section 3.1.

3.3 Protection

There are many situations where the connection of a sensor to the next unit, e.g. a microprocessor, can lead to the possibility of damage as a result of perhaps a high current or high voltage. A high current can be protected against by the incorporation in the input line of a series resistor to limit the current to an acceptable level and a fuse to break if the current does exceed a safe level. High voltages, and wrong polarity, may be protected against by the use of a Zener diode circuit (figure 3.10). Zener diodes behave like ordinary diodes up to some breakdown voltage when they become conducting. Thus to allow a maximum voltage of 5 V but stop voltages above 5.1 V getting through, a Zener diode with a voltage rating of 5.1 V might be chosen. When the voltage rises to 5.1 V the Zener diode breakdown and its resistance drops to a very low value. The result is that the voltage across the diode, and hence that outputted to the next circuit, drops. Because the Zener diode is a diode with a low resistance for current in one direction through it and a high resistance for the opposite direction, it also provides protection against wrong polarity.

In some situations it is desirable to completely isolate circuits and remove all electrical connections between them. This can be done using an *optoisolator*. This involves converting an electrical signal into an optical signal, then passing it to a detector which then converts it back into an electrical signal. Figure 3.11 shows the arrangement. The input signal is fed through an infrared light-emitting diode (LED). The infrared signal is then detected by a phototransistor. To prevent the LED having the wrong polarity or too high an applied voltage, it is likely to be protected by the Zener diode circuit shown in figure 3.10. Also, if there is alternating signal in the input a diode would be put in the input line to

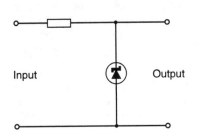

Fig. 3.10 Zener diode protection circuit

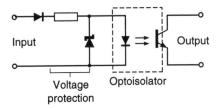

Fig. 3.11 Optoisolator

rectify it. Optoisolators are widely used with programmable logic controllers (see chapter 16) at its input/output ports to protect the microprocessor from damage.

3.4 Filtering

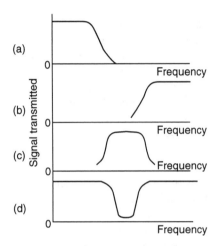

Fig. 3.12 Filters: (a) low pass, (b) high pass, (c) band pass, (d) band stop

The term *filtering* is used to describe the process of removing a certain band of frequencies from a signal and permitting others to be transmitted. The range of frequencies passed by a filter is known as the *pass band*, the range not passed as the *stop band* and the boundary between stopping and passing as the *cut-off frequency*. Filters are classified according to the frequency ranges they transmit or reject. A *low-pass filter* (figure 3.12(a)) has a pass band which allows all frequencies from 0 up to some frequency to be transmitted. A *high-pass filter* (figure 3.12(c)) has a pass band which allows all frequencies from some value up to infinity to be transmitted. A *band-pass filter* (figure 3.12(c)) allows all the frequencies within a specified band to be transmitted. A *band-stop filter* (figure 3.12(d)) stops all frequencies with a particular band from being transmitted.

The term *passive* is used to describe a filter made up using only resistors, capacitors and inductors, the term *active* being used when the filter also involves an operational amplifier. Passive filters have the disadvantage that the current that is drawn by the item that follows can change the frequency characteristic of the filter. This problem does not occur with an active filter.

Low-pass filters are very commonly used as part of signal conditioning. This is because most of the useful information being transmitted is low frequency. Since noise tends to occur at higher frequencies, a low pass filter can be used to block it off. Thus a low pass filter might be selected with a cut-off frequency of 40 Hz, thus blocking off any inference signals from the a.c. mains supply and noise in general. Figure 3.13 shows the basic forms that might be used for a passive low-pass filter and an active low-pass filter. The active filter uses a voltage-follower circuit to isolate the *RC* circuit from the next stage.

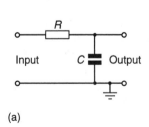

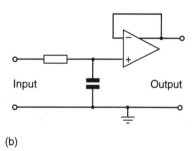

Fig. 3.13 Low-pass filters, (a) passive, (b) active

3.5 Wheatstone bridge

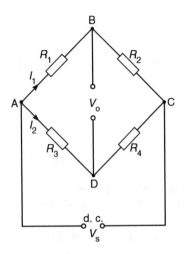

Fig. 3.14 Wheatstone bridge

The *Wheatstone bridge* can be used to convert a resistance change to a voltage change. Figure 3.14 shows the basic form of such a bridge. When the output voltage V_o is zero, then the potential at B must equal that at D. The potential difference across R_1, i.e. V_{AB}, must then equal that across R_3, i.e. V_{AD}. Thus

$$I_1R_1 = I_2R_2$$

It also means that the potential difference across R_2, i.e. V_{BC}, must equal that across R_4, i.e. V_{DC}. Since there is no current through BD then the current through R_2 must be the same as that through R_1 and the current through R_4 the same as that through R_3. Thus

$$I_1R_2 = I_2R_4$$

Dividing these two equations gives

$$\frac{R_1}{R_2} = \frac{R_3}{R_4}$$

The bridge is said to be *balanced*.

Now consider what happens when one of the resistances changes from this balanced condition. The supply voltage V_s is connected between points A and C and thus the potential drop across the resistor R_1 is the fraction $R_1/(R_1 + R_2)$ of the supply voltage. Hence

$$V_{AB} = \frac{V_sR_1}{R_1 + R_2}$$

Similarly, the potential difference across R_3 is

$$V_{AD} = \frac{V_sR_3}{R_3 + R_4}$$

Thus the difference in potential between B and D, i.e. the output potential difference V_o, is

$$V_o = V_{AB} - V_{AD} = V_s\left(\frac{R_1}{R_1 + R_2} - \frac{R_3}{R_3 + R_4}\right)$$

This equation gives the balanced condition when $V_o = 0$.

Consider resistance R_1 to be a sensor which has a resistance change. A change in resistance from R_1 to $R_1 + \delta R_1$ gives a change in output from V_o to $V_o + \delta V_o$, where

$$V_o + \delta V_o = V_s\left(\frac{R_1 + \delta R_1}{R_1 + \delta R_1 + R_2} - \frac{R_3}{R_3 + R_4}\right)$$

Hence

$$(V_o + \delta V_o) - V_o = V_s\left(\frac{R_1 + \delta R_1}{R_1 + \delta R_1 + R_2} - \frac{R_1}{R_1 + R_2}\right)$$

If δR_1 is much smaller than R_1 then the above equation approximates to

$$\delta V_o \approx V_s\left(\frac{\delta R_1}{R_1 + R_2}\right)$$

With this approximation, the change in output voltage is thus proportional to the change in the resistance of the sensor. This gives the output voltage when there is no load resistance across the output. If there is such a resistance then the loading effect has to be considered.

To illustrate the above, consider a platinum resistance temperature sensor which has a resistance at 0°C of 100 Ω and forms one arm of a Wheatstone bridge. The bridge is balanced, at this temperature, with each of the other arms also being 100 Ω. If the temperature coefficient of resistance of platinum is 0.0039 K^{-1}, what will be the output voltage from the bridge per degree change in temperature if the load across the output can be assumed to be infinite? The supply voltage, with negligible internal resistance, is 6.0 V. The variation of the resistance of the platinum with temperature can be represented by

$$R_t = R_0(1 + \alpha t)$$

where R_t is the resistance at $t\,°C$, R_0 the resistance at 0°C and α the temperature coefficient of resistance. Thus

$$\text{Change in resistance } = R_t - R_0 = R_0 \alpha t$$

$$= 100 \times 0.0039 \times 1 = 0.39 \text{ W/K}$$

Since this resistance change is small compared to the 100 Ω, the approximate equation can be used. Hence

$$\delta V_o \approx V_s\left(\frac{\delta R_1}{R_1 + R_2}\right) = \frac{6.0 \times 0.39}{100 + 100} = 0.012 \text{ V}$$

3.5.1 Temperature compensation

In many measurements involving a resistive sensor the actual sensing element may have to be at the end of long leads. Not only the sensor but the resistance of these leads will be affected by changes in temperature. For example, a platinum resistance

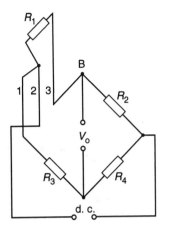

Fig. 3.15 Compensation for leads

temperature sensor consists of a platinum coil at the ends of leads. When the temperature changes, not only will the resistance of the coil change but so also will the resistance of the leads. What is required is just the resistance of the coil and so some means has to be employed to compensate for the resistance of the leads to the coil. One method of doing this is to use three leads to the coil, as shown in figure 3.15. The coil is connected into the Wheatstone bridge in such a way that lead 1 is in series with the R_3 resistor while lead 3 is in series with the platinum resistance coil R_1. Lead 2 is the connection to the power supply. Any change in lead resistance is likely to affect all three leads equally, since they are of the same material, diameter and length and held close together. The result is that changes in lead resistance occur equally in two arms of the bridge and cancels out if R_1 and R_3 are the same resistance.

The electrical resistance strain gauge is another sensor where compensation has to be made for temperature effects. The strain gauge changes resistance when the strain applied to it changes. Unfortunately, it also changes if the temperature changes. One way of eliminating the temperature effect is to use a *dummy strain gauge*. This is a strain gauge which is identical to the one under strain, the active gauge, and is mounted on the same material but is not subject to the strain. It is positioned close to the active gauge so that it suffers the same temperature changes. Thus a temperature change will cause both gauges to change resistance by the same amount. The active gauge is mounted in one arm of a Wheatstone bridge (figure 3.16) and the dummy gauge in another arm so that the effects of temperature-induced resistance changes cancel out.

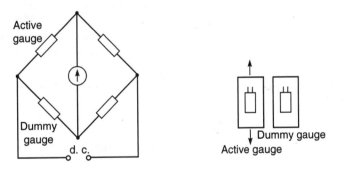

Fig. 3.16 Compensation with strain gauges

Strain gauges are often used with other sensors such as load cells (figure 2.25) or diaphragm pressure gauges (figure 2.17). In such situations, temperature compensation is still required. While dummy gauges could be used, a better solution is to use four strain gauges. Two of them are attached so that when forces are applied they are in tension and the other two in compression. The load cell in figure 3.17 shows such a mounting. The gauges that are in tension will increase in resistance while those in compression will

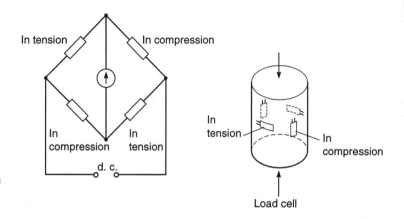

Fig. 3.17 Four active arm strain gauge bridge

decrease in resistance. As the gauges are connected as the four arms of a Wheatstone bridge (figure 3.17), then since all will be equally affected by any temperature changes the arrangement is temperature compensated. The arrangement also gives a much greater output voltage than would occur with just a single active gauge.

To illustrate this, consider a load cell with four strain gauges arranged as shown in figure 3.17 which is to be used with a four active strain gauge bridge. The gauges have a gauge factor of 2.1 and a resistance of 100 Ω. When the load cell is subject to a force, the compressive gauges suffer a strain of -1.0×10^{-5} and the tensile gauges $+0.3 \times 10^{-5}$. The supply voltage for the bridge is 6 V. The output voltage from the bridge is amplified by a differential operational amplifier circuit. What will be the ratio of the feedback resistance to that of the input resistances in the two inputs of the amplifier if the load is to produce an output of 1 mV? The change in resistance of a gauge subject to the compressive strain is given by $\Delta R/R = G\varepsilon$:

$$\text{Change in resistance} = G\varepsilon R = -2.1 \times 1.0 \times 10^{-5} \times 100$$

$$= -2.1 \times 10^{-3}\,\Omega$$

For a gauge subject to tension we have

$$\text{Change in resistance} = G\varepsilon R = 2.1 \times 0.3 \times 10^{-5} \times 100$$

$$= 6.3 \times 10^{-4}\,\Omega$$

The out-of-balance potential difference is given by (see section 3.5)

$$V_o = V_s\left(\frac{R_1}{R_1 + R_2} - \frac{R_3}{R_3 + R_4}\right)$$

$$= V_s\left(\frac{R_1(R_3+R_4)-R_3(R_1+R_2)}{(R_1+R_2)(R_3+R_4)}\right)$$

$$= V_s\left(\frac{R_1R_4-R_2R_3}{(R_1+R_2)(R_3+R_4)}\right)$$

We now have each of the resistors changing. We can, however, neglect the changes in relation to the denominators where the effect of the changes on the sum of the two resistances is insignificant. Thus

$$V_o = V_s\left(\frac{(R_1+\delta R_1)(R_4+\delta R_4)-(R_2+\delta R_2)(R_3+\delta R_3)}{(R_1+R_2)(R_3+R_4)}\right)$$

Neglecting products of δ terms and since we have an initially balanced bridge with $R_1 R_4 = R_2 R_3$, then

$$V_o = \frac{V_s R_1 R_4}{(R_1+R_2)(R_3+R_4)}\left(\frac{\delta R_1}{R_1}-\frac{\delta R_2}{R_2}-\frac{\delta R_3}{R_3}+\frac{\delta R_4}{R_4}\right)$$

Hence

$$V_o = \frac{6\times100\times100}{200\times200}\left(\frac{2\times6.3\times10^{-4}+2\times2.1\times10^{-3}}{100}\right)$$

The output is thus 3.6×10^{-5} V. This becomes the input to the differential amplifier hence, using the equation developed in section 3.2.7,

$$V_o = \frac{R_2}{R_1}(V_2-V_1)$$

$$1.0\times10^{-3} = \frac{R_2}{R_1}\times3.6\times10^{-5}$$

Thus $R_2/R_1 = 27.8$.

3.5.2 Thermocouple compensation

A thermocouple gives an e.m.f. which depends on the temperature of its two junctions (see section 2.3.7). Ideally one junction is kept at 0°C, then the temperature relating to the e.m.f. can be directly read from tables. However this is not always feasible and the cold junction is often allowed to be at the ambient temperature. To compensate for this a potential difference has to be added to the thermocouple. This must be the same as the e.m.f. that would be generated by the thermocouple with one junction at 0°C and the other at the ambient temperature. Such a potential difference can be produced by using a resistance temperature sensor in a Wheat-stone bridge. The bridge is balanced at 0°C and the output voltage from the bridge provides the correction potential difference at other temperatures.

The resistance of a metal resistance temperature sensor can be described by the relationship

$$R_t = R_0(1 + \alpha t)$$

where R_t is the resistance at t °C, R_0 the resistance at 0°C and α the temperature coefficient of resistance. Thus

$$\text{change in resistance} = R_t - R_0 = R_0 \alpha t$$

The output voltage for the bridge, taking R_1 to be the resistance temperature sensor, is given by

$$\delta V_o \approx V_s \left(\frac{\delta R_1}{R_1 + R_2} \right) = \frac{V_s R_0 \alpha t}{R_0 + R_2}$$

The thermocouple e.m.f. e is likely to vary with temperature t in a reasonably linear manner over the small temperature range being considered – from 0°C to the ambient temperature. Thus $e = at$, where a is a constant, i.e. the e.m.f. produced per degree change in temperature. Hence for compensation we must have

$$at = \frac{V_s R_0 \alpha t}{R_0 + R_2}$$

and so

$$aR_2 = R_0(V_s \alpha - a)$$

For an iron–constantan thermocouple giving 51 μV/°C, compensation can be provided by a nickel resistance element with a resistance of 10 Ω at 0°C and a temperature coefficient of resistance of 0.0067 K^{-1}, a supply voltage for the bridge of 1.0 V, and R_2 as 1304 Ω.

3.6 Digital signals

The output from most sensors tends to be in analogue form. Where a microprocessor is used as part of the measurement or control system, the analogue output from the sensor has to be converted into a digital form before it can be used as an input to the microprocessor. Likewise, most actuators operate with analogue inputs and so the digital output from a microprocessor has to be converted into an analogue form before it can be used as an input by the actuator.

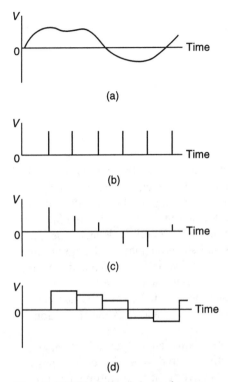

Fig. 3.18 Signals: (a) analogue,
(b) time, (c) sampled, (d) sampled
and held

Fig. 3.19 Analogue to digital
conversion

Analogue-to-digital conversion involves taking samples of the analogue signal. A clock supplies regular time signal pulses to the analogue to digital converter and every time it receives a pulse it samples the analogue signal. Figure 3.18 illustrates this. The result of the sampling is a series of narrow pulses (figure 3.18(c)). A *sample and hold* unit is then used to hold each sampled value until the next pulse occurs, with the result shown in figure 3.18(d). The sample and hold unit thus holds a sample of the analogue signal so that conversion can take place to a digital signal at an analogue-to-digital converter. Analogue-to-digital conversion thus involves a sample and hold unit followed by an analogue-to-digital converter (figure 3.19).

The relationship between the sampled and held input V_A and the output for an analogue-to-digital converter is illustrated by the graph shown in figure 3.20 for a digital output which is restricted to three bits. The binary digits of 0 and 1 are referred to as *bits*, a group of bits being a *word*. Thus the three bits give the *word length*. The position of bits in a word has the significance that the least significant bit is on the right end of the word and the most significant bit on the left. The sequence of bits in a word of n bits thus signifies

$$2^{n-1}, \dots, 2^3, 2^2, 2^1, 2^0$$

Most   Least
significant   significant
bit   bit

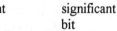

Fig. 3.20 Input–output for an
analogue to digital converter

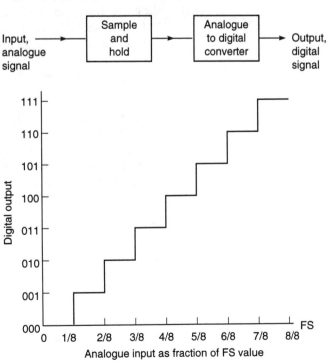

Each rise in the analogue voltage of (1/8) of the full scale (FS) value results in a further bit being generated. Bits can be added according to the basic rules:

$$0 + 0 = 0$$

$$0 + 1 = 1$$

$$1 + 1 = 10$$

The word length possible determines the *resolution* of the element, i.e. the smallest change in V_A which will result in a change in the digital output. The smallest change in digital output is one bit in the least significant bit position in the word, i.e. the far right bit. Thus with a word length of n bits the full scale analogue input V_{FS} is divided into 2^n pieces and so the minimum change in input that can be detected, i.e. the *resolution*, is $V_{FS}/2^n$. Thus if we have an analogue-to-digital converter having a word length of 10 bits and the analogue signal input range is 10 V, then the number of levels with a 10 bit word is $2^{10} = 1024$ and thus the resolution is $10/1024 = 9.8$ mV.

Consider a thermocouple giving an output of 0.5 mV/°C. What will be the word length required when its output passes through an analogue-to-digital converter if temperatures from 0 to 200°C are to be measured with a resolution of 0.5°C? The full scale output from the sensor is $200 \times 0.5 = 100$ mV. With a word length n, this voltage will be divided into $100/2^n$ mV steps. For a resolution of 0.5°C we must be able to detect a signal from the sensor of $0.5 \times 0.5 = 0.25$ mV. Thus we require

$$0.25 = \frac{100}{2^n}$$

Hence $n = 8.6$. Thus a 9-bit word length is required.

The term *conversion time* is used to specify the time it takes a converter to generate a complete digital word when supplied with the analogue input. Typically, conversion times are of the order of microseconds.

3.6.1 Analogue-to-digital converters

The input to an analogue-to-digital converter is an analogue signal and the output is a binary word that represents the level of the input signal. There are a number of forms of analogue-to-digital converter, the most common being successive approximations, flash, ramp, dual ramp and voltage-to-frequency. The reader is referred to more specialist texts, e.g. *Electronics* by D.I. Crecraft, D.A. Gorham and J.J. Sparkes (Chapman and Hall 1993) for

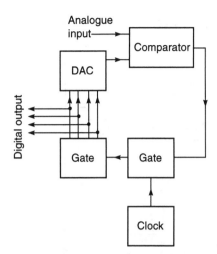

Analogue input→

Comparator

DAC

Digital output

Gate ← Gate

Clock

Fig. 3.21 Successive approximations analogue to digital converter

detailed discussions of these methods. Here we will just indicate how two of these methods operate.

Successive approximations is probably the most commonly used method. Figure 3.21 illustrates the subsystems involved. A voltage is generated by a clock emitting a regular sequence of pulses which are counted, in a binary manner, and the resulting binary word converted into an analogue voltage by a digital-to-analogue converter. This voltage rises in steps and is compared with the analogue input voltage from the sensor. When the clock-generated voltage passes the input analogue voltage the pulses from the clock are stopped from being counted by a gate being closed. The output from the counter at that time is then a digital representation of the analogue voltage. While the comparison could be accomplished by starting the count at 1, the least significant bit, and then proceeding bit by bit upwards, a faster method is by successive approximations. This involves selecting the most significant bit that is less than the analogue value, then adding successive lesser bits for which the total does not exceed the analogue value. For example, we might start the comparison with 1000. If this is too large we try 0100. If this is too small we then try 0110. If this is too large we try 0101. Because each of the bits in the word is tried in sequence, with an n-bit word it only takes n steps to make the comparison. Thus if the clock has a frequency f, the time between pulses is $1/f$. Hence the time taken to generate the word, i.e. the conversion time, is n/f.

Figure 3.22 shows the typical form of an 8-bit analogue-to-digital converter (GEC Plessey ZN439) designed for use with microprocessors and using the successive approximations method. All the active circuitry, including the clock, is contained on a single chip. Such a device has *tri-state outputs* to allow easy interfacing with a microprocessor. There are three possible outputs; a low state corresponding to logic 0, a high state corresponding to logic 1 and a high impedance state which effectively open circuits the output. An extra input, called the *enable*, is used to allow the output to give a logic output (0 or 1) or cause the output to be of high impedance. The ADC is first selected by taking the chip select pin low. The start conversion pin when taken low starts the conversion. At the end of the conversion the status output goes low. For further details of how to use the ADC the reader is referred to the manufacturer's data sheet.

The *ramp* form of analogue-to-digital converter involves an analogue voltage which is increased at a constant rate, a so-called ramp voltage, and applied to a comparator where it is compared with the analogue voltage from the sensor. The time taken for the ramp voltage to increase to the value of the sensor voltage will depend on the size of the sampled analogue voltage. When the ramp voltage starts, a gate is opened which starts a binary counter counting the regular pulses from a clock. When the two voltages

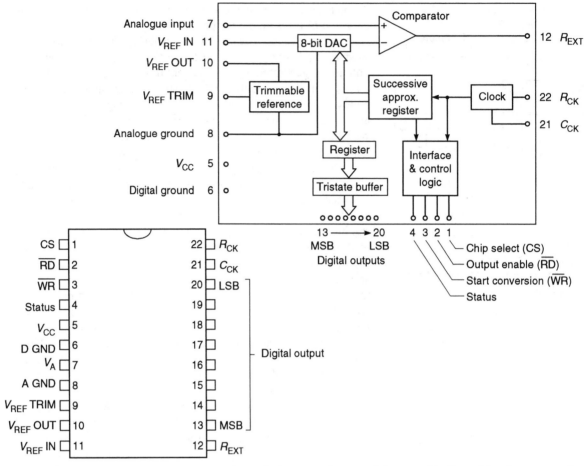

Fig. 3.22 ZN 439 ADC

are equal, the gate closes and the word indicated by the counter is the digital representation of the sampled analogue voltage. Figure 3.23 indicates the subsystems involved in the ramp form of analogue-to-digital converter.

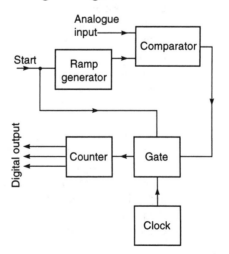

Fig. 3.23 Ramp analogue to digital converter

3.6.2 Digital-to-analogue converters

The input to a digital-to-analogue converter is a binary word; the output is an analogue signal that represents the weighted sum of the non-zero bits represented by the word. Thus, for example, an input of 0010 must give an analogue output which is twice that given by an input of 0001. Consider the situation where a microprocessor gives an output of an 8-bit word. This is fed through an 8-bit digital-to-analogue converter to a control valve. The control valve requires 6.0 V to be fully open. If the fully open state is indicated by 11111111 what will be the output to the valve for a change of 1 bit? The output voltage will be divided into 2^8 intervals. A change of 1 bit is a change in the output voltage of $6.0/2^8 = 0.023$ V.

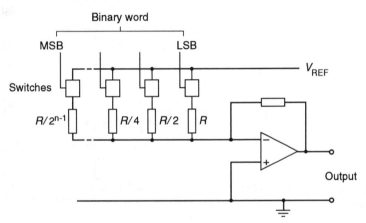

Fig. 3.24 Weighted-resistor DAC

A simple form of digital-to-analogue converter uses a summing amplifier (see section 3.2.3) to form the weighted sum of all the non-zero bits in the input word (figure 3.24). The reference voltage is connected to the resistors by means of switches which respond to binary 1. The input resistances depend on which bit in the word a switch is responding to. Hence the sum of the voltages is a weighted sum of the digits in the word. Such a system is referred to as a *weighted-resistor network*. Another, more commonly used, version uses a *R-2R ladder network* (figure 3.25). The output voltage is generated by switching sections of the ladder to either the reference voltage or 0 V according to whether there is a 1 or 0 in the digital input. Figure 3.26 shows details of the GEC Plessey ZN558D 8-bit latched input digital-to-analogue converter using a *R-2R* ladder network. After the conversion is complete, the 8-bit result is placed in an internal latch until the next conversion is complete. A latch is just a device to retain the output until a new one replaces it.

The reader is referred to more specialist texts, e.g. *Electronics* by D.I. Crecraft, D.A. Gorham and J.J. Sparkes (Chapman and Hall 1993) for more detailed discussions.

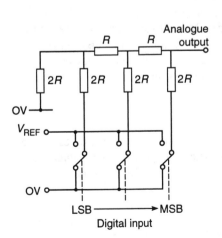

Fig. 3.25 *R-2R* ladder network DAC

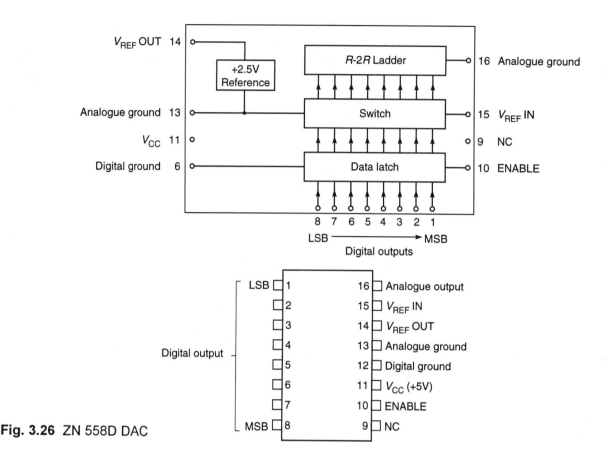

Fig. 3.26 ZN 558D DAC

3.6.3 Multiplexers

Frequently there is a need for measurements to be sampled from a number of different locations, or perhaps a number of different measurements need to be made. Rather than use a separate microprocessor for each measurement, a multiplexer can be used (figure 3.27). The multiplexer is essentially a switching device which enables each of the inputs to be sampled in turn.

3.6.4 Buffering

Frequently the signals reaching the inputs of a microprocessor are spasmodic. Sometimes the signals come in rapid bursts which may be too fast for the microprocessor to accept. Sometimes the microprocessor is engaged in some other activity and cannot accept them at the time they arrive. The output from a microprocessor may be too fast for an actuator or other device, new data being transmitted to it before it has had time to accept the previous data. For these reasons, external storage is provided for signals. This is provided in what are termed *buffers*.

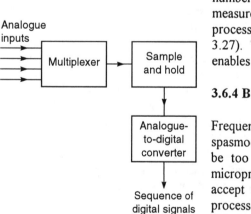

Fig. 3.27 Analogue to digital conversion using a multiplexer

For a buffer on the input, data is transferred out of it at a rate determined by the microprocessor. For a buffer on the output, data is transferred out of it at a rate determined by the peripheral device being fed. The buffer memory fills and empties to compensate for the difference between the input rate of the data into the buffer and the required output rate. The ability of the buffer to cope is determined by the size if its memory and the difference in input and output transfer rates.

3.7 Modulation

A problem that is often encountered with dealing with the transmission of low-level d.c. signals from sensors is that the gain of an operational amplifier used to amplify them may drift and so the output drifts. This problem can be overcome if the signal is alternating rather than direct. In addition, the conversion of the signal to alternating can assist in the elimination of external interference from the signal.

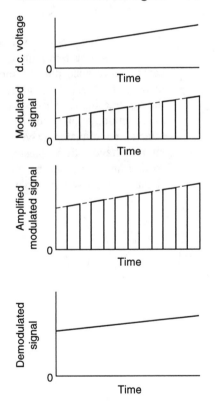

Fig. 3.28 Pulse amplitude modulation

One way this conversion can be achieved is by chopping the d.c. signal in the way suggested in figure 3.28. The output from the chopper is a chain of pulses, the heights of which are related to the d.c. level of the input signal. This process is called *pulse amplitude modulation*. After amplification and any other signal conditioning, the modulated signal can be demodulated to give a

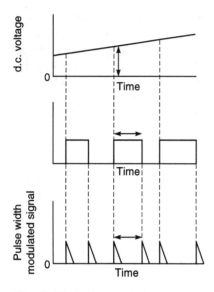

Fig. 3.29 Pulse duration modulation

d.c. output. With pulse amplitude modulation, the height of the pulses is related to the size of the d.c. voltage. An alternative to this is *pulse width modulation* where the width, i.e. duration, of a pulse depends on the size of the voltage (figure 3.29).

The above refers to d.c. signals, however it is often necessary to modulate a.c. signals. This enables data transmission at much higher frequencies and so allows the use of high-pass filters to eliminate the noise signals that usually occur at much lower frequencies. Modulation techniques used are *amplitude modulation* (figure 3.30(a)), *phase modulation* and *frequency modulation* (figure 3.30(b)). With amplitude modulation the amplitude of a carrier wave, of much higher frequency than the input, is varied according to the size of the voltage input, i.e. the wave carrying the signal from the sensor. Thus for a carrier wave that can be represented by

$$v = V \sin (\omega t + \phi)$$

the amplitude term V is varied according to the way the voltage input varies. Phase modulation involves varying the phase ϕ of the carrier wave according to the size of the voltage input. Another method is to vary the angular frequency ω according to the size of the voltage input. This is known as frequency modulation. Both phase modulation and frequency modulation produce similar effects, a modulated wave with a frequency which relates to the input voltage. After transmission, the signal can be demodulated so that an output can be obtained which is related to the original signal before it was modulated.

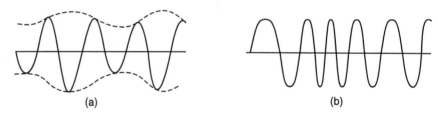

Fig. 3.30 Modulation:
(a) amplitude, (b) frequency

Problems

1 Design an operational amplifier circuit that can be used to produce an output that ranges from 0 to −5 V when the input goes from 0 to 100 mV.

2 Design a summing amplifier circuit that can be used to produce an output that ranges from −1 to −5 V when the input goes from 0 to 100 mV.

3 A differential amplifier is used with a thermocouple sensor in the way shown in figure 3.8. What values of R_1 and R_2 would give a circuit which has an output of 10 mV for a temperature difference between the thermocouple junctions of 100°C with a copper–constantan thermocouple if the thermocouple is assumed to have a constant sensitivity of 43 μV/°C?

4 The output from the differential pressure sensor used with an orifice plate for the measurement of flow rate is non-linear, the output voltage being proportional to the square of the flow rate. Determine the form of characteristic required for the element in the feedback loop of a operational amplifier signal conditioner circuit in order to linearise this output.

5 Digital signals from a sensor are polluted by noise and mains interference and are typically of the order of 100 V or more. Explain how protection can be afforded for a microprocessor to which these signals are to be inputted.

6 A platinum resistance temperature sensor has a resistance of 120 Ω at 0°C and forms one arm of a Wheatstone bridge. At this temperature the bridge is balanced with each of the other arms being 120 Ω. The temperature coefficient of resistance of the platinum is 0.0039 K^{-1}. What will be the output voltage from the bridge for a change in temperature of 20°C? The loading across the output is effectively open-circuit and the supply voltage to the bridge is from a source of 6.0 V with negligible internal resistance.

7 A diaphragm pressure gauge employs four strain gauges to monitor the displacement of the diaphragm. The four active gauges form the arms of a Wheatstone bridge, in the way shown in figure 3.16. The gauges have a gauge factor of 2.1 and resistance 120 Ω. A differential pressure applied to the diaphragm results in two of the gauges on one side of the diaphragm being subject to a tensile strain of 1.0×10^{-5} and the two on the other side a compressive strain of 1.0×10^{-5}. The supply voltage for the bridge is 10 V. What will be the voltage output from the bridge?

8 What is the resolution of an analogue-to-digital converter with a word length of 12 bits and an analogue signal input range of 100 V?

9 A sensor gives a maximum analogue output of 5 V. What word length is required for an analogue-to-digital converter if there is to be a resolution of 10 mV?

10 In monitoring the inputs from a number of thermocouples the following sequence of modules is used for each thermocouple in its interface with a microprocessor.

Protection, cold junction compensation, amplification, linearisation, sample and hold, analogue to digital converter, buffer, multiplexer.

Explain the function of each of the modules.

11 Suggest the modules that might be needed to interface the output of a microprocessor with an actuator.

4 Measurement systems

4.1 Designing measurement systems

In designing measurement systems there are a number of steps that need to be considered:

1 Identification of the *nature of the measurement* required, e.g. the variable to be measured, its nominal value, the range of values, the accuracy required, the required speed of measurement, the reliability required, the environmental conditions under which the measurement is to be made.
2 Identification of the *required output from the system*. This means considering the form of display that is required and also whether the measurement is needed as part of a control system. Thus, for example, control applications might require a 4 to 20 mA current.
3 Identification of *possible sensors*, taking into account such factors as their range, accuracy, linearity, speed of response, reliability, maintainability, life, power supply requirements, ruggedness, availability, cost. The sensor needs to match the specification arrived at in step 1 and also be amenable, with suitable signal conditioning, to give the required output arrived at in step 2.
4 Selection of appropriate *signal conditioning*. This needs to take the output from the sensor and modify the signal in such a way as to deliver the required output.

This chapter is a consideration of some examples of measurement systems and the selection of elements in measurement systems.

4.1.1 Loading

Connecting an ammeter into a circuit to make a measurement of the current changes the resistance of the circuit and so changes the current. The act of attempting to make such a measurement has modified the current that was being measured. When a voltmeter is

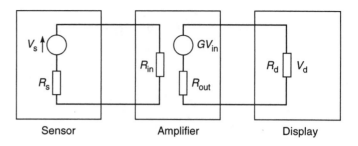

Fig. 4.1 Measurement system loading

Sensor   Amplifier   Display

connected across a resistor then we effectively have put two resistances in parallel, and if the resistance of the voltmeter is not considerably higher than that of the resistor the current through the resistor is markedly changed and so the voltage being measured is changed. The act of attempting to make the measurement has modified the voltage that was being measured. Such acts are termed *loading*.

Loading can also occur within a measurement system when the connection of one element to another modifies the characteristics of the preceding element. Consider, for example, a measurement system consisting of a sensor, an amplifier and a display element (figure 4.1). The sensor has an open-circuit output voltage of V_s and a resistance R_s. The amplifier has an input resistance of R_{in}. This is thus the load across the sensor. Hence the input voltage from the sensor is divided so that the potential difference across this load, and so the input voltage V_{in} to the amplifier, is

$$V_{in} = \frac{V_s R_{in}}{R_s + R_{in}}$$

If the amplifier has a voltage gain of G then the open-circuit voltage output from it will be GV_{in}. If the amplifier has an output resistance of R_{out} then the output voltage from the amplifier is divided so that the potential difference V_d across the display element, resistance R_d, is

$$V_d = \frac{GV_{in}R_d}{R_{out} + R_d} = \frac{GV_s R_{in} R_d}{(R_{out} + R_d)(R_s + R_{in})}$$

$$= \frac{GV_s}{\left(\dfrac{R_{out}}{R_d} + 1\right)\left(\dfrac{R_s}{R_{in}} + 1\right)}$$

Thus if loading effects are to be negligible we require $R_{out} \gg R_d$ and $R_s \gg R_{in}$.

4.2 Data presentation elements

Measurement systems consist of three elements: sensor, signal conditioner and display or data-presentation element (see section 1.3). There are a very wide range of elements that can be used for the presentation of data. They can be broadly classified into two groups: indicators and recorders. *Indicators* give an instant visual indication of the sensed variable while *recorders* record the output signal over a period of time and give automatically a permanent record. The recorder will be the most appropriate choice if the event is high speed or transient and cannot be followed by an observer, or there are large amounts of data, or it is essential to have a record of the data.

Both indicators and recorders can be subdivided into two groups of devices, *analogue* and *digital*. An example of an analogue indicator is a meter which has a pointer moving across a scale, while a digital meter would be just a display of a series of numbers. An example of an analogue recorder is a chart recorder which has a pen moving across a moving sheet of paper, while a digital recorder has the output printed out on a sheet of paper as a sequence of numbers.

This section is a brief overview of commonly used examples of such elements. For a more detailed description the reader is referred to more specialist texts such as *Jones' Instrument Technology*, volume 4 *Instrumentation Systems*, edited by B.E. Noltingk (Butterworth 1987) or *Principles of Electronic Instrumentation and Measurement* by H.M. Berlin and F.C. Getz (Merrill 1988).

4.2.1 Analogue and digital meters

The *moving coil meter* is an analogue indicator with a pointer moving across a scale. The basic instrument movement is a d.c. microammeter with shunts, multipliers and rectifiers being used to convert it to other ranges of direct current and measurement of alternating current, direct voltage, and alternating voltage. With alternating current and voltages, the instrument is restricted to between about 50 Hz and 10 kHz. The accuracy of such a meter depends on a number of factors, among which are temperature, the presence nearby of magnetic fields or ferrous materials, the way the meter is mounted, bearing friction, inaccuracies in scale marking during manufacture, etc. In addition there are errors involved in reading the meter, e.g. parallax errors when the position of the pointer against the scale is read from an angle other than directly at right angles to the scale and errors arising from estimating the position of the pointer between scale markings. The overall accuracy is generally of the order of ±0.1 to ±5 %. The time taken for a moving coil meter to reach a steady deflection is typically in the region of a few seconds. The low resistance of the meter can present loading problems.

The *digital voltmeter* gives its reading in the form of a sequence of digits. Such a form of display eliminates parallax and interpolation errors and can give accuracies as high as ±0.005%.

Fig. 4.2 Principle of digital voltmeter

The digital voltmeter is essentially just a sample and hold unit feeding an analogue-to-digital converter with its output counted by a counter (figure 4.2). It has a high resistance, of the order of 10 MΩ, and so loading effects are less likely than with the moving coil meter with its lower resistance. Thus, if a digital voltmeter specification includes the statement 'sample rate approximately 5 readings per second' then this means that every 0.2 s the input voltage is sampled. It is the time taken for the instrument to process the signal and give a reading. Thus, if the input voltage is changing at a rate which results in significant changes during 0.2 s then the voltmeter reading can be in error.

For details of the 'mechanics' of meters the reader is referred to such texts as *Electrical and Electronic Measurement and Testing* by W. Bolton (Longman 1992) or *Electronic Instruments and Measurement Techniques* by F.F. Mazda (Cambridge University Press 1987).

4.2.2 Analogue chart recorders

The *galvanometric type* of chart recorder (figure 4.3) works on the same principle as the moving coil meter movement. The coil is suspended between two fixed points by a suspension wire. When a current passes through the coil a torque acts on it, causing it to rotate and twist the suspension. The coil rotates to an angle at which the torque is balanced by the opposing torque resulting from the twisting of the suspension. The rotation of the coil results in a pen being moved across a chart.

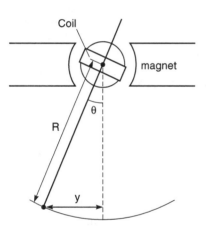

Fig. 4.3 Galvanometric type of chart recorder

If R is the length of the pointer and θ the angular deflection of the coil, then the displacement y of the pen is $y = R \sin \theta$. Since θ is proportional to the current i through the coil, then y is proportional to sin i. This is a non-linear relationship. However, if the angular deflections are restricted to less than ±10°, then the relationship is reasonably linear, the non-linearity error being less than 0.5%. A greater problem is however the fact that the pen moves in an arc rather than a straight line and thus curvilinear paper (figure 4.4) has to be used for the plotting. With such forms of chart there are difficulties in interpolation for points between the lines.

Figure 4.5 illustrates the general principles of the *potentiometric recorder*. Such a recorder is sometimes referred to as a *closed-loop recorder* or a *closed-loop servo recorder*. The position of the pen is monitored by means of a slider which moves along a linear potentiometer. The position of the slider determines

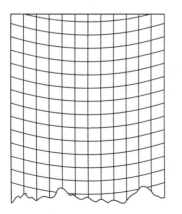

Fig. 4.4 Curvilinear chart paper

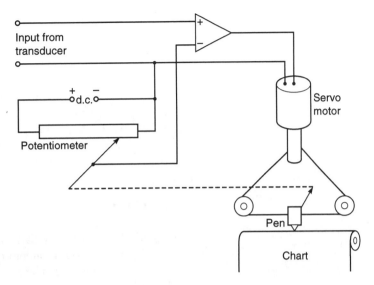

Fig. 4.5 Potentiometric recorder

the potential applied to an operational amplifier. The operational amplifier subtracts the slider signal, which is obtained from the input signal, from the sensor/signal conditioner. The output from the amplifier is thus a signal that is related to the difference between the pen and sensor signals. This signal is used to operate a servo motor which in turn controls the movement of the pen across the chart. The pen thus ends up moving to a position where there is no difference between the pen and sensor signals. The pen is thus made to track the sensor signal.

Potentiometric recorders have high input resistances, higher accuracies than galvanometric recorders (about ±0.1% of full-scale reading) but considerably slower response times. Response times are typically of the order of 1 to 2 s and can only be used for d.c. signals or very low frequencies, up to about 2 Hz. They can thus only be used for slowly changing signals. Because of friction, a minimum current is required to operate the motor and thus there is an error due to the recorder not responding to a small input signal. This error is known as the *dead band*. Typically it is about ±0.3% of the range of the instrument. Thus, for a range of 5 mV the dead band amounts to about ±0.015 mV.

4.2.3 Cathode-ray oscilloscope

The cathode-ray oscilloscope is a voltage-measuring instrument which is capable of displaying extremely high frequency signals. A general-purpose instrument can respond to signals up to about 10 MHz while more specialist instruments can respond up to about 1 GHz. Double-beam oscilloscopes enable two separate traces to be observed simultaneously on the screen while storage oscilloscopes enable the trace to remain on the screen after the input signal has ceased, only being removed by a deliberate action of erasure. Permanent records of traces can be made with special-purpose cameras that attach directly to the oscilloscope.

For details of the 'mechanics' of cathode-ray oscilloscopes the reader is referred to such texts as *Electrical and Electronic Measurement and Testing* by W. Bolton (Longman 1992), *Principles of Electronic Instrumentation and Measurement* by H.M. Berlin and F.C. Getz (Merrill 1988) or *Electronic Instruments and Measurement Techniques* by F.F. Mazda (Cambridge University Press 1987).

4.2.4 Monitors

Output data is increasingly being presented using a television type of display, termed a monitor. This builds up a picture on a cathode-ray tube screen by moving the spot formed by the electron beam in a series of horizontal scan lines, one after another down the screen. The image is built up by varying the intensity of the spot on the screen as each line is scanned. To reduce the effects of flicker two scans down the screen are used to trace a complete picture. On the first scan all the odd-numbered lines are traced out and on the second the even-numbered lines are traced. This technique is called interlaced scanning. A text character or a diagram is produced on the screen by selectively lighting dots. Thus, for example, the display might show a diagrammatic representation of a plant and values of quantities being measured numerically indicated at the relevent locations on the diagram.

4.2.5 Magnetic media

The *magnetic tape recorder* can be used to record both analogue and digital signals. It consists of a recording head which responds to the input signal and produces corresponding magnetic patterns on magnetic tape, a replay head to give an output by converting the magnetic patterns on the tape to electrical signals, a tape transport system which moves the magnetic tape in a controlled way under the heads, and signal-conditioning elements such as amplifiers and filters.

The recording head consists of a core of ferromagnetic material which has a non-magnetic gap (figure 4.6). When electrical signals are fed to the coil which is wound round the core, magnetic flux is produced in the core. The proximity of the magnetic tape to the non-magnetic gap means that the magnetic flux readily follows a path through the core and that part of the magnetic tape in the region of the gap. The magnetic tape is a flexible plastic base coated with a ferromagnetic powder. When there is magnetic flux passing through a region of the tape it

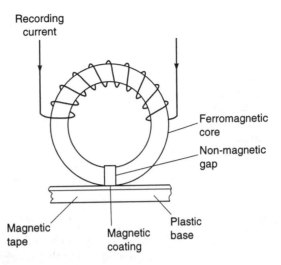

Fig. 4.6 Basis of magnetic recording head

becomes permanently magnetised. Hence a magnetic record is produced of the electrical input signal.

The replay head has a similar construction to that of the recording head. When a piece of magnetised tape bridges the non-magnetised gap then magnetic flux is induced in the core. Flux changes in the core induce e.m.f.s in the coil wound round the core. Thus the output from the coil is an electrical signal which is related to the magnetic record on the tape.

If the input signal to the recording head is sinusoidal with a frequency f then a sinusoidal variation in the magnetisation is produced along the tape. The time interval for one cycle is $1/f$ and so if the tape is moving through the head with a constant velocity v then the distance along the tape taken by one cycle is v/f. This distance is called the *recorded wavelength*. The minimum size this recorded wavelength can have is the gap width. At this wavelength the average magnetic flux across the gap is the average of one cycle and thus has a zero value. The size of the gap and the tape velocity thus determine the upper limit to the frequency response of the recorder. Typically tape velocities range between about 23 and 1500 mm/s with a gap width of 5 μm. This gives an upper frequency limit in the range of 4.6 to 300 kHz. Recorders generally have more than one recording head, the heads being spaced across the tape, with each producing a magnetisation track. Thus several different signals can be recorded simultaneously.

To illustrate the above, let us determine the maximum frequency that can be used for a magnetic tape recorder if it has a gap of width 5 μm and is used with a tape speed of 95.5 mm/s. The recorded wavelength is $v/f = 95.5 \times 10^{-3}/f$. The minimum recorded wavelength is the gap width. Thus $5 \times 10^{-6} = 95.5 \times 10^{-3}/f$ and so $f = 19.1$ kHz.

Digital recording involves the recording of signals as a coded combinations of bits. A commonly used method is the

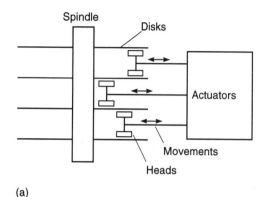

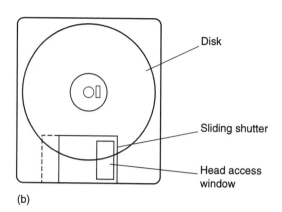

Fig. 4.7 Computer disks: (a) hard, (b) floppy

non-return-to-zero method. With this system the flux is recorded on the tape such that no change in flux represents 0 and a change in flux 1. Digital recording has the advantages over analogue recording of higher accuracy and relative insensitivity to tape speed.

Digital data can also be stored, in a similar way, on magnetic disks. These can be in the form of, so-termed, *hard disks* or *floppy disks*. Hard disks and floppy disks operate in much the same way. The digital data is stored on the disk surface along concentric circles called tracks, a single disk having many such tracks. A single read/write head is used for each disk surface and the heads are moved, by means of mechanical actuator, backwards and forwards to access different tracks. The disk is spun by the drive and the read/write heads read or write data into a track. Figure 4.7 shows the basic forms of hard disks and floppy disks. The 3½ inch floppy disk used in the personal computer has 135 tracks per inch and can store 1.4 Mbytes of data.

4.2.6 Digital printers

Analogue chart recorders give records in the form of a continuous trace. Digital printers give records in the form of numbers, letters or special characters. Such printers are known as *alphanumeric printers*. There are a number of versions of such printers, a commonly used one being the *dot-matrix printer*. With such a printer, the print head consists of either 9 or 24 pins in a vertical line. Each pin is controlled by an electromagnet which when turned on propels the pin onto the inking ribbon. This impacts a small blob of ink onto the paper behind the ribbon. A character is formed by moving the print head across the paper and firing the appropriate pins.

4.2.7 Data loggers

A *data logger* consists essentially of a multiplexer, a sample and hold element and an analogue-to-digital converter (figure 4.8).

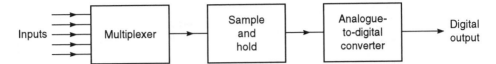

Fig. 4.8 Principle of data logger

Such a unit can monitor the inputs from a large number of sensors. Inputs from individual sensors, after suitable signal conditioning, are fed into the multiplexer. The mutiplexer is used to select one signal which is then fed to the sample and hold element. This holds the signal long enough for the analogue-to-digital conversion to occur. The output from the unit is thus a digital signal for one of the sensors. The multiplexer can be switched to each sensor in turn and so the output consists of a sequence of samples from each sensor. Scanning of the inputs can be selected to be just a sampling of a single channel, a single scan of all channels, a continuous scan of all channels, and periodic scan of all channels, say every 1, 5, 15, 30 or 60 minutes. The output from the system might be displayed on a digital meter that indicates the output and channel number, or be used to give a permanent record with a digital printer, a magnetic tape recorder or a magnetic disk memory.

Typically a data logger may handle 20 to 100 inputs, though some may handle considerably more, perhaps 1000. It might have a sample and conversion time of 10 µs and be used to make perhaps 1000 readings per second. The accuracy is typically about 0.01% of full scale input and linearity is about ± 0.005% of full scale input. Cross-talk is typically 0.01% of full scale input on any one input. The term *cross-talk* is used to describe the interference that can occur when one sensor is being sampled as a result of signals from other sensors.

4.3 Indicators

Many display systems use light indicators to indicate on–off status or give alphanumeric displays. Such light indicators might be neon lamps, incandescent lamps, liquid crystal displays (LCDs) or light-emitting diodes (LEDs). Neon lamps need high voltages and low currents and can be powered directly from the mains voltage but can only be used to give a red light. Incandescent lamps can be used with a wide range of voltages but need a comparatively high current. They emit white light so can be used with appropriate lenses to generate any required colour. Their main advantage is their brightness. Liquid crystal displays do not produce any light of their own by rely on reflected light. They require an alternating voltage of about 12 to 30 V but no current. Light-emitting diodes require low voltages and low currents and are cheap. The most

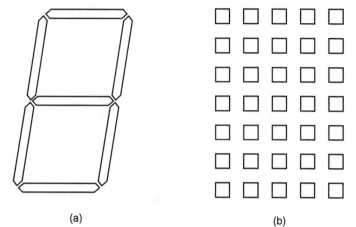

Fig. 4.9 Displays: (a) 7-segment, (b) 5 × 7 dot matrix

(a)   (b)

commonly used LEDs can give red, yellow or green colours. With microprocessor-based systems, LEDs are the most common form of indicator used.

LEDs are often used to give numeric and alphabetic displays. Two basic types, segmented and dot matrix, are used. The 7-segment display (figure 4.9(a)) is a common form. By illuminating different segments of the display the full range of numbers and a small range of alphabetical characters can be formed. The 5 × 7 dot matrix display enables a full range of numbers and alphabetical characters to be produced.

4.3.1 Alarm indicators

A wide variety of alarm systems are used with measurement and control systems. Commonly met ones are:

1 Temperature alarms which respond when the temperature reaches a particular value or falls to some other value. These may be based on the use of a resistance element or thermocouple to sense the temperature.
2 Current alarms which respond when the current reaches a particular value or falls below some other value.
3 Voltage alarms which respond when the voltage reaches a particular value or falls below some other value.
4 Weight alarms which respond when the weight in a container reaches a particular value or falls below some other value. These generally use load cells with electrical resistance strain gauges.

Alarm indicators take an analogue input from some sensor, possibly via a signal conditioner, and turn it into an on–off signal for some indicator. Figure 4.10 shows the basic form of alarm systems. The input is compared with the alarm set point. The

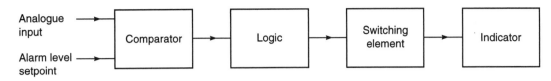

Fig. 4.10 An alarm system

comparator takes two inputs and gives an output when, for example, input A is greater than input B. Thus when the set point is exceeded a logic 0 or 1 signal passes to the logic unit which then gives an output which triggers the switching unit and switches on, or off, an indicator. The indicator can take a variety of forms, e.g. a bell, a horn, a klaxon, a coloured light, a flashing light, a backlighted display (the light comes on behind a message on a screen).

4.4 Measurement systems

The following examples illustrate the points involved in the design of measurement systems for particular applications.

Example 1

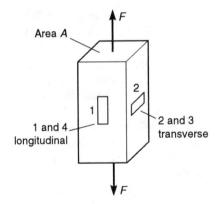

Fig. 4.11 Example

A link-type load cell, of the form shown in figure 4.11, has four strain gauges attached to its surface. Two of the gauges are in the longitudinal axis direction and two in a transverse direction. When the link is subject to tensile forces, the axial gauges will be in tension and the transverse gauges in compression. Design a complete measurement system which could be used to monitor the forces acting on the load cell. The sensitivity required is an output of about 30 mV when the stress applied to the link is 500 MPa. The strain gauges may be assumed to have gauge factors of 2.0 and resistances of 100 Ω. Use tables to find any other values required.

When a load F is applied to the link then, since the elastic modulus E is stress/strain and stress is force per unit area, the longitudinal axis strain ε_1 is F/AE and the transverse strain ε_t is $-\nu F/AE$, where A is the cross-sectional area and ν is Poisson's ratio for the link material. The responses of the strain gauges (see section 2.3.1) to these strains are

$$\frac{\delta R_1}{R_1} = \frac{\delta R_4}{R_4} = G\varepsilon_1 = \frac{GF}{AE}$$

$$\frac{\delta R_3}{R_3} = \frac{\delta R_2}{R_2} = G\varepsilon_t = -\frac{\nu GF}{AE}$$

The output voltage from the Wheatstone bridge (see the example given in section 3.5.1 for a load cell) is given by

$$V_{\circ} = \frac{V_s R_1 R_4}{(R_1 + R_2)(R_3 + R_4)} \left(\frac{\delta R_1}{R_1} - \frac{\delta R_2}{R_2} - \frac{\delta R_3}{R_3} + \frac{\delta R_4}{R_4} \right)$$

With $R_1 = R_2 = R_3 = R_4 = R$, and with $\delta R_1 = \delta R_4$ and $\delta R_2 = \delta R_3$, then

$$V_{\circ} = \frac{V_s}{2R}(\delta R_1 - \delta R_2) = \frac{V_s GF}{2AE}(1 + v)$$

Suppose we consider steel for the link. Then tables give E as about 210 GPa and v about 0.30. Thus with a stress ($= F/A$) of 500 MPa we have, for strain gauges with a gauge factor of 2.0,

$$V_{\circ} = 3.09 \times 10^{-3} V_s$$

For a bridge voltage of 10 V this would be an output voltage of 30.9 mV. Thus no amplification is required. The output can be displayed on a high-resistance voltmeter, high resistance to avoid loading problems. A digital voltmeter might thus be suitable.

Example 2

A measurement system is required which will set of an alarm when the temperature of a liquid rises above 40°C. The liquid is normally at 30°C. The output from the system must be a 1 V signal to operate the alarm. Suggest a possible measurement system.

  Since the output is to be electrical and a reasonable speed of response is likely to be required, an obvious possibility is an electrical resistance element. To generate a voltage output the resistance element could be used with a Wheatstone bridge. The output voltage will probably be less than 1 V for a change from 30°C to 40°C but a differential amplifier could be used to enable the required voltage to be obtained.

  Suppose a nickel element is used. Nickel has a temperature coefficient of resistance of 0.0067 K^{-1}. Thus if the resistance element is taken as being 100 Ω at 0°C then its resistance at 30°C will be

$$R_{30} = R_0(1 + \alpha t) = 100(1 + 0.0067 \times 30) = 120.1 \ \Omega$$

and at 40°C

$$R_{40} = 100(1 + 0.0067 \times 40) = 126.8 \ \Omega$$

Thus there is a change in resistance of 6.7 Ω. If this element forms one arm of a Wheatstone bridge which is balanced at 30°C, then the output voltage $V_{\circ}$ is given by (see section 3.5)

$$\delta V_{\circ} = \frac{V_s \delta R_1}{R_1 + R_2}$$

With the bridge balanced at 30°C and, say, all the arms the same value and a supply voltage of 4 V, then

$$\delta V_o = \frac{4 \times 6.7}{126.8 + 120.1} = 0.109 \text{ V}$$

To amplify this to 1 V we can use a differential amplifier (see section 3.2.5)

$$V_o = \frac{R_2}{R_1}(V_2 - V_1)$$

$$1 = \frac{R_2}{R_1} \times 0.109$$

Hence $R_2/R_1 = 9.17$ and so if we use an input resistance of 1 kΩ the feedback resistance must be 9.17 kΩ.

Example 3

A potentiometer is to be used to monitor the angular position of a pulley wheel. Suggest the items that might be needed to enable there to be an output to a recorder of 10 mV per degree if the potentiometer has a full scale angular rotation of 320°.

When the supply voltage V_s is connected across the potenti-ometer we will need to safeguard it and the wiring against possible high currents and so a resistance R_s can be put in series with the potentiometer R_p. The total voltage drop across the potentiometer is thus $V_s R_p/(R_s + R_p)$. For an angle θ with a potentiometer having a full scale angular deflection of θ_F we will obtain an output from the potentiometer of

$$V_\theta = \frac{\theta}{\theta_F} \frac{V_s R_p}{R_s + R_p}$$

Suppose we consider a potentiometer with a resistance of 4 kΩ and let R_s be 2 kΩ. Then for 1 mV per degree we have

$$0.01 = \frac{1}{320} \frac{4 V_s}{4 + 2}$$

Hence we would need a supply voltage of 4.8 V.

To prevent loading of the potentiometer by the resistance of the recorder, a voltage follower circuit can be used. Thus the circuit might be of the form shown in figure 4.12.

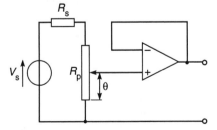

Fig. 4.12 Example

4.5 Testing and calibration

Testing a measurement system installation falls into three stages:

1 *Pre-installation testing* This is the testing of each instrument for correct calibration and operation prior to it being installed.
2 *Piping and cabling testing* In the case of pneumatic lines this involves, prior to the connection of the instruments, blowing through with clear, dry, air prior to connection and pressure testing to ensure they are leak free. With process piping, all the piping should be flushed through and tested prior to the connection of instruments. With instrument cables, all should be checked for continuity and insulation resistance prior to the connection of any instruments.
3 *Precommissioning* This involves testing that the installation is complete, all instrument components are in full operational order when interconnected and all control room panels or displays function.

Calibration consists of comparing the output of a measurement system and its subsystems against standards of known accuracy. The standards may be other instruments which are kept specially for calibration duties or some means of defining standard values. In many companies some instruments and items such as standard resistors and cells are kept in a company standards department and used solely for calibration purposes. The relationship between the calibration of an instrument in everyday use and national standards is likely to be:

1 National standards are used to calibrate standards for calibration centres.
2 Calibration centre standards are used to calibrate standards for instrument manufacturers.
3 Standardised instruments from instrument manufacturers are used to provide in-company standards.
4 In-company standards are used to calibrate process instruments.

There is a simple traceability chain from the instrument used in a process back to national standards. For a more detailed discussion of calibration the reader is referred to *Measurement and Calibration for Quality Assurance* by A.S. Morris (Prentice Hall 1991).

The following are some examples of calibration procedures that might be used in-company:

1 *Voltmeters* These can be checked against standard voltmeters or standard cells giving standard e.m.f.s.
2 *Ammeters* These can be checked against standard ammeters.
3 *Gauge factor of strain gauges* This can be checked by taking

a sample of gauges from a batch and applying measured strains to them when mounted on some test piece. The resistance changes can be measured and hence the gauge factor computed.

4   *Wheatstone bridge circuits*   The output from a Wheatstone bridge can be checked when a standard resistance is introduced into one of the arms.

5   *Load cells*   For low-capacity load cells, dead-weight loads using standard weights can be used.

6   *Pressure sensors*   Pressure sensors can be calibrated by using a dead-weight tester (figure 4.13). The calibration pressures are generated by adding standard weights W to the piston tray. After the weights are placed on the tray, a screw-driven plunger is forced into the hydraulic oil in the chamber to lift the piston-weight assembly. The calibration pressure is then W/A, where A is the cross-sectional area of the piston. Alternatively the dead-weight tester can be used to calibrate a pressure gauge and this gauge can be used for the calibration of other gauges.

7   *Temperature sensors*   These can be calibrated by immersion in a melt of a pure metal or water. The temperature of the substance is then slowly reduced and a temperature–time record obtained. When the substance changes state from liquid to solid, the temperature remains constant. Its value can be looked up from tables and hence an accurate reference temperature for calibration obtained. Alternatively, the temperature at which a liquid boils can be used. However, the boiling point depends on the atmospheric pressure and corrections have to be applied if it differs from the standard atmospheric pressure. Alternatively, in-company the readings given by the measurement system can be compared with those of a standard thermometer.

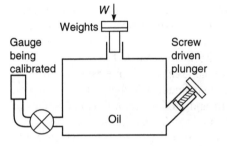

Fig. 4.13 Dead-weight calibration for pressure gauges

Problems

1   Explain the significance of the following terms taken from the specifications of display systems:
  (a) Closed-loop servo recorder: dead band ±0.2% of span.
  (b) Magnetic tape recorder: tape speed can be varied from 1524 mm/s to 23.8 mm/s in 6 calibrated steps, when signal frequency band varies from 300–300 000 to100–4650 Hz.
  (c) Data logger: number of inputs 100, cross-talk on any one input 0.01% of full scale input.
  (d) Double-beam oscilloscope: vertical deflection with two identical channels, bandwidth d.c. to 15 MHz, deflection factor of 10 mV/div to 20 V/div in 11 calibrated steps, time base of 0.5 μs/div to 0.5 s/div in 19 calibrated steps.

2   Explain the problems of loading when a measurement system is being assembled from a sensor, signal conditioner and display.

3   Suggest a display unit that could be used to give:

(a) A permanent record of the output from a thermocouple.

(b) A display which enables the oil pressure in a system to be observed.

(c) A record to be kept of the digital output from a micro-processor.

(d) The transient voltages resulting from monitoring of the loads on an aircraft during simulated wind turbulence.

4   A cylindrical load cell, of the form shown in figure 4.7, has four strain gauges attached to its surface. Two of the gauges are in the circumferential direction and two in the longitudinal axis direction. When the cylinder is subject to a compressive load, the axial gauges will be in compression while the circumferential ones will be in tension. If the material of the cylinder has a cross-sectional area A and an elastic modulus E, then a force F acting on the cylinder will give a strain acting on the axial gauges of $-F/AE$ and on the circumferential gauges of $+vF/AE$, where v is Poisson's ratio for the material. Design a complete measurement system, using load cells, which could be used to monitor the mass of water in a tank. The tank itself has a mass of 20 kg and the water when at the required level 40 kg. The mass is to be monitored to an accuracy of ±0.5 kg. The strain gauges have a gauge factor of 2.1 and are all of the same resistance of 120.0 Ω. For all other items, specify what your design requires. If you use mild steel for the load cell material, then the tensile modulus may be taken as 210 GPa and Poisson's ratio 0.30.

5   Design a complete measurement system involving the use of a thermocouple to determine the temperature of the water in a boiler and give a visual indication on a meter. The temperature will be in the range 0 to 100°C and is required to an accuracy of ±1 % of full scale reading. Specify the materials to be used for the thermocouple and all other items necessary. In advocating your design you must consider the problems of cold junction and non-linearity. You will probably need to consult thermocouple tables. The following data is taken from such tables, the cold junction being at 0°C, and may be used as a guide.

Materials	e.m.f. in mV at				
	20°C	40°C	60°C	80°C	100°C
Copper−constantan	0.789	1.611	2.467	3.357	4.277
Chromel−constantan	1.192	2.419	3.683	4.983	6.317
Iron−constantan	1.019	2.058	3.115	4.186	5.268
Chromel−alumel	0.798	1.611	2.436	3.266	4.095
Platinum−10% Rh, Pt	0.113	0.235	0.365	0.502	0.645

6   Design a measurement system which could be used to monitor the temperatures, of the order of 100°C, in positions scattered over a number of points in a plant and present the results on a control panel.

7   A suggested design for the measurement of liquid level in a vessel involves a float which in its vertical motion bends a cantilever. The degree of bending of the cantilever is then taken as a measure of the liquid level. When a force F is applied to the free end of a cantilever of length L, the strain on its surface a distance x from the clamped end is given by

$$\text{Strain} = \frac{6(L-x)}{wt^2E}$$

where w is the width of the cantilever, t its thickness and E the elastic modulus of the material. Strain gauges are to be used to monitor the bending of the cantilever with two strain gauges being attached longitudinally to the upper surface and two longitudinally to the lower surface. The gauges are then to be incorporated into a four-gauge Wheatstone bridge and the output voltage, after possible amplification, then taken as a measure of the liquid level. Determine the specifications required for the components of this system if there is to be an output of 10 mV per 10 cm change in level.

8   Design a static pressure measurement system based on a sensor involving a 40 mm diameter diaphragm across which there is to be a maximum pressure difference of 500 MPa. For a diaphragm where the central deflection y is much smaller than the thickness t of the diaphragm,

$$y \approx \frac{3r^2P(1-v^2)}{16Et^3}$$

where r is the radius of the diaphragm, P the pressure difference, E the modulus of elasticity and v Poisson's ratio. Explain how the deflection y will be converted into a signal that can be displayed on a meter.

9   Suggest the elements that might be considered for the measurement systems to be used to:
(a) Monitor the pressure in an air pressure line and present the result on a dial, no great accuracy being required.
(b) Continuously monitor and record the temperature of a room with an accuracy of ±1°C.
(c) Monitor the weight of lorries passing over a weighing platform.
(d) Monitor the angular speed of rotation of a shaft.

5 Pneumatic and hydraulic actuation systems

5.1 Actuation systems

Actuation systems are the elements of control systems which are responsible for transforming the output of a microprocessor or control system into a controlling action on a machine or device. Thus, for example, we might have an electrical output from the controller which has to be transformed into a linear motion to move a load. Another example might be where an electrical output from the controller has to be transformed into an action which controls the amount of liquid passing along a pipe.

In this chapter pneumatic and hydraulic actuation systems are discussed. In chapter 6 mechanical actuator systems are discussed and in chapter 7 electrical actuation systems. For a more detailed consideration of pneumatic and hydraulic systems the reader is referred to more specialist books such as *Hydraulics and Pneumatics* by A. Parr (Newnes 1991).

5.2 Pneumatic and hydraulic systems

Pneumatic signals are often used to control final control elements, even when the control system is otherwise electrical. This is because such signals can be used to actuate large valves and other high power control devices. The main drawback with pneumatic systems is, however, the compressibility of air. Hydraulic signals can be used for even higher power control devices but are more expensive than pneumatic systems and there are hazards associated with oil leaks which do not occur with air leaks.

Generally the signals required by a pneumatic control device are in the region of 20 to 100 kPa gauge pressure, i.e. pressure above the atmospheric pressure. Figure 5.1 shows the principle of a current to pressure converter that can be used to convert a current output of say 4 to 20 mA from the controller to a pneumatic pressure signal of 20 to 100 kPa to operate an actuator. The input current passes through coils mounted on a core which is then attracted towards a magnet, the extent of the attraction depending on the size of the current. The movement of the core causes movement of the lever about its pivot and so the movement of a flapper

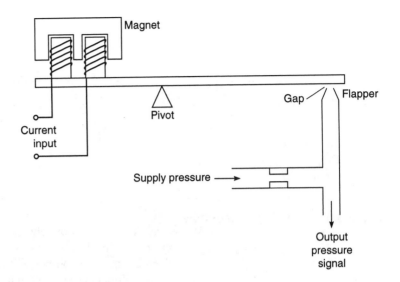

Fig. 5.1 Current to pressure converter

above a nozzle. The position of the flapper in relation to the nozzle determines the size of the air pressure in the system.

5.2.1 Power supplies

With a hydraulic system a source of pressurised oil is required. Typically this is provided by a pump driven by an electric motor. The pump pumps oil from a sump through a non-return valve and an accumulator to the system, from which it returns to the sump. Figure 5.2 illustrates the arrangement. A pressure relief valve is included, this being to release the pressure if it rises above a safe

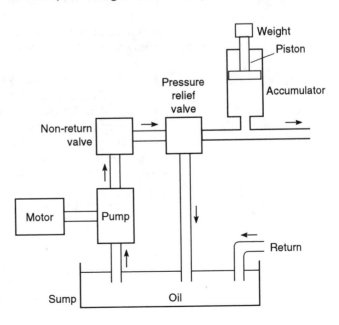

Fig. 5.2 Hydraulic power supply

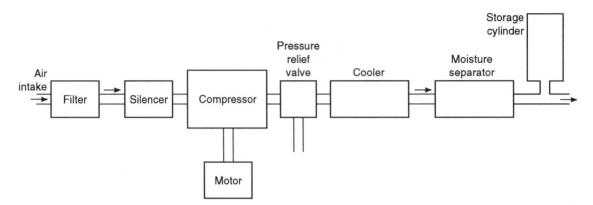

Fig. 5.3 Pneumatic power supply

level. The non-return valve is to prevent the oil being back driven to the pump. The accumulator is to smooth out any short-term fluctuations in the output oil pressure. Essentially the accumulator is just a container in which the oil is held under pressure against an external force, figure 5.2 showing one form involving a piston. If the oil pressure rises then the piston moves to increase the volume the oil can occupy and so reduces the pressure. If the oil pressure falls then the piston moves in to reduce the volume occupied by the oil and so increase its pressure.

With a pneumatic power supply (figure 5.3) an electric motor drives an air compressor. The air inlet to the compressor is likely to be filtered and via a silencer to reduce the noise level. A pressure relief valve provides protection against the pressure in the system rising above a safe level. Since the air compressor increases the temperature of the air a cooling system is likely to follow. Also there is likely to be a moisture separator to remove moisture from the air. A storage cylinder acts in a similar manner to the accumulator in the hydraulic system and smoothes out any short-term pressure fluctuations.

5.3 Control valves

Pneumatic and hydraulic systems use control valves to direct and regulate the flow of fluid through a system. Valves can be considered to be either infinite position valves or finite position valves. *Infinite position valves* can take up any position between the fully open and fully closed positions. *Finite position valves* are either completely open or completely closed, i.e. they are on/off devices. The connections to the valves are through, what are termed, *ports*.

Figure 5.4 shows a basic, finite position, control valve system with four ports. The load is connected to the ports A and B, the pressure supply from the pump or compressor to port P and, in the case of a hydraulic valve, the fluid is returned to the tank through port T. In the case of a pneumatic system valve, the return air would be vented from this port. In figure 5.4(a), the pressure supply through port P is connected to port B and A to the vent

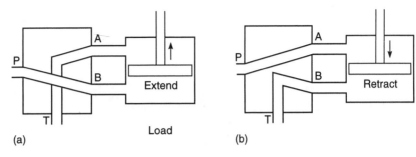

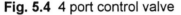

Fig. 5.4 4 port control valve

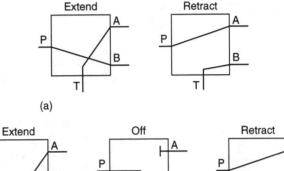

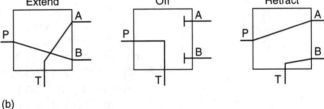

Fig. 5.5 Valves: (a) 4/2, (b) 4/3

port T. In figure 5.4(b), the pressure supply through port P is connected to port A and B to the vent port T. The control of such a valve might be just that the load ram is driven to one end or other of its stroke. The valve is then said to have two control *positions*. However, we could have a three-position valve with the control being such that the ram is driven to one end or other of its stroke or be in an intermediate off position.

Finite position valves are specified in terms of the number of ports and number of positions they have. Thus a 4/2 valve is one with four ports and two positions. A 4/3 valve is one with four ports and three positions. Figure 5.5 indicates the possible valve actions for these valves. With the 4/3 valve in the off position, ports P and T are connected and ports A and B are blocked. Other possibilities could be for the off position to have all the ports blocked or A, B and T connected.

5.3.1 Valve symbols

The symbol used for control valves consists of a square for each of its switching positions. Thus a two-position valve will have two

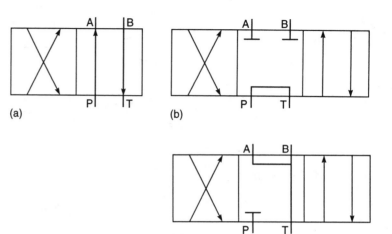

Fig. 5.6 Valve connections: (a) 4/2, (b) 4/3 with P and T connected in off position, (c) 4/3 with A, B and T connected in off position

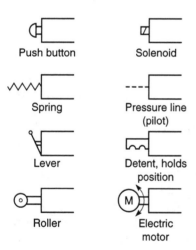

Fig. 5.7 Control methods

squares, a three-position valve three squares. Arrow-headed lines are used to indicate the directions of flow in each of the positions. The ports are labelled P for pressure supply, T normally for hydraulic return port, R and S normally for pneumatic systems for pneumatic exhaust ports. Figure 5.6 shows examples of such symbols. Figure 5.6(a) shows the 4/2 valve described in figure 5.5(a). The first square indicates one position for the valve, the ram extended state. The second square indicates the other position for the valve, the retracted state. Figure 5.6(b) shows the 4/3 valve described in figure 5.5(b) with A and B closed and P connected to T. Figure 5.6(c) shows the 4/3 valve with A, B and T connected and P closed.

Figure 5.7 shows some of the symbols which are used to indicate the various ways the valves can be operated. More than one of these symbols might be used with the valve symbol. Thus figure 5.8(a) shows how, with a 4/2 valve, a push-button can be depressed to cause the ram to extend. The push-button movement gives the state indicated by the symbols used in the square to which it is attached. When the push-button is released, the spring pushes the valve back to its initial position and the ram retracts. The spring movement gives the state indicated by the symbols used in the square to which it is attached. Figure 5.8(b) shows a solenoid-operated version of the same valve. When the current through the solenoid is switched on, the ram extends. When the

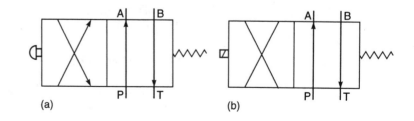

Fig. 5.8 Valves with controls indicated

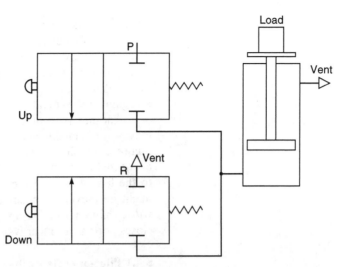

Fig. 5.9 Pneumatic lift system

current ceases, the spring pushes the valve back to its original position and the ram retracts.

Figure 5.9 shows a simple example of an application of valves in a pneumatic lift system. Two push-button 2/2 valves are used. When the button on the up valve is pressed, the load is lifted. When the button on the down valve is pressed, the load is lowered. Note that with pneumatic systems an open arrow is used to indicate a vent to the atmosphere.

5.4 Types of valves

This section is just a brief indication of types of control valves, only simple forms of each type being discussed.

Figure 5.10 shows the basic form of a simple 2/2 *poppet*

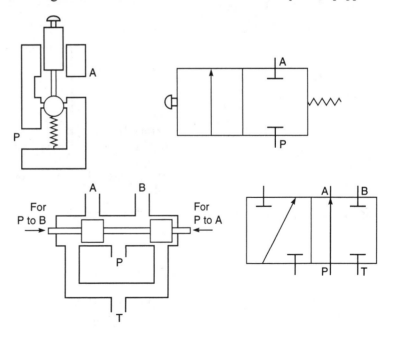

Fig. 5.10 2/2 poppet valve

Fig. 5.11 4/2 shuttle valve

valve which is normally in the closed condition. In poppet valves, balls, discs or cones are used in conjunction with valve seats to control the flow. In the figure a ball is shown. When the push-button is depressed, the ball is pushed out of its seat and flow occurs. When the button is released, the spring forces the ball back up against its seat and so closes off the flow.

Another type of control valve is the *spool valve*. A spool moves horizontally within the valve body to control the flow. Figure 5.11 shows the form for a 4/2 valve. As shown, B is connected to P with A and T are closed. When the spool is moved to the left, A is connected to P with B and T closed. The spool might be moved by push-button, lever or solenoid. *Rotary spool valves* have a rotating spool which, when it rotates, opens and closes ports in a similar way.

5.4.1 Pilot-operated valves

The force required to move the ball or shuttle in a valve can often be too large for manual or solenoid operation. To overcome this problem a *pilot-operated system* is used where one valve is used to control a second valve. Figure 5.12 illustrates this. The pilot valve is small capacity and can be operated manually or by a solenoid. It is used to allow the main valve to be operated by the system pressure. Pilot ports on the main valve are denoted by the letters Z, Y, X, etc. The pilot pressure line is indicated by dashes. The pilot and main valves can be operated by two separate valves but they are often combined in a single housing.

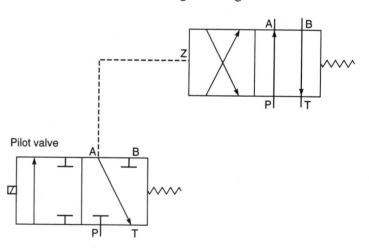

Fig. 5.12 Pilot-operated system

5.4.2 Directional valves

Figure 5.13 shows a simple *directional valve* and its symbol. Free flow can only occur in one direction through the valve, that which

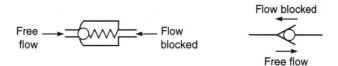

Fig. 5.13 Directional valve

results in the ball being pressed against the spring. Flow in the other direction is blocked by the spring forcing the ball against its seat.

5.4.3 Pressure relief valves

Figure 5.14 shows a *pressure relief valve* which has one orifice which is normally closed. When the inlet pressure overcomes the force exerted by the spring, the valve opens and permits flow to occur. This can be used as just a pressure relief valve to safeguard a system against excessive pressures. However, it can be used as a *sequential valve* and allow flow to occur to some part of the system when the pressure has risen to the required level. Figure 5.15 shows a system where such a sequential valve is used. When the 4/3 valve first operates, the ram in cylinder 1 moves. While this is happening the pressure is too low to operate the sequence valve. When the ram of cylinder 1 reaches the end stop, then the pressure in the system rises and, at an appropriate level, triggers the sequence valve to open and so start the ram in cylinder 2 in motion.

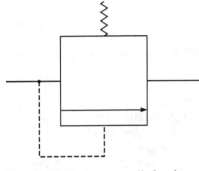

Fig. 5.14 Pressure relief valve

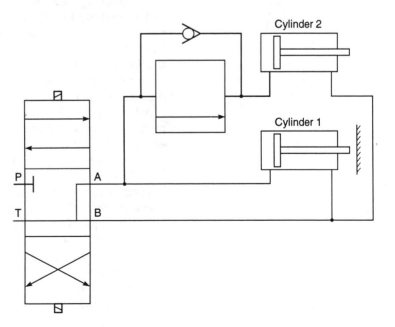

Fig. 5.15 Sequential system

5.5 Actuators

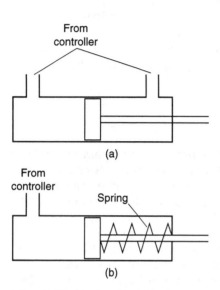

From
controller

(a)

From
controller

Spring

(b)

Fig. 5.16 Hydraulic cylinders:
(a) double acting, (b) single acting
with spring return

The *hydraulic* or *pneumatic cylinder* is an example of a linear actuator. The principles and form are the same for both hydraulic and pneumatic versions, differences being purely a matter of size as a consequence of the higher pressures used with hydraulics. The hydraulic cylinder consists of a hollow cylindrical tube along which a piston/ram can slide. There are a number of types, figure 5.16 showing two common types. The term *double acting* is used when the control pressures are applied to each side of the piston. A difference in pressure between the two sides then results in motion of the piston, the piston being able to move in either direction along the cylinder. The term *single acting* is used when the control pressure is applied to just one side of the piston, a spring often being used to provide the opposition to the movement of the piston. The piston is only moved in one direction along the cylinder by the signal from the controller. The linear motion produced by the hydraulic cylinder is often used to operate a spool valve.

The choice of cylinder is determined by the force required to move the load and the speed required. Hydraulic cylinders are capable of much larger forces than pneumatic cylinders. However, pneumatic cylinders are capable of greater speeds. The force produced by a cylinder is equal to the cross-sectional area of the cylinder multiplied by the working pressure in the cylinder. A cylinder for use with a working pneumatic pressure of 500 kPa and having a diameter of 50 mm will thus give a force of 982 N. A hydraulic cylinder with the same diameter and a working pressure of 15 000 kPa will give a force of 29.5 kN.

If the flow rate of hydraulic liquid into a cylinder is a volume of Q per second, then the volume swept out by the piston in a time of 1 second must be Q. But for a piston of cross-sectional area A this is a movement through a distance of v in 1 second, where we have $Q = Av$. Thus the speed v of a hydraulic cylinder is equal to the flow rate of liquid Q through the cylinder divided by the cross-sectional area A of the cylinder. Thus for a hydraulic cylinder of diameter 50 mm and a hydraulic fluid flow of 7.5×10^{-3} m³/s the speed is 3.8 m/s. The speed of a pneumatic cylinder cannot be calculated in this way since its speed depends on the rate at which air can be vented ahead of the advancing piston. A valve to adjust this can be used to regulate the speed.

To illustrate the above consider the problem of a hydraulic cylinder to be used to move a workpiece in a manufacturing operation through a distance of 250 mm in 15 s. If a force of 50 kN is required to move the workpiece, what is the required working pressure and hydraulic liquid flow rate if a cylinder with a piston diameter of 150 mm is available? The cross-sectional area of the piston is $\tfrac{1}{4}\pi \times 0.150^2 = 0.0177$ m². The force produced by the cylinder is equal to the product of the cross-sectional area of the cylinder and the working pressure. Thus the **working pressure** is $50 \times 10^3/0.0177 = .28$ MPa. The speed of a hydraulic cylinder is equal to the flow rate of liquid through the cylinder divided by the cross-sectional area of the cylinder. Thus the required flow rate is $(0.250/15) \times 0.0177 = 2.95 \times 10^{-4}$ m^3/s.

5.5.1 Rotary actuator

Fluid pressure can also be used to produce rotary motion. Figure 5.17(a) shows a simplified diagram of one form, a *vane type rotary actuator*. Fluid pressure at the inlet causes the vane to rotate. By using spring-loaded vanes, as illustrated in figure 5.7(b), a continuous rotation can be produced and a *hydraulic/pneumatic motor* produced.

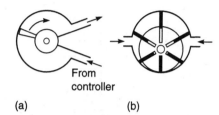

5.5.2 Diaphragm actuator

A common form of pneumatic actuator is that associated with process control valves, the *diaphragm actuator*. Essentially it consists of a diaphragm with the input pressure signal from the controller on one side and atmospheric pressure on the other, this difference in pressure being termed the *gauge pressure*. The diaphragm is made of rubber which is sandwiched in its centre between two circular steel discs. The effect of changes in the input pressure is thus to move the central part of the diaphragm, as illustrated in figure 5.18. This movement is communicated to the final control element by a shaft which is attached to the diaphragm. The force F acting on the shaft is the force that is acting on the diaphragm and is thus the gauge pressure P multiplied by the diaphragm area A. A restoring force is provided by a spring. Thus if the shaft moves through a distance x, and assuming the compression of the spring is proportional to the force, i.e. $F = kx$ with

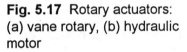

Fig. 5.17 Rotary actuators: (a) vane rotary, (b) hydraulic motor

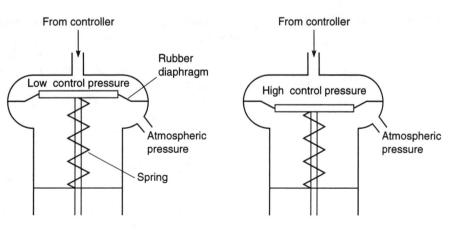

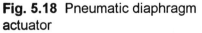

Fig. 5.18 Pneumatic diaphragm actuator

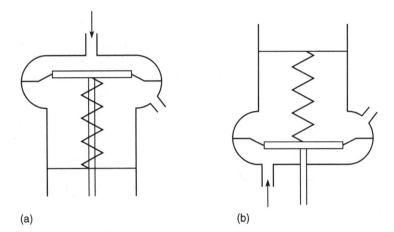

Fig. 5.19 Actuators: (a) direct, (b) reverse

(a)   (b)

k being a constant, then $kx = PA$ and thus the displacement of the shaft is proportional to the gauge pressure.

With the spring in the position shown in figure 5.18 (figure 5.19(a)) the actuator is referred to as a *direct actuator*. By mounting the spring on the other side of the diaphragm to the shaft a *reverse actuator* is produced (figure 5.19(b)).

To illustrate the above consider the problem of a diaphragm actuator to be used to open a control valve if a force of 500 N must be applied to the valve. What diaphragm area is required for a control gauge pressure of 100 kPa? The force F applied to the diaphragm of area A by a pressure P is given by $P = F/A$. Hence $A = 500/(100 \times 10^3) = 0.005$ m^2.

This form of actuator is often referred to as a *linear actuator* since the signal from the controller is converted into linear motion. This linear motion is generally used to operate a process control valve.

5.6 Process control valves

Pneumatic process control valves are frequently the final control element actuated by the movement of actuators. Figure 5.20 shows a cross-section of a valve for the control of rate of flow of a fluid. The pressure change in the actuator causes the diaphragm to move and so consequently the valve stem. The result of this is a movement of the inner-valve plug within the valve body. The plug restricts the fluid flow and so its position determines the flow rate. The figure shows a direct actuator being used; another form can operate with a reverse actuator.

There are many forms of valve body and plug. Figure 5.21 shows some forms of valve bodies. The term *single seated* is used for a valve where there is just one path for the fluid through the valve and so just one plug is needed to control the flow. The term *double seated* is used for a valve where the fluid on entering the valve splits into two streams, each stream passing through an orifice controlled by a plug. There are thus two plugs with such a

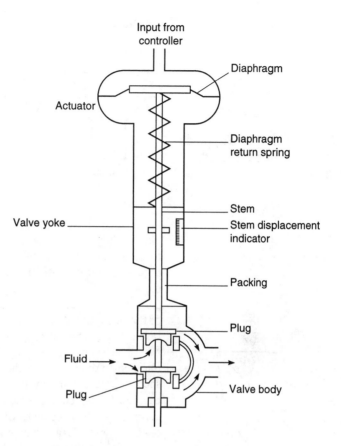

Input from controller

Diaphragm

Actuator

Diaphragm return spring

Stem

Valve yoke

Stem displacement indicator

Packing

Plug

Fluid

Plug

Valve body

Fig. 5.20 Diaphragm-operated control valve

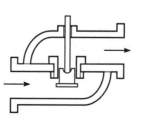
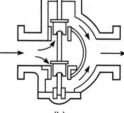
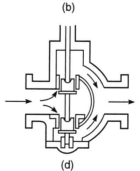

(a)

(b)

(c)

(d)

Fig. 5.21 Valve bodies:
(a) single-seated normally open,
(b) double-seated normally open,
(c) single-seated normally closed,
(d) double-seated normally closed

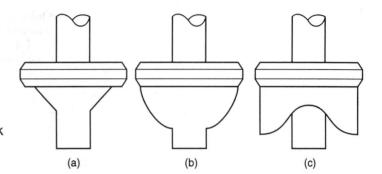

Fig. 5.22 Plug shapes: (a) quick opening, (b) linear contoured, (c) equal percentage

(a)   (b)   (c)

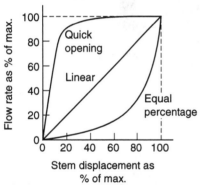

Fig. 5.23 Flow characteristics with different plugs

valve. A single-seated valve has the advantage that is can be closed more tightly than a double seated one but the disadvantage that the force on the plug due to the flow is much higher and so the diaphragm in the actuator has to exert considerably higher forces on the stem. This can result in problems in accurately positioning the plug. Double-seated valves have thus an advantage here. The form of the body also determines whether an increasing air pressure will result in the valve opening or closing.

The shape of the plug determines the relationship between the stem movement and the effect on the flow rate. Figure 5.22 shows three commonly used types and figure 5.23 how the percentage by which the volumetric rate of flow is related to the percentage displacement of the valve stem. With the *quick-opening* type a large change in flow rate occurs for a small movement of the valve stem. Such a plug is used where on/off control of flow rate is required.

With the *linear-contoured* type, the change in flow rate is proportional to the change in displacement of the valve stem, i.e.

Change in flow rate = k (change in stem displacement)

where k is a constant. If Q is the flow rate at a valve stem displacement S and Q_{max} is the maximum flow rate at the maximum stem displacement S_{max}, then we have

$$\frac{Q}{Q_{max}} = \frac{S}{S_{max}}$$

or percentage change in the flow rate equals the percentage change in the stem displacement.

To illustrate the above consider the problem of an actuator which has a stem movement at full travel of 30 mm. It is mounted on a linear plug valve which has a minimum flow rate of 0 and a maximum flow rate of 40 m³/s. What will be the flow rate when the stem movement is (a) 10 mm, (b) 20 mm? Since the percentage flow rate is the same as the percentage stem displacement, **then:**

(a) a percentage stem displacement of 33% gives a percentage flow rate of 33%, i.e. 13 m³/s; (b) a percentage stem displacement of 67% gives a percentage flow rate of 67%, i.e. 27 m³/s.

With the *equal percentage* type of plug, equal percentage changes in flow rate occur for equal changes in the valve stem position, i.e.

$$\frac{\Delta Q}{Q} = k\,\Delta S$$

where ΔQ is the change in flow rate at a flow rate of Q and ΔS the change in valve position resulting from this change. If we write this expression for small changes and then integrate it we obtain

$$\int_{Q_{min}} \frac{1}{Q}\,dQ = k \int_{S_{min}}^{S} dS$$

Hence

$$\ln Q - \ln Q_{min} = k(S - S_{min})$$

If we consider the flow rate Q_{max} which is given by S_{max} then

$$\ln Q_{max} - \ln Q_{min} = k(S_{max} - S_{min})$$

Eliminating k from these two equations gives

$$\frac{\ln Q - \ln Q_{min}}{\ln Q_{max} - \ln Q_{min}} = \frac{S - S_{min}}{S_{max} - S_{min}}$$

$$\ln \frac{Q}{Q_{min}} = \frac{S - S_{min}}{S_{max} - S_{min}} \ln \frac{Q_{max}}{Q_{min}}$$

and so

$$\frac{Q}{Q_{min}} = \left(\frac{Q_{max}}{Q_{min}}\right)^{(S-S_{min})/(S_{max}-S_{min})}$$

The term *rangeability R* is used for the ratio Q_{max}/Q_{min}.

To illustrate the above, consider the problem of an actuator which has a stem movement at full travel of 30 mm. It is mounted with a control valve having an equal percentage plug and which has a minimum flow rate of 2 m³/s and a maximum flow rate of 24 m³/s. What will be the flow rate when the stem movement is (a) 10 mm, (b) 20 mm? Using the equation

$$\frac{Q}{Q_{min}} = \left(\frac{Q_{max}}{Q_{min}}\right)^{(S-S_{min})/(S_{max}-S_{min})}$$

we have for (a) $Q = 2 \times (24/2)^{10/30} = 4.6$ m³/s and for (b) $Q = 2 \times (24/2)^{20/30} = 10.5$ m³/s

The relationship between the flow rate and the stem displacement is the inherent characteristic of a valve. It is only realised in practice if the pressure losses in the rest of the pipework, etc., are negligible compared with the pressure drop across the valve itself. If there are large pressure drops in the pipework so that, for example, less than half the pressure drop occurs across the valve then a linear characteristic might become almost a quick-opening characteristic. The linear characteristic is thus widely used when a linear response is required and most of the system pressure is dropped across the valve. The effect of large pressure drops in the pipework with an equal percentage valve is to make it more like a linear characteristic. For this reason, if a linear response is required when only a small proportion of the system pressure is dropped across the valve, then an equal percentage value might be used.

5.6.1 Control valve sizing

The term *control valve sizing* is used for the procedure of determining the correct size of valve body. The equation relating the rate of flow of liquid Q through a wide open valve to its size is

$$Q = A_V \sqrt{\frac{\Delta P}{\rho}}$$

where A_V is the valve flow coefficient, ΔP the pressure drop across the valve and ρ the density of the fluid. This equation is sometimes written, with the quantities in SI units, as

$$Q = 2.37 \times 10^{-5} C_V \sqrt{\frac{\Delta P}{\rho}}$$

where C_V is the valve flow coefficient. Alternatively it may be found written as

$$Q = 0.75 \times 10^{-6} C_V \sqrt{\frac{\Delta P}{G}}$$

where G is the specific gravity or relative density. These last two forms of the equation derive from its original specification in terms of US gallons. Table 5.1 shows some typical values of A_V, C_V and valve size.

To illustrate the above, consider the problem of determining the valve size for a valve that is required to control the flow of water when the maximum flow required is 0.012 m³/s and the permissible pressure drop across the valve at this flow rate is 300 kPa. Using the equation

Table 5.1 Flow coefficients and valve sizes

Flow coefficients	Valve size (mm)							
	480	640	800	960	1260	1600	1920	2560
C_V	8	14	22	30	50	75	110	200
$A_V \times 10^{-5}$	19	33	52	71	119	178	261	474

$$Q = A_V \sqrt{\frac{\Delta P}{\rho}}$$

then, since the density of water is 1000 kg/m³,

$$A_V = Q \sqrt{\frac{\rho}{\Delta P}} = 0.012 \sqrt{\frac{1000}{300 \times 10^3}} = 69.3 \times 10^{-5}$$

Thus, using table 5.1, the valve size is 960 mm.

5.7 Hydraulic amplifier

Figure 5.24 shows the combination of a *spool valve* with a hydraulic cylinder, the combination often being referred to as a

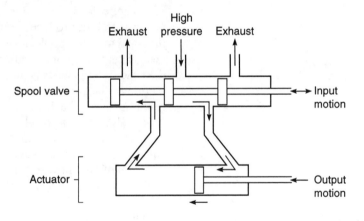

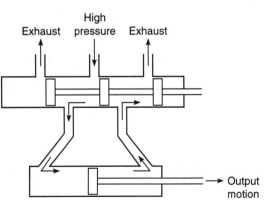

Fig. 5.24 Spool valve with hydraulic cylinder

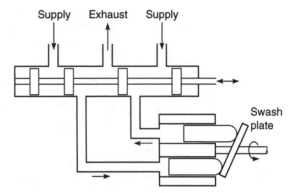

Fig. 5.25 Hydraulic motor

hydraulic amplifier. The spool valve has a constant high-pressure supply which is switched to either side of the actuator piston by the input motion of the spool valve shaft. The force acting on the cylinder piston will be the product of the cylinder piston area and the pressure difference between the high pressure and the exhaust. This force can be much larger than the force which moves the spool valve shaft. Such an arrangement might be used as the final control element to apply the high forces needed to move the workpiece for a machine tool.

Figure 5.25 shows the use of a spool valve with an axial piston motor to produce a *hydraulic motor*. The input motion to the shaft of the spool valve results in the high-pressure supply being alternately connected to first one piston of the motor and then the other. The movement of these pistons results in angular motion of the swash plate and hence rotation of the shaft. The net result is that a backwards-and-forwards motion applied to the spool valve shaft has resulted in angular motion of the motor shaft.

The spool valve can also operate as a *directional control valve*. Thus the spool valve part of figure 5.25 is controlling the direction of fluid flow to the motor, or any other device, as a result of the motion of the spool valve shaft.

Problems

1 Describe the basic details of (a) a poppet valve, (b) a shuttle valve.
2 Explain the principle of a pilot-operated valve.
3 Explain how a sequential valve can be used to initiate an operation only when another operation has been completed.
4 Draw the symbols for (a) a pressure relief valve, (b) a 2/2 valve which has actuators of a push-button and a spring, (c) a 4/2 valve, (d) a directional valve.
5 Explain how a diaphragm pneumatic actuator operates.
6 A force of 400 N is required to open a process control valve. What area of diaphragm will be needed with a diaphragm actuator to open the valve with a control gauge pressure of 70 kPa?

7 A pneumatic system is operated at a pressure of 1000 kPa. What diameter cylinder will be required to move a load requiring a force of 12 kN?

8 A hydraulic cylinder is to be used to move a workpiece in a manufacturing operation through a distance of 50 mm in 10 s. A force of 10 kN is required to move the workpiece. Determine the required working pressure and hydraulic liquid flow rate if a cylinder with a piston diameter of 100 mm is available.

9 An actuator has a stem movement which at full travel is 40 mm. It is mounted with a linear plug process control valve which has a minimum flow rate of 0 and a maximum flow rate of 0.20 m³/s. What will be the flow rate when the stem movement is (a) 10 mm, (b) 20 mm?

10 An actuator has a stem movement which at full travel is 40 mm. It is mounted on a process control valve with an equal percentage plug and which has a minimum flow rate of 0.2 m³/s and a maximum flow rate of 4.0 m³/s. What will be the flow rate when the stem movement is (a) 10 mm, (b) 20 mm?

11 What is the process control valve size required for a valve that is required to control the flow of water when the maximum flow required is 0.002 m³/s and the permissible pressure drop across the valve at this flow rate is 100 kPa? The density of water is 1000 kg/m³.

12 Explain how a spool valve can be used to produce a large force from a smaller force.

6 Mechanical actuation systems

6.1 Mechanical systems

This chapter is a consideration of what are termed *mechanisms*. Mechanisms are devices which can be considered to be motion converters in that they transform motion from one form to some required form. They might, for example, transform linear motion into rotational motion, or motion in one direction into a motion in a direction at right angles, or perhaps a linear reciprocating motion into rotary motion, as in the internal combustion engine where the reciprocating motion of the pistons are converted into rotation of the crank and hence drive shaft.

Mechanical elements can include the use of linkages, cams, gears, rack-and-pinion, chains, belt drives, etc. For example, the rack-and-pinion can be used to convert rotational motion to linear motion. Parallel shaft gears might be used to reduce a shaft speed. Bevel gears might be used for the transmission of rotary motion through 90°. A toothed belt or chain drive might be used to transform rotary motion in one plane to motion in another. Cams and linkages can be used to obtain motions which are prescribed to vary in a particular manner. This chapter is a consideration of the basic characteristics of a range of such mechanisms.

Note that a mechanism is principally concerned with transformation of motion while the term *machine* is used for a system that transmits or modifies the action of a force or torque to do useful work. A machine is thus defined as a system of elements which are arranged to transmit motion and energy from one form to some required form while a mechanism is defined as a system of elements which are arranged to transmit motion and energy from one form to some required form. A mechanism can therefore be thought of as a machine which is not required to transmit energy but merely to reproduce exactly the motions that take place in an actual machine.

The term *kinematics* is used for the study of motion without regard to forces. When we consider just the motions without any consideration of the forces or energy involved then we are carrying out a kinematic analysis of the mechanism. This chapter is an introduction to a consideration of kinematics and mechanisms. For more detail the reader is referred to general texts for mechanical engineers, such as *Mechanical Science* by W. Bolton (Blackwell Scientific Publications 1993), or more specialist texts on the principles of machines, such as *Design of Machinery* by R.L. Norton (McGraw-Hill 1992).

6.1.1 Types of motion

A rigid body can have a very complex motion which might seem difficult to describe. However, the motion of any rigid body can be considered to be a combination of translational and rotational motions. By considering the three dimensions of space, a translation motion can be considered to be a movement along one or more of the three axes. A rotation can be defined as a rotation about one or more of the axes.

For example, think of the motion required for you to pick up a pencil from a table. This might involve your hand moving at a particular angle towards the table, rotation of the hand, and then all the movement associated with opening your fingers and moving them to the required positions to grasp the pencil. However, we can break down all these motions into combinations of translational and rotational motions. Such an analysis is particularly relevant if we are not moving a human hand to pick up the pencil but instructing a robot to carry out the task. Then it really is necessary to break down the motion into combinations of translational and rotational motions. Among the sequence of control signals might be such groupings of signals as those to instruct joint 1 to rotate by 20° and link 2 to be extended by 4 mm for translational motion.

6.2 Kinematic chains

When we consider the movements of a mechanism without any reference to the forces involved, we can treat the mechanism as being composed of a series of individual links. Each part of a mechanism which has motion relative to some other part is termed a *link*. A link need not necessarily be a rigid body but it must be a resistant body which is capable of transmitting the required force with negligible deformation. For this reason is it usually taken as being represented by a rigid body which has two or more joints, which are points of attachment to other links. Each link is capable of moving relative to its neighbouring links. Levers, cranks, connecting rods and pistons, sliders, pulleys, belts and shafts are all examples of links. A sequence of joints and links is known as a *kinematic chain*. For a kinematic chain to transmit motion, one link must be fixed. Movement of one link will then produce predictable relative movements of the others. It is possible to

obtain from one kinematic chain a number of different mechanisms by having a different link as the fixed one.

As an illustration of a kinematic chain, consider a motor car
engine where the reciprocating motion of a piston is transformed
into rotational motion of a crankshaft on bearings mounted in a
fixed frame (figure 6.1(a)). We can represent this as being four
connected links (figure 6.1(b)). Link 1 is the crankshaft, link 2 the
connecting rod, link 3 the fixed frame and link 4 the slider, i.e.
piston, which moves relative to the fixed frame (see section 6.2.2
for further discussion).

The designs of many machines are based on two kinematic
chains, the four-bar chain and the slider–crank chain.

6.2.1 The four-bar chain

The four-bar chain consists of four links connected to give four
joints about which turning can occur. Figure 6.2 shows a number
of forms of the four-bar chain produced by altering the relative
lengths of the links. In figure 6.2(a), link 3 is fixed and the relative
lengths of the links are such that links 1 and 4 can oscillate but not
rotate. The result is a *double-lever mechanism*. By shortening link
4 relative to link 1, then link 4 can rotate (figure 6.2(b)) with link 1
oscillating and the result is termed a *lever–crank mechanism*. With
links 1 and 4 the same length and both able to rotate (figure
6.2(c)), then the result is a *double-crank mechanism*. By altering
which link is fixed, other forms of mechanism can be produced.

Figure 6.3 illustrates how such a mechanism can be used to

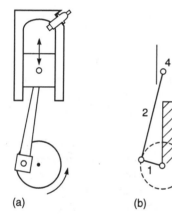

Fig. 6.1 Simple engine
mechanism

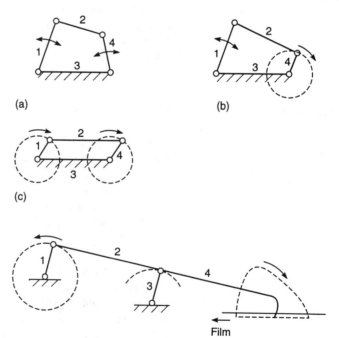

(a)

(b)

(c)

Fig. 6.2 Examples of four-bar
chains

Fig. 6.3 Camera film advance
mechanism

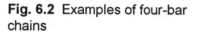

Film

advance the film in a cine camera. As link 1 rotates so the end of link 4 locks into a sprocket of the film, pulls it forward before releasing and moving up and back to lock into the next sprocket.

6.2.2 The slider–crank mechanism

This form of mechanism consists of a crank, a connecting rod and a slider and is the type of mechanism described in figure 6.1 which showed the simple engine mechanism. With that configuration, link 3 is fixed, i.e. there is no relative movement between the centre of rotation of the crank and the housing in which the piston slides. Link 1 is the crank that rotates, link 2 the connecting rod and link 4 the slider which moves relative to the fixed link. When the piston moves backwards and forwards, i.e. link 4 moves backwards and forwards, then the crank, link 1, is forced to rotate. Hence the mechanism transforms an input of backwards and forwards motion into rotational motion.

Figure 6.4 shows another form of this type of mechanism, a *quick-return mechanism*. It consists of a rotating crank, link AB, which rotates round a fixed centre, an oscillating lever CD, which is caused to oscillate about C by the sliding of the block at B along CD as AB rotates, and a link DE which causes E to move backwards and forwards. E might be the ram of a machine and have a cutting tool attached to it. The ram will be at the extremes of its movement when the positions of the crank are AB_1 and AB_2. Thus as the crank moves anticlockwise from B_1 to B_2 the ram makes a complete stroke, the cutting stroke. When the crank continues its movement from B_2 anticlockwise to B_1 then the ram again makes a complete stroke in the opposite direction, the return stroke. With the crank rotating at constant speed, then, because the angle of crank rotation required for the cutting stroke is greater than the angle for the return stroke, the cutting stroke takes more time than the return stroke. Hence the term, quick-return for the mechanism.

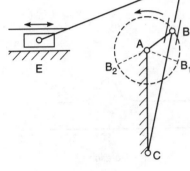

Fig. 6.4 Quick-release mechanism

6.3 Cams

A *cam* is a body which rotates or oscillates and in doing so imparts a reciprocating or oscillatory motion to a second body, called the *follower*, with which it is in contact (figure 6.5). As the cam rotates so the follower is made to rise, dwell and fall, the lengths of times spent at each of these positions depending on the shape of the cam. The rise section of the cam is the part that drives the follower upwards, its profile determining how quickly the cam follower will be lifted. The fall section of the cam is the part that lowers the follower, its profile determining how quickly the cam follower will fall. The dwell section of the cam is the part that allows the follower to remain at the same level for a significant period of time. The dwell section of the cam is where it is circular with a radius that does not change.

The cam shape required to produce a particular motion of the

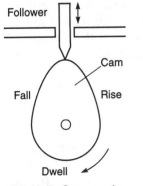

Fig. 6.5 Cam and cam follower

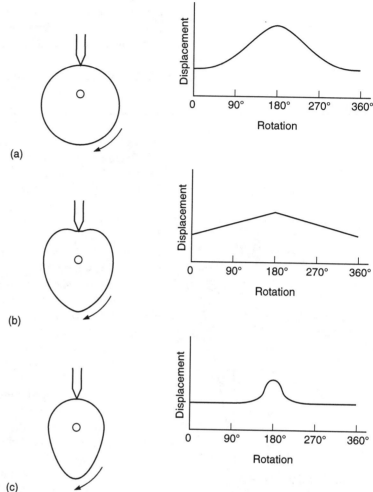

Fig. 6.6 Cams: (a) eccentric,
(b) heart-shaped, (c) pear-shaped

follower will depend on the shape of the cam and the type of follower used. Figure 6.6 shows the type of follower displacement diagrams that can be produced with different shaped cams and either point or knife followers. The radial distance from the axis of rotation of the cam to the point of contact of the cam with the follower gives the displacement of the follower with reference to the axis of rotation of the cam. The figures show how these radial distances, and hence follower displacements, vary with the angle of rotation of the cams. The eccentric cam (figure 6.6(a)) is a circular cam with an offset centre of rotation. It produces an oscillation of the follower which is simple harmonic motion and is often used with pumps. The heart-shaped cam (figure 6.6(b)) gives a follower displacement which increases at a constant rate with time before decreasing at a constant rate with time, hence a uniform speed for the follower. The pear-shaped cam (figure 6.6(c)) gives a follower motion which is stationary for about half a

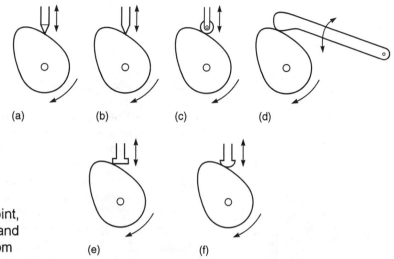

(a) (b) (c) (d)

(e) (f)

Fig. 6.7 Cam followers: (a) point, (b) knife, (c) roller, (d) sliding and oscillating, (e) flat, (f) mushroom

revolution of the cam and rises and falls symmetrically in each of the remaining quarter revolutions. Such a pear-shaped cam is used for engine valve control. The dwell holds the valve open while the petrol/air mixture passes into the cylinder. The longer the dwell, i.e. the greater the length of the cam surface with a constant radius, the more time is allowed for the cylinder to be completely charged with flammable vapour.

Figure 6.7 shows a number of examples of different types of cam followers. Roller followers are essentially ball or roller bearings. They have the advantage of lower friction than a sliding contact but can be more expensive. Flat-faced followers are often used because they are cheaper and can be made smaller than roller followers. Such followers are widely used with engine valve cams. While cams can be run dry, they are often used with lubrication and may be immersed in an oil bath.

6.4 Gear trains

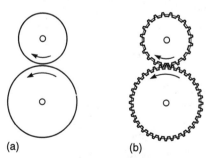

(a) (b)

Fig. 6.8 Transfer of rotatory motion: (a) rolling cylinders, (b) meshed gear wheels

Gear trains are mechanisms which are very widely used to transfer and transform rotational motion. They are used when a change in speed or torque of a rotating device is needed. For example, the car gearbox enables the driver to match the speed and torque requirements of the terrain with the engine power available.

Rotary motion can be transferred from one shaft to another by a pair of rolling cylinders (figure 6.8(a)), however there is a possibility of slip. The transfer of the motion between the two cylinders depends on the frictional forces between the two surfaces in contact. Slip can be prevented by the addition of meshing teeth to the two cylinders and the result is then a pair of meshed gear wheels (figure 6.8(b)).

Gears can be used for the transmission of rotary motion between parallel shafts (figure 6.9(a)) and for shafts which have

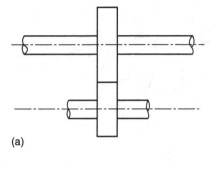

(a)

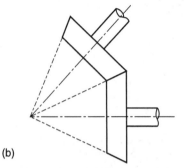

(b)

Fig. 6.9 Gear axes: (a) parallel, (b) inclined to one another

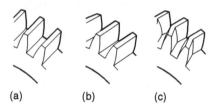

(a)   (b)   (c)

Fig. 6.10 Teeth: (a) axial, (b) helical, (c) double helical

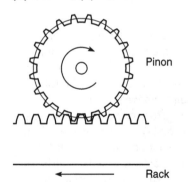

Pinon

Rack

Fig. 6.11 Rack and pinion

axes inclined to one another (figure 6.9(b)). The term *bevel gears* is used when the lines of the shafts intersect, as illustrated in figure 6.9(b). When two gears are in mesh, the larger gear wheel is often called the *spur* or *crown wheel* and the smaller one the *pinion*. Gears for use with parallel shafts may have axial teeth with the teeth cut along axial lines parallel to the axis of the shaft (figure 6.10(a)). Such gears are then termed *spur gears*. Alternatively they may have helical teeth with the teeth being cut on a helix (figure 6.10(b)) and are then termed *helical gears*. Helical gears have the advantage that there is a gradual engagement of any individual tooth and consequently there is a smoother drive and generally prolonged life of the gears. However, the inclination of the teeth to the axis of the shaft results in an axial force component on the shaft bearing. This can be overcome by using double helical teeth (figure 6.10(c)). Another form of gear is the *rack and pinion* (figure 6.11). This transforms either linear motion to rotational motion or rotational motion to linear motion.

Consider two meshed gear wheels A and B (as in figure 6.8(b)). If there are 40 teeth on wheel A and 80 teeth on wheel B, then wheel A must rotate through two revolutions in the same time as wheel B rotates through one. Thus the angular velocity ω_A of wheel A must be twice that ω_B of wheel B, i.e.

$$\frac{\omega_A}{\omega_B} = \frac{\text{number of teeth on B}}{\text{number of teeth on A}} = \frac{80}{40} = 2$$

Since the number of teeth on a wheel is proportional to its diameter, we can write:

$$\frac{\omega_A}{\omega_B} = \frac{\text{number of teeth on B}}{\text{number of teeth on A}} = \frac{d_B}{d_A}$$

Thus for the data we have been considering, wheel B must have twice the diameter of wheel A. The term *gear ratio* is used for the ratio of the angular speeds of a pair of intermeshed gear wheels. Thus the gear ratio for this example is 2.

6.4.1 Gear trains

The term *gear train* is used to describe a series of intermeshed gear wheels. The term *simple gear train* is used for a system where each shaft carries only one gear wheel, as in figure 6.12(a). For such a gear train, the overall gear ratio is the ratio of the angular velocities at the input and output shafts and is thus ω_A/ω_C.

$$G = \frac{\omega_A}{\omega_C}$$

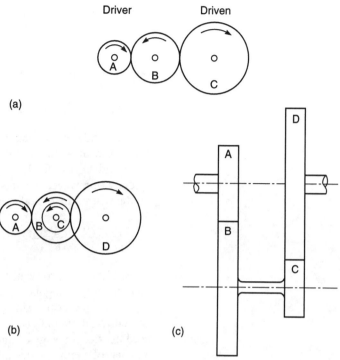

Fig. 6.12 Gear trains: (a) simple, (b) and (c) compound

Consider a simple gear train consisting of wheels A, B and C, as in figure 6.12(a), with A having 9 teeth and C having 27 teeth. Then, as the angular velocity of a wheel is inversely proportional to the number of teeth on the wheel, the gear ratio is 27/9 = 3. The effect of wheel B is purely to change the direction of rotation of the output wheel compared with what it would have been with just the two wheels A and C intermeshed. The intermediate wheel, B, is termed the *idler wheel*.

We can rewrite this equation for the overall gear ratio G as

$$G = \frac{\omega_A}{\omega_C} = \frac{\omega_A}{\omega_B} \times \frac{\omega_B}{\omega_C}$$

But ω_A/ω_B is the gear ratio for the first pair of gears and ω_B/ω_C the gear ratio for the second pair of gears.

The term *compound gear train* is used to describe a gear train when two wheels are mounted on a common shaft. Figure 6.12(b) and (c) shows two examples of such a compound gear train. The gear train in figure 6.12(c) enables the input and output shafts to be in line. An alternative way of achieving this is the epicyclic gear train discussed in the next section.

When two gear wheels are mounted on the same shaft they have the same angular velocity. Thus, for both of the compound gear trains in figure 6.12(b) or (c), $\omega_B = \omega_C$. The overall gear ratio G is thus

$$G = \frac{\omega_A}{\omega_D} = \frac{\omega_A}{\omega_B} \times \frac{\omega_B}{\omega_C} \times \frac{\omega_C}{\omega_D} = \frac{\omega_A}{\omega_B} \times \frac{\omega_C}{\omega_D}$$

For the arrangement shown in figure 6.12(c), for the input and output shafts to be in line we must also have

$$r_A + r_B = r_D + r_C$$

Consider a compound gear train of the form shown in figure 6.12(b), with A, the first driver, having 15 teeth, B 30 teeth, C 18 teeth and D, the final driven wheel, 36 teeth. Since the angular velocity of a wheel is inversely proportional to the number of teeth on the wheel, the overall gear ratio is

$$G = \frac{30}{15} \times \frac{36}{18} = 4$$

Thus, if the input to wheel A is an angular velocity of 160 rev/min, then the output angular velocity of wheel D is 160/4 = 40 rev/min.

A simple gear train of spur, helical or bevel gears is usually limited to an overall gear ratio of about 10. This is because of the need to keep the gear train down to a manageable size if the number of teeth on the pinion is to be kept above a minimum number which is usually about 10 to 20. Higher gear ratios can, however, be obtained with compound gear trains (or epicyclic gears). This is because the gear ratio is the product of the individual gear ratios of parallel gear sets.

6.4.2 Epicyclic gear trains

In the *epicyclic gear train* one or more wheels is carried on an arm which can rotate about the main axis of the train. Such wheels are called *planets* and the wheel around which the planets revolve is the *sun*. Figure 6.13 shows such a system with the centres of rotation of the sun wheel S and the planet wheel P linked by an arm. Two inputs are required for such a system. Typically, the arm and the sun gear will both be driven and the output taken from the rotation of the planet wheel. In order to determine the amount of this rotation, a technique that can be used is to first imagine the arm to be fixed while S rotates through +1 revolution. This causes wheel P to rotate through $-t_S/t_P$ revolutions, where t_S is the number of teeth on the sun wheel and t_P the number of teeth on the planet wheel. Then we imagine the gears to be locked solid and give a rotation of −1 revolution to all the wheels and the arm about the axis through S. If we had taken S to be fixed with the arm rotating, then the result is just the sum of the above two operations. Thus while the arm rotates through −1 revolution, the planet gear rotates through $-(1 + t_S/t_P)$ revolutions.

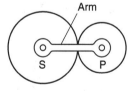

Fig. 6.13 Simple epicyclic train

Operation	Rotation		
	Arm	S	P
A Fix arm and rotate S by +1 rev.	0	+1	$-t_S/t_P$
B Give all −1 rev.	−1	−1	−1
Adding **A** and **B**	−1	0	$-(1 + t_S/t_P)$

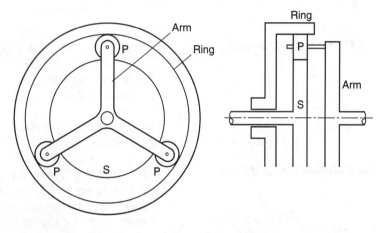

Fig. 6.14 Epicyclic train with ring output

It is difficult to get a useful output from the orbiting planet as its axis of rotation is moving. A more useful form is shown in figure 6.14. This has a ring (often termed the annulus) gear R added. This has internal teeth and there are three planets which mesh with it and can rotate about pins through the arms emanating from the centre and the axis of the sun. There are usually three or four planets. We can use the same technique as above to determine the relative motion of the wheels and arm. Consider the arm to be fixed and the ring is rotated through +1 revolution. This causes the sun to rotate through $-t_R/t_S$ and the planets through $+t_R/t_P$. Then we consider the gears to be locked solid and all given −1 rotation about the axis through the sun. Now if we had taken the ring to be fixed with the arm rotating, then the result is the same as the sum of the above two operations. Thus −1 revolution of the arm, with the ring fixed, results in a revolution for the sun of $(-1 - t_R/t_S)$ and for the planets of $(-1 + t_R/t_P)$.

By fixing different parts of the epicyclic gear, different gear ratios can be obtained. The above discussion related to a fixed ring; we could, however, have had the ring rotate and kept the arm or the sun fixed. Epicyclic gears are the basis of most car automatic gearboxes.

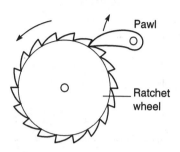

Fig. 6.15 Ratchet and pawl

Operation	Rotation			
	Arm	Ring	S	P
A Fix arm and rotate arm by +1 rev.	0	+1	$-t_R/t_S$	$+t_R/t_P$
B Give all −1 rev.	−1	−1	−1	−1
Adding **A** and **B**	−1	0	$-1 - t_R/t_S$	$-1 + t_R/t_P$

6.5 Ratchet mechanisms

Figure 6.15 shows the basic form of a ratchet mechanism. It consists of a wheel, called a *ratchet*, with saw-shaped teeth which engage with an arm called a *pawl*. The arm is pivoted and can move back and forth to engage the wheel. The shape of teeth are such that rotation can occur in only one direction. Thus the mechanism allows motion in only one direction and prevents motion in the reverse direction.

6.6 Belt and chain drives

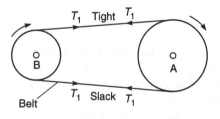
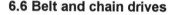

Fig. 6.16 Belt drive

Belt drives are essentially just a pair of rolling cylinders, as described in figure 6.8(a) and section 6.4, with the motion of one cylinder being transferred to the other by a belt (figure 6.16). Belt drives use the friction that develops between the pulleys attached to the shafts and the belt around the arc of contact in order to transmit a torque. Since the transfer relies on frictional forces then slip can occur. The transmitted torque is due to the differences in tension that occur in the belt during operation. This difference results in a tight side and a slack side for the belt. If the tension on tight side is T_1, and that on the slack side T_2, then with pulley A in figure 6.16 as the driver

$$\text{Torque on A} = (T_1 - T_2)r_A$$

where r_A is the radius of pulley A. For the driven pulley B we have

$$\text{Torque on B} = (T_1 - T_2)r_B$$

where r_B is the radius of pulley B. Since the power transmitted is the product of the torque and the angular velocity, and since the angular velocity is v/r_A for pulley A and v/r_B for pulley B, where v is the belt speed, then for either pulley we have

$$\text{Power} = (T_1 - T_2)v$$

As a method of transmitting power between two shafts, belt drives have the advantage that the length of the belt can easily be adjusted to suit a wide range of shaft-to-shaft distances and the system is automatically protected against overload because slipping occurs if the loading exceeds the maximum tension that can be sustained by frictional forces. If the distances between shafts is large, a belt drive is more suitable than gears, but over small distances gears are to be preferred. Different size pulleys can be used to give a gearing effect. However, the gear ratio is limited to about 3 because of the need to maintain an adequate arc of contact between the belt and the pulleys.

Slip can be prevented by the use of chains which lock into teeth on the rotating cylinders to give the equivalent of a pair of intermeshing gear wheels. A chain drive has the same relationship for gear ratio as a simple gear train. The drive mechanism used with a bicycle is an example of a chain drive.

Problems

1   Explain the terms (a) mechanism, (b) kinematic chain.
2   Explain what is meant by the four-bar chain.
3   For the mechanism shown in figure 6.17, the arm AB rotates at a constant rate. B and E are sliders moving along CD and AF. Describe the behaviour of this mechanism.

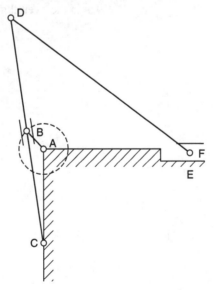

Fig. 6.17 Problem 3

4   Describe how the displacement of the cam follower shown in figure 6.18 will vary with the angle of rotation of the cam.
5   A circular cam of diameter 100 mm has an eccentric axis of rotation which is offset 30 mm from the centre. When used with a knife follower with its line of action passing through the centre of rotation, what will be the difference between the maximum and minimum displacements of the follower?
6   Design a cam–follower system to give constant follower speeds over follower displacements varying from 40 to 100 mm.

Fig. 6.18 Problem 4

7 Design a mechanical system which can be used to:
   (a) Operate a sequence of microswitches in a timed sequence.
   (b) Move a tool at a steady rate in one direction and then quickly move it back to the beginning of the path.
   (c) Transform a rotation into a linear back-and-forth movement with simple harmonic motion.
   (d) Transform a rotation through some angle into a linear displacement.
   (e) Transform a rotation of a shaft into rotation of another, parallel, shaft some distance away.
   (f) Transform a rotation of one shaft into rotation of another, close, shaft which is at right angles to it.

8 A three-planet epicyclic gear, of the form shown in figure 6.14, has a sun with 65 teeth and a ring with 125 teeth. What is the angular speed of the arm when the ring is fixed and the sun rotates at 120 rev/min?

9 A compound gear train consists of the final driven wheel with 15 teeth which meshes with a second wheel with 90 teeth. On the same shaft as the second wheel is a wheel with 15 teeth. This meshes with a fourth wheel, the first driver, with 60 teeth. What is the overall gear ratio?

7 Electrical actuation systems

7.1 Electrical systems

In any discussion of electrical systems used as actuators for control, the discussion has to include:

1 Switching devices, such as mechanical switches, diodes, thyristors, and transistors, where the control signal is applied to a switch which switches on or off some electrical device, perhaps a heater or a motor.
2 Solenoid type devices where a current through a solenoid is used to actuate a soft iron core, as for example the solenoid-operated hydraulic/pneumatic valve where a control current through a solenoid is used to actuate a hydraulic/pneumatic flow.
3 Drive systems, such as d.c. and a.c. motors, where a current through a motor is used to produce rotation.

This chapter is an overview of such devices and their characteristics.

7.2 Switches

Switches are frequently items in many electrical actuation systems. Switches make or break connections in an electrical circuit. This may be to switch on electric motors, switch on heating elements, actuate solenoid valves controlling hydraulic or pneumatic cylinders, control a set of sequential actions (see the description of the washing machine in section 1.5), etc. The switching action may be obtained as a result of the mechanical action of an actuator, electromechanically by a current in a solenoid, e.g. a relay, or electronically by means of a solid-state device.

A problem that occurs with mechanical switches is *switch bounce*. When a mechanical switch is switched to close the contacts, we have one contact being moved towards the other. It hits the other and, because the contacting elements are elastic, bounces. It may bounce a number of times before finally settling to its closed state after, typically, some 20 ms. Each of the contacts during this bouncing time can register as a separate contact. Thus, to a microprocessor, it might appear that perhaps two or more separate switch actions have occurred. Similarly, when a mechanical switch is opened, bouncing can occur. To overcome this problem either hardware or software can be used. With software, the microprocessor is programmed to detect if the switch is closed and then wait, say, 20 ms. After checking that bouncing has ceased and the switch is in the same closed position, the next part of the program can take place.

7.2.1 Mechanical switches

Mechanical switches consist of one or more pairs of contacts which can be mechanically closed or opened and in doing so make or break electrical circuits. Mechanically actuated switches may be operated manually by an operator or automatically by liquid level, temperature, flow, pressure, a cam or some other object. Liquid level actuated switches open or close their contacts at a particular liquid level. Temperature-actuated switches open or close their contacts at a particular temperature. Flow-actuated switches open or close their contacts at a particular flow rate. Pressure-actuated switches open or close their contacts at a particular pressure. Limit switches open or close their contacts when actuated by a physical object, such as a cam or the work-piece.

7.2.2 Relays

The electrical relay offers a simple on/off switching action in response to a control signal. Figure 7.1 illustrates the principle. When a current flows through the coil of wire a magnetic field is produced. This pulls a movable arm that forces the contact to open or close. This might then be used to supply a current to a motor or perhaps an electric heater in a temperature control system.

Time-delay relays are control relays that have a delayed switching action. The time delay is usually adjustable and can be initiated when a current flows through the relay coil or when it ceases to flow through the coil.

7.2.3 Solid-state switches

There are a number of solid-state devices which can be used to electronically switch circuits. These include diodes, thyristors, triacs and transistors.

The *diode* has the characteristic shown in figure 7.2 and so allows a significant current in one direction only. A diode can thus be regarded as a 'directional valve', only passing a current when forward biased, i.e. with the anode being positive with respect to the cathode. If the diode is sufficiently reverse biased, i.e. a very

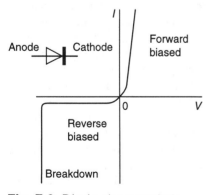

Fig. 7.1 Relay principle

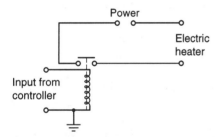

Fig. 7.2 Diode characteristic

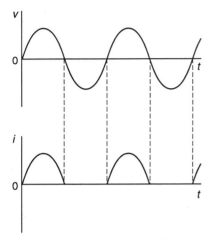

Fig. 7.3 Half-wave rectification

high voltage, it will break down. If an alternating voltage is applied across a diode, it can be regarded as only switching on when the direction of the voltage is such as to forward bias it and being off in the reverse biased direction. The result is that the current through the diode is half-rectified to become just the current due to the positive halves of the input voltage (figure 7.3).

The *thyristor* can be regarded as a diode which has a gate controlling the conditions under which the diode can be switched on. Figure 7.4 shows the thyristor characteristic. With the gate current zero, the thyristor passes negligible current when reverse biased (unless sufficiently reverse biased, hundreds of volts, when it breaks down). When forward biased the current is also negligible until the forward breakdown voltage is exceeded. When this occurs the voltage across the diode falls to a low level, about 1 to 2 V, and the current is then only limited by the external resistance in a circuit. Thus, for example, if the forward breakdown is at 300 V then when this voltage is reached the thyristor switches on and the voltage across it drops to 2 V. If the thyristor is in series with a resistance of, say, 20 Ω then before breakdown we have a very high resistance in series with the 20 Ω and so virtually all the 300 V is across the thyristor and there is negligible current. When forward breakdown occurs, the voltage across the thyristor drops to 2 V and so there is now 300 − 2 = 298 V across the 20 Ω resistor, hence the current rises to 298/20 = 14.9 A. When once switched on the thyristor remains on until the forward current is reduced to below a level of a few milliamps. The voltage at which forward breakdown occurs is determined by the current entering the gate, the higher the current the lower the breakdown voltage.

The *triac* is similar to the thyristor and is equivalent to a pair of thyristors connected in reverse parallel on the same chip. The triac can be turned on in either the forward or reverse direction. Figure 7.5 shows the characteristic.

Figure 7.6 shows the type of effect that occurs when a sinusoidal alternating voltage is applied across a thyristor and a triac. Forward breakdown occurs when the voltage reaches the breakdown value and then the voltage across the device remains low. As an example of how such devices can be used for control

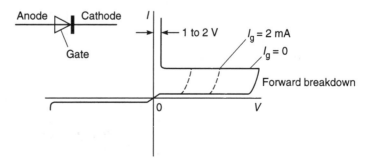

Fig. 7.4 Thyristor characteristic

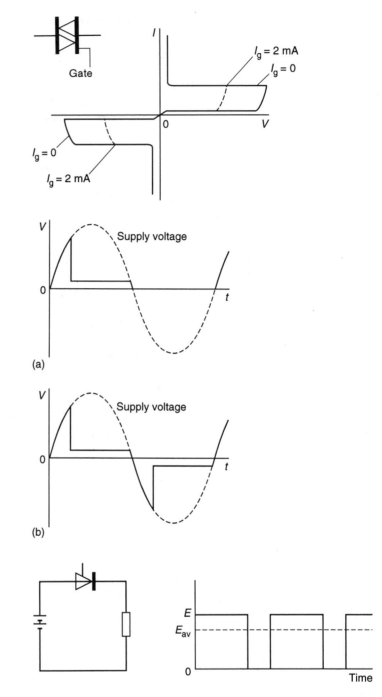

Fig. 7.5 Triac characteristic

Fig. 7.6 Voltage control:
(a) thyristor, (b) triac

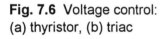

Fig. 7.7 Thyristor d.c. control

purposes, figure 7.7 illustrates how a thyristor could be used to control a d.c. voltage. In this the thyristor is operated as a switch by using the gate to switch the device on or off. By using an alternating signal to the gate, the supply voltage can be chopped

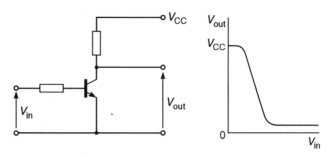

Fig. 7.8 Transistor switch circuit

and so the average value of the output d.c. voltage can be varied and hence controlled.

Bipolar transistors can be used as switches, as illustrated in figure 7.8. When there is no input voltage V_{in} then virtually the entire V_{CC} voltage appears at the output. When the input voltage is made sufficiently high the transistor switches so that very little of the V_{CC} voltage appears at the output.

For more information on solid-state switches, the reader is referred to specialist texts such *Advanced Industrial Electronics* by N. Morris (McGraw-Hill 1974) or *Electronics* by D.I. Crecraft, D.A. Gorham and J.J. Sparkes (Chapman and Hall 1993).

7.3 Solenoids

Solenoids can be used to provide electrically operated actuators. Figure 7.9 shows the basic principle. When a current passes through the solenoid a magnetic field is produced which exerts a force on the soft iron plunger, pulling it towards the coil. Thus changing the current changes the force acting on the plunger and so its movement. This movement may be magnified by a system of levers.

Solenoid valves are operated in a similar manner and are used to control fluid flow in hydraulic or pneumatic systems (see section 5.4). When a current passes through a coil a soft iron core is pulled into the coil and, in doing so, can open or close ports to allow the flow of a fluid.

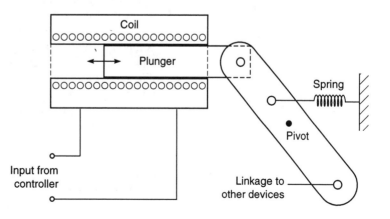

Fig. 7.9 A solenoid actuator

7.4 Motors

Electric motors are frequently used as the final control element in positional or speed-control systems. Motors can be classified into two main categories: d.c. motors and a.c. motors. The basic principles involved are:

1   A force is exerted on a conductor in a magnetic field when a current passes through it. For a conductor of length L carrying a current I in a magnetic field of flux density B at right angles to the conductor, the force F equals BIL.

2   When a conductor moves in a magnetic field then an e.m.f. is induced across it. The induced e.m.f. e is equal to the rate at which the magnetic flux Φ swept through by the conductor changes (Faraday's law), i.e. $e = -d\Phi/dt$. The minus sign is because the e.m.f. is in such a direction as to oppose the change producing it (Lenz's law). For this reason it is often referred to as a back e.m.f.

7.4.1 D.C. motors

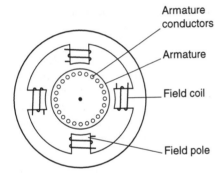

Armature conductors

Armature

Field coil

Field pole

Fig. 7.10 D.C. motor

In the d.c. motor, coils of wire are mounted in slots on a cylinder of magnetic material called the *armature*. The armature is mounted on bearings and is free to rotate. It is mounted in the magnetic field produced by *field poles*. These may be, for small motors, permanent magnets or electromagnets with their magnetism produced by a current through the *field coils*. Figure 7.10 shows the basic principle of a four pole d.c. motor. The ends of each armature coil are connected to adjacent segments of a segmented ring called the commutator with electrical contacts made to the segments through carbon contacts called brushes. As the armature rotates, the commutator reverses the current in each coil as it moves between the field poles. This is necessary if the forces acting on the coil are to remain acting in the same direction and so the rotation continue. To overcome the problems of brush wear, d.c. brushless motors are now available. These have a rotating magnet with stationary armature coils (see 7.4.3). The direction of rotation of the d.c. motor can be reversed by reversing either the armature current or the field current.

D.C. motors with field coils are classified as series, shunt, compound and separately excited according to how the field windings and armature windings are connected. Figure 7.11 illustrates these different types. With the *series wound motor* the armature and fields coils are in series. Such a motor exerts the highest starting torque and has the greatest no-load speed. With light loads there is a danger that a series wound motor might run at too high a speed. Reversing the polarity of the supply to the coils has no effect on the direction of rotation of the motor; it will continue rotating in the same direction since both the field and armature currents have been reversed. With the *shunt wound*

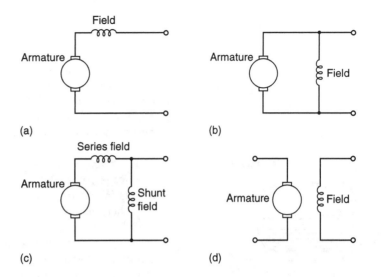

Fig. 7.11 Excitation of motors: (a) series, (b) shunt, (c) compound, (d) separate

motor the armature and field coils are in parallel. It provides the lowest starting torque, a much lower no-load speed and has good speed regulation. Because of this almost constant speed regardless of load, shunt wound motors are very widely used. To reverse the direction of rotation, either the armature or field supplied must be reversed. For this reason, the separately excited windings are preferable for such a situation. The *compound motor* has two field windings, one in series with the armature and one in parallel. Compound wound motors aim to get the best features of the series and shunt wound motors, namely a high starting torque and good speed regulation. The *separately excited motor* has separate control of the armature and field currents and can be considered to be a special case of the shunt wound motor. Figure 7.12 indicates the torque–speed characteristics of the above motors.

The speed of such d.c. motors can be changed by either changing the armature current or the field current. Generally it is the armature current that is varied. However, because fixed voltage supplies are often used, a variable voltage is often obtained by an electronic circuit. This might be a circuit using thyristors, as illustrated in figure 7.7, to control the average d.c. voltage by varying the time for which a d.c. voltage is switched on. Another possibility, when the input supply is a.c., is to convert the a.c. into d.c. and then control its level by means of the thyristor circuit.

The choice of d.c. motor will depend on its application. For example, with a robot manipulator, the robot wrist might use a series-wound motor because the speed decreases as the load increases. A shunt-wound motor would be used where a constant speed was required, regardless of the load. For more details of d.c. motors, the reader is referred to specialist texts such as *Electric Machines and Drives* by J.D. Edwards (Macmillan 1991) or *Electrical Machines and Drive Systems* by C.B. Gray (Longman 1989).

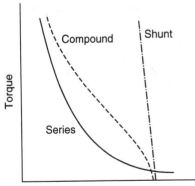

Fig. 7.12 Torque–speed characteristics

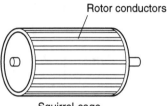

Rotor conductors

Squirrel-cage

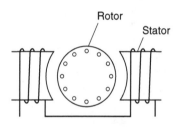

Rotor

Stator

Fig. 7.13 Single-phase induction motor

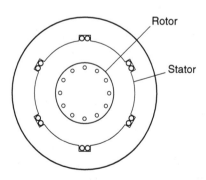

Rotor

Stator

Fig. 7.14 Three-phase induction motor

7.4.2 A.C. motors

Alternating current motors can be classified into two groups, single phase and poly phase with each group being further sub-divided into induction and synchronous motors. Single-phase motors tend to be used for low power requirements while poly-phase motors are used for higher powers. Induction motors tend to be cheaper than synchronous motors and are thus very widely used.

The *single-phase squirrel-cage induction motor* consists of a squirrel-cage rotor, this being copper or aluminium bars that fit into slots in end rings to form complete electrical circuits (figure 7.13). There are no external electrical connections to the rotor. The basic motor consists of this rotor with a stator having a set of windings. When an alternating current passes through the stator windings an alternating magnetic field is produced. As a result of electromagnetic induction, e.m.f.s are induced in the conductors of the rotor and currents flow in the rotor. Initially, when the rotor is stationary, the forces on the current-carrying conductors of the rotor in the magnetic field of the stator are such as to result in no net torque. The motor is not self-starting. However, if the rotor is given an initial start, it will experience forces and continue to rotate in the direction in which it was started. A number of methods are used to make the motor self-starting and give this initial impetus to start it. The rotor rotates at a speed determined by the frequency of the alternating current applied to the stator. For a constant frequency supply to a two-pole single-phase motor the magnetic field will alternate at this frequency. This speed of rotation of the magnetic field is termed the *synchronous speed*. The rotor will never quite match this frequency of rotation, typically differing from it by about 1 to 3%. This difference is termed *slip*. Thus for a 50 Hz supply the speed of rotation of the rotor will be almost 50 revolutions per second.

The *three-phase induction motor* (figure 7.14) is similar to the single-phase induction motor but has a stator with three windings located 120° apart, each winding being connected to one of the three lines of the supply. Because the three phases reach their maximum currents at different times, the magnetic field can be considered to rotate round the stator poles, completing one rotation in one full cycle of the current. The rotation of the field is much smoother than with the single-phase motor. The three-phase motor has a great advantage over the single-phase motor of being self-starting. The direction of rotation is reversed by interchanging any two of the line connections, this changing the direction of rotation of the magnetic field.

Synchronous motors have stators similar to those described above for induction motors but a rotor which is a permanent magnet (figure 7.15). The magnetic field produced by the stator

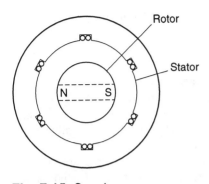

Fig. 7.15 Synchronous three-phase motor

rotates and so the magnet rotates with it. With one pair of poles per phase of the supply, the magnetic field rotates through 360° in one cycle of the supply and so the frequency of rotation with this arrangement is the same as the frequency of the supply. Synchronous motors are used when a precise speed is required. They are not self-starting and some system has to be employed to start them.

A.C. motors have the great advantage over d.c. motors of being cheaper, more rugged, reliable and maintenance free. However speed control is generally more complex than with d.c. motors and as a consequence a speed-controlled d.c. drive generally works out cheaper than a speed-controlled a.c. drive, though the price difference is steadily dropping as a result of technological developments and the reduction in price of solid-state devices. Speed control of a.c. motors is based around the provision of a variable frequency supply, since the speed of such motors is determined by the frequency of the supply. The torque developed by an a.c. motor is constant when the ratio of the applied stator voltage to frequency is constant. Thus to maintain a constant torque at the different speeds when the frequency is varied the voltage applied to the stator has also to be varied. With one method, the a.c. is first rectified to d.c. by a *converter* and then *inverted* back to a.c. again but at a frequency that can be selected (figure 7.16). Another method that is often used for operating slow-speed motors is the *cycloconverter*. This converts a.c. at one frequency·directly to a.c. at another frequency without the intermediate d.c. conversion.

For more details of a.c. motors, the reader is referred to specialist texts such as *Electric Machines and Drives* by J.D. Edwards (Macmillan 1991) or *Electrical Machines and Drive Systems* by C.B. Gray (Longman 1989).

Fig. 7.16 Variable speed a.c. motor

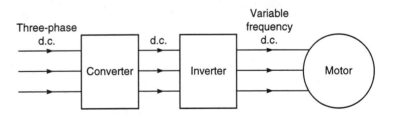

7.4.3 Brushless permanent magnet d.c. motors

Brushless permanent magnet d.c. motors are becoming increasingly used in situations where high performance coupled with reliability and low maintenance are essential. Essentially they consist of a sequence of stator coils and a permanent magnet rotor. Figure 7.17 shows the basic form of such a motor. The rotor is a ferrite or ceramic permanent magnet.

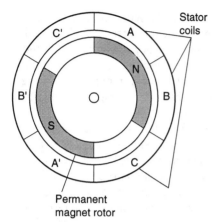

Fig. 7.17 Brushless permanent magnet d.c. motor

A current-carrying conductor in a magnetic field experiences a force; likewise, as a consequence of Newton's third law of motion, the magnet will also experience an opposite and equal force. With the conventional d.c. motor the magnet is fixed and the current-carrying conductors made to move. With the brushless d.c. permanent magnet d.c. motor the reverse is the case, the current-carrying conductors are fixed and the magnet moves. The current to the stator coils is electronically switched, typically using transistors, in sequence round the coils, the switching being controlled by the position of the rotor so that there are always forces acting on the magnet causing it to rotate in the same direction.

For more details of brushless d.c. motors, the reader is referred to specialist texts such as *Electric Machines and Drives* by J.D. Edwards (Macmillan 1991), *Electrical Machines and Drive Systems* by C.B. Gray (Longman 1989) or *Brushless Permanent-Magnet and Reluctance Motor Drives* by T.J.E. Miller (Oxford University Press 1989).

7.5 Stepping motors

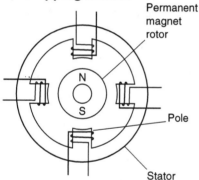

Fig. 7.18 Permanent magnet stepping motor

The *stepping motor* is a device that produces rotation through equal angles, the so-called *steps*, for each digital pulse supplied to its input. Thus, for example, if with such a motor 1 pulse produces a rotation of 6° then 60 pulses will produce a rotation through 360°. A stepper motor consists of a stator with a number of poles, four such poles being shown in figure 7.18 and six poles in figure 7.19. Each pole is wound with a field winding, the coils on opposite pairs of poles being in series. Current is supplied from a d.c. source to the windings through switches. If, when the rotor is a permanent magnet (figure 7.18), a pair of stator poles has a current switched to it, the rotor will move to line up with it. If the next pair of stator windings are then switched on, the rotor will move a step to line up with it. Figure 7.19 shows another form of rotor, termed the *variable reluctance* type. With this form the rotor is made of soft steel and is cylindrical with four poles, i.e. less poles than on the stator. When an opposite pair of windings has current switched to them, a magnetic field is produced with lines of force which pass from the stator poles through the nearest set of poles on the rotor. Since lines of force can be considered to be rather like elastic thread and always trying to shorten themselves, the rotor will move until the rotor and stator poles line up. This is termed the position of minimum reluctance.

Solid-state electronics is used to switch the d.c. supply between the pairs of stator windings. Thus the arrangement might be of the form shown in figure 7.20. The input pulses may come from a microprocessor.

The following are some of the terms commonly used in specifying stepping motors:

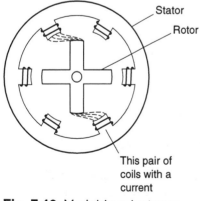

Fig. 7.19 Variable reluctance stepping motor

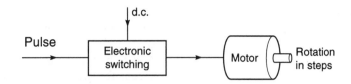

Fig. 7.20 Drive system for a stepping motor

1 *Holding torque*   This is the maximum torque that can be applied to a powered motor without moving it from its rest position and causing spindle rotation.
2 *Pull-in torque*   This is the maximum torque against which a motor will start, for a given pulse rate, and reach synchronism without losing a step.
3 *Pull-out torque*   This is the maximum torque that can be applied to a motor, running at a given stepping rate, without losing synchronism.
4 *Pull-in rate*   This is the maximum switching rate at which a loaded motor can start without losing a step.
5 *Pull-out rate*   This is the switching rate at which a loaded motor will remain in synchronism as the switching rate is reduced.
6 *Slew range*   This is the range of switching rates between pull-in and pull-out within which the motor runs in synchronism but cannot start up or reverse.

Figure 7.21 shows the general characteristics of a stepping motor.

For more details of stepping motors, the reader is referred to specialist texts such as *Stepping Motors and their Microprocessor Controls* by T. Kenjo (Oxford University Press 1984), *Electric Machines and Drives* by J.D. Edwards (Macmillan 1991) or *Electrical Machines and Drive Systems* by C.B. Gray (Longman 1989).

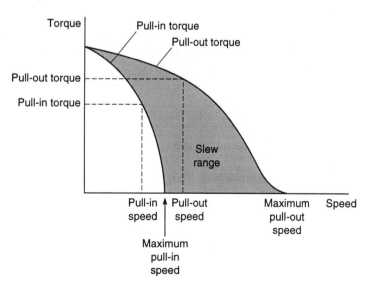

Fig. 7.21 Stepper motor characteristics

Problems

1 Explain how a thyristor can be used to control the level of a d.c. voltage by chopping the output from a constant voltage supply.

2 A d.c. motor is required to have (a) a high torque at low speeds for the movement of large loads, (b) a torque which is almost constant regardless of speed. Suggest suitable forms of motor.

3 Suggest possible motors, d.c. or a.c., which can be considered for applications where (a) cheap, constant torque operation is required, (b) high controlled speeds are required, (c) low speeds are required, (d) maintenance requirements have to be minimised.

4 Explain the principles of operation of the variable reluctance stepping motor.

5 Explain the principle of the brushless d.c. permanent magnet motor.

8 Basic system models

8.1 Mathematical models

Consider the following situation. A microprocessor switches on a motor. How will the rotation of the motor shaft vary with time? The speed will not immediately assume the full-speed value but will only attain that speed after some time. Consider another situation. A hydraulic system is used to open a valve which allows water into a tank to restore the water level to that required. How will the water level vary with time? The water level will not immediately assume the required level but will only attain that level after some time. This chapter, and chapters 9, 10 and 11, are about determining how systems behave with time when subject to some disturbance.

In order to understand the behaviour of systems, *mathematical models* are needed. These mathematical models are equations which describe the relationship between the input and output of a system. They can be used to enable forecasts to be made of the behaviour of a system under specific conditions. The basis for any mathematical model is provided by the fundamental physical laws that govern the behaviour of the system. In this chapter a range of systems will be considered, including mechanical, electrical, thermal and fluid examples.

Like a child building houses, cars, cranes, etc., from a number of basic building blocks, systems can be made up from a range of building blocks. Each building block is considered to have a single property or function. Thus to take a simple example, an electrical circuit system may be made up from building blocks which rep- resent the behaviour of resistors, capacitors and inductors. The resistor building block is assumed to have purely the property of resistance, the capacitor purely that of capacitance and the inductor purely that of inductance. By combining these building blocks in different ways a variety of electrical circuit systems can be built up and the overall input–output relationships obtained for the system by combining in an appropriate way the relationships for the building blocks. Thus a mathematical model for the system can be obtained. A system built up in this way is called a *lumped parameter* system. This is because each parameter, i.e. property or function, is considered independently.

There are similarities in the behaviour of building blocks used in mechanical, electrical, thermal and fluid systems. This chapter is about the basic building blocks and their combination to produce mathematical models for physical, real, systems. Chapter 9 looks at more complex models.

A more detailed discussion will be found in *Dynamic Modelling and Control of Engineering Systems* by J. Lowen Shearer and Bohdan T. Kulakowski (Macmillan 1990).

8.2 Mechanical system building blocks

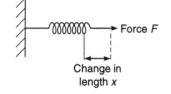

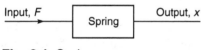

Fig. 8.1 Spring

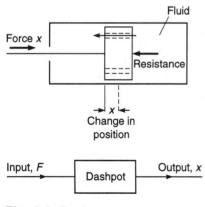

Fig. 8.2 Dashpot

The models used to represent mechanical systems have the basic building blocks of springs, dashpots and masses. *Springs* represent the stiffness of a system, *dashpots* the forces opposing motion, i.e. frictional or damping effects, and *masses* the inertia or resistance to acceleration. The mechanical system does not have to be really made up of springs, dashpots and masses but have the properties of stiffness, damping and inertia. All these building blocks can be considered to have a force as an input and a displacement as an output.

The stiffness of a *spring* is described by the relationship between the forces F used to extend or compress a spring and the resulting extension or compression x (figure 8.1). In the case of a spring where the extension or compression is proportional to the applied forces, i.e. a linear spring,

$$F = kx$$

where k is a constant. The bigger the value of k the greater the forces have to be to stretch or compress the spring and so the greater the stiffness. The object applying the force to stretch the spring is also acted on by a force, the force being that exerted by the stretched spring (Newton's third law). This force will be in the opposite direction and equal in size to the force used to stretch the spring, i.e. kx.

The *dashpot* building block represents the type of forces experienced when we endeavour to push an object through a fluid or move an object against frictional forces. The faster the object is pushed the greater becomes the opposing forces. The dashpot which is used pictorially to represent these damping forces which slow down moving objects consists of a piston moving in a closed cylinder (figure 8.2). Movement of the piston requires the fluid on one side of the piston to flow through or past the piston. This flow produces a resistive force. In the ideal case, the damping or resistive force F is proportional to the velocity v of the piston. Thus

$$F = cv$$

where c is a constant. The larger the value of c the greater the damping force at a particular velocity. Since velocity is the rate of change of displacement x of the piston, i.e. $v = dx/dt$, then

$$F = c\frac{dx}{dt}$$

Thus the relationship between the displacement x of the piston, i.e. the output, and the force as the input is a relationship depending on the rate of change of the output.

The *mass* building block (figure 8.3) exhibits the property that the bigger the mass the greater the force required to give it a specific acceleration. The relationship between the force F and the acceleration a is (Newton's second law) $F = ma$, where the constant of proportionality between the force and the acceleration is the constant called the mass m. Acceleration is the rate of change of velocity, i.e. dv/dt, and velocity v is the rate of change of displacement x, i.e. $v = dx/dt$. Thus

$$F = ma = m\frac{dv}{dt} = m\frac{d(dx/dt)}{dt} = m\frac{d^2x}{dt^2}$$

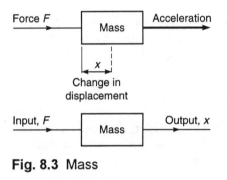

Fig. 8.3 Mass

Energy is needed to stretch the spring, accelerate the mass and move the piston in the dashpot. However, in the case of the spring and the mass we can get the energy back but with the dashpot we cannot. The spring when stretched stores energy, the energy being released when the spring springs back to its original length. The energy stored when there is an extension x is $\frac{1}{2}kx^2$. Since $F = kx$ this can be written as

$$E = \frac{1}{2}\frac{F^2}{k}$$

There is also energy stored in the mass when it is moving with a velocity v, the energy being referred to as kinetic energy, and released when it stops moving.

$$E = \frac{1}{2}mv^2$$

However, there is no energy stored in the dashpot. It does not return to its original position when there is no force input. The dashpot dissipates energy rather than storing it, the power P dissipated depending on the velocity v and being given by

$$P = cv^2$$

8.2.1 Rotational systems

The spring, dashpot and mass are the basic building blocks for mechanical systems where forces and straight line displacements are involved without any rotation. If there is rotation then the equivalent three building blocks are a torsional spring, a rotary damper and the moment of inertia, i.e. the inertia of a rotating mass. With such building blocks the inputs are torque and the outputs angle rotated. With a *torsional spring* the angle θ rotated is proportional to the torque T. Hence

$$T = k\theta$$

With the *rotary damper* a disc is rotated in a fluid and the resistive torque T is proportional to the angular velocity ω, and since angular velocity is the rate at which angle changes, i.e. $d\theta/dt$,

$$T = c\omega = c\frac{d\theta}{dt}$$

The *moment of inertia* building block exhibits the property that the greater the moment of inertia I the greater the torque needed to produce an angular acceleration α.

$$T = I\alpha$$

Thus, since angular acceleration is the rate of change of angular velocity, i.e. $d\omega/dt$, and angular velocity is the rate of change of angular displacement, then

$$T = I\frac{d\omega}{dt} = I\frac{d(d\theta/dt)}{dt} = I\frac{d^2\theta}{dt^2}$$

The torsional spring and the rotating mass store energy; the rotary damper just dissipates energy. The energy stored by a torsional spring when twisted through an angle θ is $\frac{1}{2}k\theta^2$ and since $T = k\theta$ this can be written as

$$E = \frac{1}{2}\frac{T^2}{k}$$

The energy stored by a mass rotating with an angular velocity ω is the kinetic energy E, where

$$E = \frac{1}{2}I\omega^2$$

Table 8.1 Mechanical building blocks

Building block	Describing equation	Energy stored or power dissipated
Translational		
Spring	$F = kx$	$E = \dfrac{1}{2}\dfrac{F^2}{k}$
Dashpot	$F = c\dfrac{dx}{dt}$	$P = cv^2$
Mass	$F = m\dfrac{d^2x}{dt^2}$	$E = \dfrac{1}{2}mv^2$
Rotational		
Spring	$T = k\theta$	$E = \dfrac{1}{2}\dfrac{T^2}{k}$
Rotational damper	$T = c\dfrac{d\theta}{dt}$	$P = c\omega^2$
Moment of inertia	$T = I\dfrac{d^2\theta}{dt^2}$	$E = \dfrac{1}{2}I\omega^2$

The power P dissipated by the rotatory damper when rotating with an angular velocity ω is

$$P = c\omega^2$$

Table 8.1 summarises the equations defining the character-istics of the mechanical building blocks when there is, in the case of straight line displacements (termed translational), a force input F and a displacement x output and, in the case of rotation, a torque T and angular displacement θ.

8.2.2 Building up a mechanical system

Many systems can be considered to be essentially a mass, a spring and dashpot combined in the way shown in figure 8.4. To evaluate the relationship between the force and displacement for the system the procedure to be adopted is to consider just one mass, and just the forces acting on that body. A diagram of the mass and just the forces acting on it is called a *free-body diagram*. When several forces act concurrently on a body, their single equivalent resultant can be found by vector addition. If the forces are all acting along the same line or parallel lines, this means that the resultant or net force acting on the block is the algebraic sum. Thus for the mass in figure 8.4, if we consider just the forces acting on that block then the net force applied to the mass is the applied force F minus the

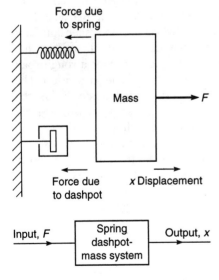

Fig. 8.4 Spring–dashpot–mass system

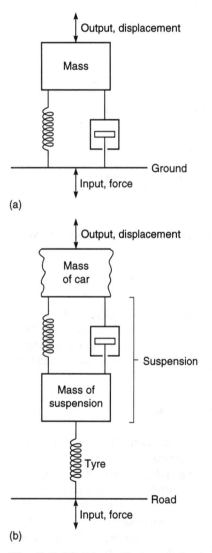

(a)

(b)

Fig. 8.5 Mathematical models: (a) a machine mounted on the ground, (b) the wheel of a car moving along a road

force resulting from the stretching or compressing of the spring and minus the force from the damper. Thus

Net force applied to mass $m = F - kx - cv$

where v is the velocity with which the piston in the dashpot, and hence the mass, is moving. This net force is the force applied to the mass to cause it to accelerate. Thus

Net force applied to mass $= ma$

Hence

$$F - kx - c\frac{dx}{dt} = m\frac{d^2x}{dt^2}$$

or, when rearranged,

$$m\frac{d^2x}{dt^2} + c\frac{dx}{dt} + kx = F$$

This equation, called a *differential equation*, describes the relationship between the input of force F to the system and the output of displacement x. It is a *second-order* differential equation, a first-order differential equation would only have dx/dt.

There are many systems which can be built up from suitable combinations of the spring, dashpot and mass building blocks. Figure 8.5(a) shows the model for a machine mounted on the ground and could be used as a basis for studying the effects of ground disturbances on the displacements of a machine bed. Figure 8.5(b) shows a model for the wheel and its suspension for a car or truck and can be used for the study of the behaviour that could be expected of the vehicle when driven over a rough road and hence as a basis for the design of the vehicle suspension. The procedure to be adopted for the analysis of such models is just the same as outlined above for the simple spring–dashpot–mass model. A free-body diagram is drawn for each mass in the system, such diagrams showing each mass independently and just the forces acting on it. Then for each mass the resultant of the forces acting on it is then equated to the product of the mass and the acceleration of the mass.

Similar models can be constructed for rotating systems. To evaluate the relationship between the torque and angular displacement for the system the procedure to be adopted is to consider just one rotational mass block, and just the torques acting on that body. When several torques act on a body simultaneously, their single equivalent resultant can be found by addition in which the direction of the torques is taken into account. Thus a system

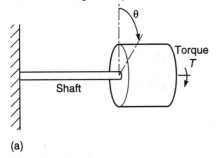

Angular displacement

θ

Torque
T

Shaft

(a)

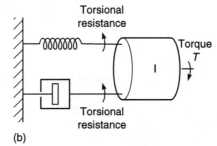

Torsional
resistance

Torque
T

I

Torsional
resistance

(b)

Fig. 8.6 Rotating a mass on the end of a shaft: (a) physical situations, (b) the building block model

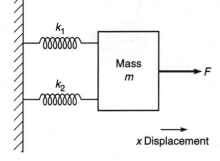

k_1

Mass
m

k_2

F

x Displacement

Fig. 8.7 Example

involving a torque being used to rotate a mass on the end of a shaft (figure 8.6(a)) can be considered to be represented by the rotational building blocks shown in figure 8.6(b). This is a comparable situation to that analysed above (figure 8.4) for linear displacements and yields a similar equation

$$I\frac{d^2\theta}{dt^2} + c\frac{d\theta}{dt} + k\theta = T$$

To illustrate the above, consider the development of the equations in the following examples.

Derive the differential equation describing the relationship between the input of the force F and the output of displacement x for the system shown in figure 8.7.

The net force applied to the mass is F minus the resisting forces exerted by each of the springs. Since these are k_1x and k_2x, then

$$\text{Net force} = F - k_1x - k_2x$$

Since the net force causes the mass to accelerate, then

$$\text{Net force} = m\frac{d^2x}{dt^2}$$

Hence

$$m\frac{d^2x}{dt^2} + (k_1 + k_2)x = F$$

Derive the differential equation describing the motion of the mass m_1 in figure 8.8 when a force F is applied.

The first step is to consider just the mass m_1 and the forces acting on it (figure 8.9). These are the forces exerted by the two springs. The force exerted by the lower spring is as a result of that spring being stretched. The amount it is stretched is $(x_1 - x_2)$. Thus

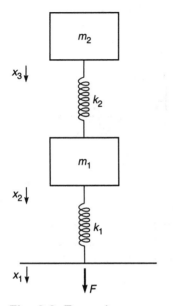

Fig. 8.8 Example

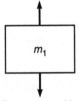

Force exerted by
upper spring

m_1

Force exerted by
lower spring

Fig. 8.9 Example

the force is $k_1(x_1 - x_2)$. The force exerted by the upper spring is due to it being stretched by $(x_2 - x_3)$ and so is $k_2(x_3 - x_2)$. Thus the net force acting on the mass is

$$\text{Net force} = k_1(x_2 - x_1) - k_2(x_3 - x_2)$$

This net force will cause the mass to have an acceleration. Hence

$$m\frac{d^2x}{dt^2} = k_1(x_2 - x_1) - k_2(x_3 - x_2)$$

But the force causing the extension of the lower spring is F. Thus

$$F = k_1(x_2 - x_1)$$

Hence the equation can be written as

$$m\frac{d^2x}{dt^2} + k_2(x_3 - x_2) = F$$

A motor is used to rotate a load. Devise a model and obtain the differential equation for it.

The model used can be that described by figure 8.6(a) and thus the model is that described by figure 8.6(b). The differential equation is thus

$$I\frac{d^2\theta}{dt^2} + c\frac{d\theta}{dt} + k\theta = T$$

8.3 Electrical system building blocks

The basic building blocks of electrical systems are inductors, capacitors and resistors. For an *inductor* the potential difference v across it at any instant depends on the rate of change of current (di/dt) through it:

$$v = L\frac{di}{dt}$$

where L is the inductance. The direction of the potential difference is in the opposite direction to the potential difference used to drive the current through the inductor, hence the term back e.m.f. The equation can be rearranged to give

$$i = \frac{1}{L}\int v\,dt$$

For a *capacitor*, the potential difference across it depends on the charge q on the capacitor plates at the instant concerned.

$$v = \frac{q}{C}$$

where C is the capacitance. Since the current i to or from the capacitor is the rate at which charge moves to or from the capacitor plates, i.e. $i = dq/dt$, then the total charge q on the plates is given by

$$q = \int i\, dt$$

and so

$$v = \frac{1}{C} \int i\, dt$$

Alternatively, since $v = q/C$ then

$$\frac{dv}{dt} = \frac{1}{C} \frac{dq}{dt} = \frac{1}{C} i$$

and so

$$i = C \frac{dv}{dt}$$

For a *resistor*, the potential difference v across it at any instant depends on the current i through it.

$$v = Ri$$

where R is the resistance.

Both the inductor and capacitor store energy which can then be released at a later time. A resistor does not store energy but just dissipates it. The energy stored by an inductor when there is a current i is

$$E = \tfrac{1}{2}Li^2$$

The energy stored by a capacitor when there is a potential difference v across it is

$$E = \tfrac{1}{2}Cv^2$$

The power P dissipated by a resistor when there is a potential difference v across it is

Table 8.2 Electrical building blocks

Building block	Describing equation	Energy stored or power dissipated
Inductor	$i = \dfrac{1}{L} \int v \, dt$	$E = \dfrac{1}{2} L i^2$
Capacitor	$i = C \dfrac{dv}{dt}$	$E = \dfrac{1}{2} C v^2$
Resistor	$i = \dfrac{v}{R}$	$P = \dfrac{v^2}{R}$

$$P = iv = \frac{v^2}{R}$$

Table 8.2 summarises the equations defining the characteristics of the electrical building blocks when the input is current and the output is potential difference. Compare them with the equations given in table 8.1 for the mechanical system building blocks.

8.3.1 Building up a model for an electrical system

The equations describing how the electrical building blocks can be combined are *Kirchhoff's laws*. These can be expressed as:

Law 1: the total current flowing towards a junction is equal to the total current flowing from that junction, i.e. the algebraic sum of the currents at the junction is zero.

Law 2: in a closed circuit or loop, the algebraic sum of the potential differences across each part of the circuit is equal to the applied e.m.f.

A convenient way of using law 1 is called *node analysis* since the law is applied to each principal node of a circuit, a node being a point of connection or junction between building blocks or circuit elements and a principal node being one where three or more branches of the circuit meet. A convenient way of using law 2 is called *mesh analysis* since the law is applied to each mesh, a mesh being a closed path or loop which contains no other loop.

To illustrate the use of these two methods of analysis to generate relationships, consider the circuit shown in figure 8.10. All the components are resistors for this illustrative example. With node analysis a principal node, point A on the figure, is picked and the voltage at the node given a value v_A with reference to some other principal node that has been picked as the reference. In this case it is convenient to pick node B as the reference. We then

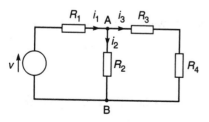

Fig. 8.10 Node analysis

consider all the currents entering and leaving node A and thus, according the Kirchhoff's first law,

$$i_1 = i_2 + i_3$$

The current entering through R_1 is i_1 and since the potential difference across R_1 is $(v_A - v)$, then $i_1R_1 = v_A - v$. The current through R_2 is i_2 and since the potential difference across R_2 is v_A then $i_2R_2 = v_A$. The current i_3 passes through R_3 in series with R_4 and there is a potential difference of v_A across the combination. Hence $i_3(R_3 + R_4) = v_A$. Thus equating the currents gives

$$\frac{v - v_A}{R_1} = \frac{v_A}{R_2} + \frac{v_A}{R_3 + R_4}$$

To illustrate the use of mesh analysis for the circuit in figure 8.10 we assume there are currents circulating in each mesh in the way shown in figure 8.11. Then Kirchhoff's second law is applied to each mesh. Thus for the mesh with current i_1 circulating, since the current through R_1 is i_1 and that through R_2 is $(i_1 - i_2)$, then

$$v = i_1R_1 + (i_1 - i_2)R_2$$

Similarly for the mesh with current i_2 circulating, since there is no source of e.m.f., then

$$0 = i_2R_3 + i_2R_4 + (i_2 - i_1)R_2$$

We thus have two simultaneous equations which can be solved to obtain the two mesh currents and hence the currents through each branch of the circuit. In general, when the number of nodes in a circuit is less than the number of meshes it is easier to employ nodal analysis.

Now consider a simple electrical system consisting of a resistor and capacitor in series, as shown in figure 8.12. Applying Kirchhoff's second law to the circuit loop gives

$$v = v_R + v_C$$

where v_R is the potential difference across the resistor and v_C that across the capacitor. Since it is just a single loop the current i through all the circuit elements will be the same. If the output from the circuit is the potential difference across the capacitor, v_C, then since $v_R = iR$ and $i = C(dv_C/dt)$,

$$v = RC\frac{dv_C}{dt} + v_C$$

This gives the relationship between the output v_C and the input v

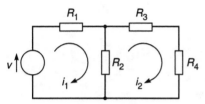

Fig. 8.11 Mesh analysis

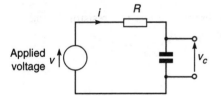

Fig. 8.12 Resistor–capacitor system

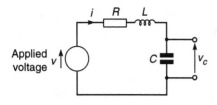

Fig. 8.13 Resistor–inductor–capacitor system

and is a first order differential equation.

Figure 8.13 shows a resistor–inductor–capacitor system. If Kirchhoff's second law is applied to this circuit loop,

$$v = v_R + v_L + v_C$$

where v_R is the potential difference across the resistor, v_L that across the inductor and v_C that across the capacitor. Since there is just a single loop the current i will be the same through all circuit elements. If the output from the circuit is the potential difference across the capacitor, v_C, then since $v_R = iR$ and $v_L = L(di/dt)$

$$v = iR + L\frac{di}{dt} + v_C$$

But $i = C(dv_C/dt)$ and so

$$\frac{di}{dt} = C\frac{d(dv_C/dt)}{dt} = C\frac{d^2 v_C}{dt^2}$$

Hence

$$v = RC\frac{dv_C}{dt} + LC\frac{d^2 v_C}{dt^2} + v_C$$

This is a second-order differential equation.

To illustrate the above, consider the relationship between the output, the potential difference across the inductor of v_L, and the input v for the circuit shown in figure 8.14.

Applying Kirchhoff's second law to the circuit loop,

$$v = v_R + v_L$$

where v_R is the potential difference across the resistor R and v_L that across the inductor. Since $v_R = iR$,

$$v = iR + v_L$$

Since

$$i = \frac{1}{L}\int v_L \, dt$$

then

$$v = \frac{R}{L}\int v_L \, dt + v_L$$

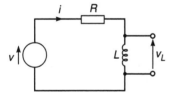

Fig. 8.14 Example

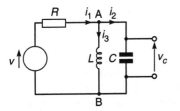

Fig. 8.15 Example

As another example, consider the relationship between the output, the potential difference v_C across the capacitor, and the input v for the circuit shown in figure 8.15.

Using nodal analysis, node B is taken as the reference node and node A taken to be at a potential of v_A relative to B. Applying Kirchhoff's law 1 to node A gives

$$i_1 = i_2 + i_3$$

But

$$i_1 = \frac{v - v_A}{R}$$

$$i_2 = \frac{1}{L} \int v_A \, dt$$

$$i_3 = C \frac{dv_A}{dt}$$

Hence

$$\frac{v - v_A}{R} = \frac{1}{L} \int v_A \, dt + C \frac{dv_A}{dt}$$

But $v_C = v_A$. Hence, with some rearrangement,

$$v = RC \frac{dv_C}{dt} + v_C + \frac{R}{L} \int v_C \, dt$$

The same answer could have been obtained by mesh analysis.

8.3.2 Electrical and mechanical analogies

The building blocks for electrical and mechanical systems have many similarities. For example, the electrical resistor does not store energy but dissipates it with the current i through the resistor being given by $i = v/R$, where R is a constant, and the power P dissipated by $P = v^2/R$. The mechanical analogue of the resistor is the dashpot. It also does not store energy but dissipates it with the force F being related to the velocity v by $F = cv$, where c is a constant, and the power P dissipated by $P = cv^2$. Both these sets of equations have similar forms. Comparing them, and taking the current as being analogous to the force, then the potential difference is analogous to the velocity and the dashpot constant c to the reciprocal of the resistance, i.e. $(1/R)$. These analogies between current and force, potential difference and velocity, hold for the other building blocks with the spring being analogous to inductance and mass to capacitance.

The analogy between current and force is the one most often used. However, another set of analogies can be drawn between potential difference and force.

8.4 Fluid system building blocks

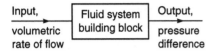

Fig. 8.16 Example

In fluid flow systems there are three basic building blocks which can be considered to be the equivalent of electrical resistance, capacitance and inductance. For such systems (figure 8.16) the input, the equivalent of the electrical current, is the volumetric rate of flow q, and the output, the equivalent of electrical potential difference, is pressure difference $(p_1 - p_2)$. Fluid systems can be considered to fall into two categories: *hydraulic*, where the fluid is a liquid and is deemed to be incompressible; and *pneumatic*, where it is a gas which can be compressed and consequently show a density change.

Hydraulic resistance is the resistance to flow which occurs as a result of a liquid flowing through valves or changes in a pipe diameter (figure 8.17). The relationship between the volume rate of flow of liquid q through the resistance element and the resulting pressure difference $(p_1 - p_2)$ is

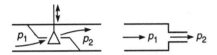

Fig. 8.17 Hydraulic resistance examples

$$p_1 - p_2 = Rq$$

where R is a constant called the hydraulic resistance. The bigger the resistance the bigger the pressure difference for a given rate of flow. This equation, like that for the electrical resistance and Ohm's law, assumes a linear relationship. Such hydraulic linear resistances occur with orderly flow through capillary tubes and porous plugs but non-linear resistances occur with flow through sharp-edged orifices or if flow is turbulent.

Hydraulic capacitance is the term used to describe energy storage with a liquid where it is stored in the form of potential energy. A height of liquid in a container (figure 8.18), i.e. a so-called pressure head, is one form of such a storage. For such a capacitance, the rate of change of volume V in the container, i.e. dV/dt, is equal to the difference between the volumetric rate at which liquid enters the container q_1 and the rate at which it leaves q_2,

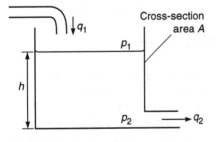

Fig. 8.18 Hydraulic capacitance

$$q_1 - q_2 = \frac{dV}{dt}$$

But $V = Ah$, where A is the cross-sectional area of the container and h the height of liquid in it. Hence

$$q_1 - q_2 = \frac{d(Ah)}{dt} = A\frac{dh}{dt}$$

But the pressure difference between the input and output is p, where $p = h\rho g$ with ρ being the liquid density and g the acceleration due to gravity. Thus, if the liquid is assumed to be incompressible, i.e. its density does not change with pressure,

$$q_1 - q_2 = A\frac{d(p/\rho g)}{dt} = \frac{A}{\rho g}\frac{dp}{dt}$$

The hydraulic capacitance C is defined as being

$$C = \frac{A}{\rho g}$$

Thus

$$q_1 - q_2 = C\frac{dp}{dt}$$

Integration of this equation gives

$$p = \frac{1}{C}\int(q_1 - q_2)\, dt$$

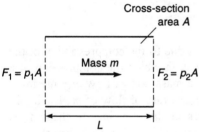

Cross-section area A

Mass m

$F_1 = p_1 A$ $F_2 = p_2 A$

L

Fig. 8.19 Hydraulic inertance

Hydraulic inertance is the equivalent of inductance in electrical systems or a spring in mechanical systems. To accelerate a fluid and so increase its velocity a force is required. Consider a block of liquid of mass m (figure 8.19). The net force acting on the liquid is

$$F_1 - F_2 = p_1 A - p_2 A = (p_1 - p_2)A$$

where $(p_1 - p_2)$ is the pressure difference and A the cross-sectional area. This net force causes the mass to accelerate with an acceleration a, and so

$$(p_1 - p_2)A = ma$$

But a is the rate of change of velocity dv/dt, hence

$$(p_1 - p_2)A = m\frac{dv}{dt}$$

But the mass of liquid concerned has a volume of AL, where L is the length of the block of liquid or the distance between the points in the liquid where the pressures p_1 and p_2 are measured. If the liquid has a density ρ then $m = AL\rho$ and so

$$(p_1 - p_2)A = AL\rho\frac{dv}{dt}$$

But the volume rate of flow $q = Av$, hence

$$(p_1 - p_2)A = L\rho \frac{dq}{dt}$$

$$p_1 - p_2 = I \frac{dq}{dt}$$

where the hydraulic inertance I is defined as

$$I = \frac{L\rho}{A}$$

With *pneumatic systems* the three basic building blocks are, as with hydraulic systems, resistance, capacitance and inertance. However, gases differ from liquids in being compressible, i.e. a change in pressure causes a change in volume and hence density. *Pneumatic resistance* R is defined in terms of the mass rate of flow $\dot{m}$ (note that the dot above the m is used to indicate that the symbol refers to the mass rate of flow and not just the mass) and the pressure difference $(p_1 - p_2)$ as

$$p_1 - p_2 = R\dot{m}$$

Pneumatic capacitance C is due to the compressibility of the gas, and is comparable to the way in which the compression of a spring stores energy. If there is a mass rate of flow $\dot{m}_1$ entering a con- tainer of volume V and a mass rate of flow of $\dot{m}_2$ leaving it, then the rate at which the mass in the container is changing is $(\dot{m}_1 - \dot{m}_2)$. If the gas in the container has a density ρ then the rate of change of mass in the container is

$$\text{Rate of change of mass in container} = \frac{d(\rho V)}{dt}$$

But, because a gas can be compressed, both ρ and V can vary with time. Hence

$$\text{Rate of change of mass in container} = \rho \frac{dV}{dt} + V \frac{d\rho}{dt}$$

Since $(dV/dt) = (dV/dp)(dp/dt)$ and, for an ideal gas, $pV = mRT$ with consequently $p = (m/V)RT = \rho RT$ and $d\rho/dt = (1/RT)(dp/dt)$, then

$$\text{rate of change of mass in container} = \rho \frac{dV}{dp}\frac{dp}{dt} + \frac{V}{RT}\frac{dp}{dt}$$

where R is the gas constant and T the temperature, assumed to be constant, on the Kelvin scale. Thus

$$\dot{m}_1 - \dot{m}_2 = \left(\rho \frac{dV}{dp} + \frac{V}{RT} \right) \frac{dp}{dt}$$

The pneumatic capacitance due to the change in volume of the container C_1 is defined as

$$C_1 = \rho \frac{dV}{dp}$$

and the pneumatic capacitance due to the compressibility of the gas C_2 as

$$C_2 = \frac{V}{RT}$$

Hence

$$\dot{m}_1 - \dot{m}_2 = (C_1 + C_2) \frac{dp}{dt}$$

or

$$p_1 - p_2 = \frac{1}{C_1 + C_2} \int (\dot{m}_1 - \dot{m}_2) \, dt$$

Pneumatic inertance is due to the pressure drop necessary to accelerate a block of gas. According to Newton's second law the net force is $ma = d(mv)/dt$. Since the force is provided by the pressure difference $(p_1 - p_2)$, then if A is the cross-sectional area of the block of gas being accelerated

$$(p_1 - p_2)A = \frac{d(mv)}{dt}$$

But m, the mass of the gas being accelerated, is ρLA with ρ being the gas density and L the length of the block of gas being accelerated. But the volume rate of flow $q = Av$, where v is the velocity. Thus

$$mv = \rho LA \frac{q}{A} = \rho Lq$$

and so

$$(p_1 - p_2)A = L \frac{d(\rho q)}{dt}$$

Table 8.3 Hydraulic and pneumatic building blocks

Building blocks	Describing equation	Energy stored or power dissipated
Hydraulic		
Inertance	$q = \dfrac{1}{L}\displaystyle\int (p_1 - p_2)\,dt$	$E = \dfrac{1}{2}Iq^2$
Capacitance	$q = C\dfrac{d(p_1 - p_2)}{dt}$	$E = \dfrac{1}{2}C(p_1 - p_2)^2$
Resistance	$q = \dfrac{p_1 - p_2}{R}$	$P = \dfrac{1}{R}(p_1 - p_2)^2$
Pneumatic		
Inertance	$\dot{m} = \dfrac{1}{L}\displaystyle\int (p_1 - p_2)\,dt$	$E = \dfrac{1}{2}I\dot{m}^2$
Capacitance	$\dot{m} = C\dfrac{d(p_1 - p_2)}{dt}$	$E = \dfrac{1}{2}C(p_1 - p_2)^2$
Resistance	$\dot{m} = \dfrac{p_1 - p_2}{R}$	$P = \dfrac{1}{R}(p_1 - p_2)^2$

But $\dot{m} = \rho q$ and so

$$p_1 - p_2 = \frac{L}{A}\frac{d\dot{m}}{dt}$$

$$p_1 - p_2 = I\frac{d\dot{m}}{dt}$$

with the pneumatic inertance I being

$$I = \frac{L}{A}$$

Table 8.3 shows the basic characteristics of the fluid building blocks, both hydraulic and pneumatic. For hydraulics the volumetric rate of flow and for pneumatics the mass rate of flow are analogous to the electrical current in an electric system. For both hydraulics and pneumatics the pressure difference is analogous to the potential difference in electrical systems. Compare the table with table 8.2. Hydraulic and pneumatic inertance and capacitance are both energy storage elements; hydraulic and pneumatic resistance are both energy dissipators.

8.4.1 Building up a model for a fluid system

Figure 8.20 shows a simple hydraulic system, a liquid entering and leaving a container. Such a system can be considered to consist of

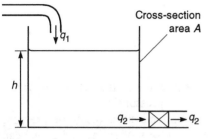

Fig. 8.20 A fluid system

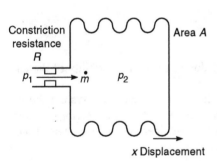

x Displacement

Fig. 8.21 A pneumatic system

a capacitor, the liquid in the container, with a resistor, the valve. Inertance can be neglected since flow rates change only very slowly. For the capacitor we can write

$$q_1 - q_2 = C\frac{dp}{dt}$$

The rate at which liquid leaves the container q_2 equals the rate at which it leaves the valve. Thus for the resistor

$$p_1 - p_2 = Rq_2$$

The pressure difference $(p_1 - p_2)$ is the pressure due to the height of liquid in the container and is thus equal to $h\rho g$. Thus substituting for q_2 in the first equation gives

$$q_1 - \frac{h\rho g}{R} = C\frac{d(h\rho g)}{dt}$$

and, since $C = A/\rho g$,

$$q_1 = A\frac{dh}{dt} + \frac{\rho g h}{R}$$

This equation describes how the height of liquid in the container depends on the rate of input of liquid into the container.

A bellows is an example of a simple pneumatic system (figure 8.21). Resistance is provided by a constriction which restricts the rate of flow of gas into the bellows and capacitance is provided by the bellows itself. Inertance can be neglected since the flow rate changes only slowly.

The mass flow rate $\dot{m}$ into the bellows is given by

$$p_1 - p_2 = R\dot{m}$$

where p_1 is the pressure prior to the constriction and p_2 the pressure after the constriction, i.e. the pressure in the bellows. All the gas that flows into the bellows remains in the bellows, there being no exit from the bellows. The capacitance of the bellows is given by

$$\dot{m}_1 - \dot{m}_2 = (C_1 + C_2)\frac{dp_2}{dt}$$

But $\dot{m}_1$ is the mass flow rate $\dot{m}$ given by the equation for the resistance and since there is no exit of gas from the bellows $\dot{m}_2$ is zero. Thus

$$\frac{p_1 - p_2}{R} = (C_1 + C_2)\frac{dp_2}{dt}$$

Hence

$$p_1 = R(C_1 + C_2)\frac{dp_2}{dt} + p_2$$

This equation describes how the pressure in the bellows p_2 varies with time when there is an input of a pressure p_1.

The bellows expands or contracts as a result of pressure changes inside it. Bellows are just a form of spring and so we can write $F = kx$ for the relationship between the force F causing an expansion or contraction and the resulting displacement x, where k is the spring constant for the bellows. But the force F depends on the pressure p_2, with $p_2 = F/A$ where A is the cross-sectional area of the bellows. Thus $p_2A = F = kx$. Hence substituting for p_2 in the above equation gives

$$p_1 = R(C_1 + C_2)\frac{k}{A}\frac{dx}{dt} + \frac{k}{A}x$$

This equation, a first-order differential equation, describes how the extension or contraction x of the bellows changes with time when there is an input of a pressure p_1. The pneumatic capacitance due to the change in volume of the container C_1 is $\rho\, dV/dp_2$ and since $V = Ax$, C_1 is $\rho A\, dx/dp_2$. But for the bellows $p_2A = kx$, thus

$$C_1 = \rho A \frac{dx}{d(kx/A)} = \frac{\rho A^2}{k}$$

C_2, the pneumatic capacitance due to the compressibility of the air, is $V/RT = Ax/RT$.

The following illustrates how, for the hydraulic system shown in figure 8.22, relationships can be derived which describe how the heights of the liquids in the two containers will change with time. With this model inertance is neglected.

Container 1 is a capacitor and thus

$$q_1 - q_2 = C_1 \frac{dp}{dt}$$

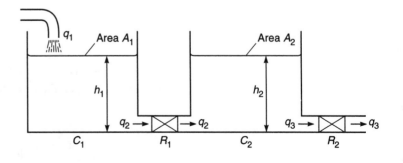

Fig. 8.22 Example

where $p = h_1 \rho g$ and $C_1 = A_1/\rho g$ and so

$$q_1 - q_2 = A_1 \frac{dh_1}{dt}$$

The rate at which liquid leaves the container q_2 equals the rate at which it leaves the valve R_1. Thus for the resistor,

$$p_1 - p_2 = R_1 q_2$$

The pressures are $h_1 \rho g$ and $h_2 \rho g$. Thus

$$(h_1 - h_2)\rho g = R_1 q_2$$

Using the value of q_2 given by this equation and substituting it into the earlier equation gives

$$q_1 - \frac{(h_1 - h_2)\rho g}{R_1} = A_1 \frac{dh_1}{dt}$$

This equation describes how the height of the liquid in container 1 depends on the input rate of flow.

For container 2 a similar set of equations can be derived. Thus for the capacitor C_2

$$q_2 - q_3 = C_2 \frac{dp}{dt}$$

where $p = h_2 \rho g$ and $C_2 = A_2/\rho g$ and so

$$q_2 - q_3 = A_2 \frac{dh_2}{dt}$$

The rate at which liquid leaves the container q_3 equals the rate at which it leaves the valve R_2. Thus for the resistor,

$$p_2 - 0 = R_2 q_3$$

This assumes that the liquid exits into the atmosphere. Thus, using the value of q_3 given by this equation and substituting it into the earlier equation gives

$$q_2 - \frac{h_2 \rho g}{R_2} = A_2 \frac{dh_2}{dt}$$

Substituting for q_2 in this equation using the value given by the equation derived for the first container gives

$$\frac{(h_1 - h_2)\rho g}{R_1} - \frac{h_2 \rho g}{R_2} = A_2 \frac{dh_2}{dt}$$

This equation describes how the height of liquid in container 2 changes.

8.5 Thermal system building blocks

There are only two basic building blocks for thermal systems: resistance and capacitance. There is a net flow of heat between two points if there is a temperature difference between them. The electrical equivalent of this is that there is only a net current i between two points if there is a potential difference v between them, the relationship between the current and potential difference being $i = v/R$, where R is the electrical resistance between the points. A similar relationship can be used to define *thermal resistance* R. If q is the rate of flow of heat and $(T_1 - T_2)$ the temperature difference, then

$$q = \frac{T_2 - T_1}{R}$$

The value of the resistance depends on the mode of heat transfer. In the case of conduction through a solid, for unidirectional conduction

$$q = Ak \frac{T_1 - T_2}{L}$$

where A is the cross-sectional area of the material through which the heat is being conducted and L the length of material between the points at which the temperatures are T_1 and T_2. k is the thermal conductivity. Hence, with this mode of heat transfer,

$$R = \frac{L}{Ak}$$

When the mode of heat transfer is convection, as with liquids and gases, then

$$q = Ah(T_2 - T_1)$$

where A is the surface area across which there is the temperature difference and h is the coefficient of heat transfer. Thus, with this mode of heat transfer,

$$R = \frac{1}{Ah}$$

Thermal capacitance is a measure of the store of internal

energy in a system. Thus, if the rate of flow of heat into a system is q_1 and the rate of flow out is q_2, then

Rate of change of internal energy = $q_1 - q_2$

An increase in internal energy means an increase in temperature. Since

Internal energy change = mc × change in temperature

where m is the mass and c the specific heat capacity, then

Rate of change of internal energy = mc × rate of change of temperature

Thus

$$q_1 - q_2 = mc\frac{\mathrm{d}T}{\mathrm{d}t}$$

where $\mathrm{d}T/\mathrm{d}t$ is the rate of change of temperature. This equation can be written as

$$q_1 - q_2 = C\frac{\mathrm{d}T}{\mathrm{d}t}$$

where C is the thermal capacitance and so $C = mc$. Table 8.5 gives a summary of the thermal building blocks.

Table 8.4 Thermal building blocks

Building block	Describing equation	Energy stored
Capacitance	$q_1 - q_2 = C\dfrac{\mathrm{d}T}{\mathrm{d}t}$	$E = CT$
Resistance	$q = \dfrac{T_1 - T_2}{R}$	

8.5.1 Building up a model for a thermal system

Consider a thermometer at temperature T which has just been inserted into a liquid at temperature T_L (figure 8.23). If the thermal resistance to heat flow from the liquid to the thermometer is R then,

$$q = \frac{T_L - T}{R}$$

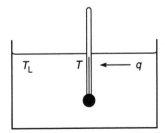

Fig. 8.23 A thermal system

where q is the net rate of heat flow from liquid to thermometer. The thermal capacitance C of the thermometer is given by the equation

$$q_1 - q_2 = C\frac{dT}{dt}$$

Since there is only a net flow of heat from the liquid to the thermometer, $q_1 = q$ and $q_2 = 0$. Thus

$$q = C\frac{dT}{dt}$$

Substituting this value of q in the earlier equation gives

$$C\frac{dT}{dt} = \frac{T_L - T}{R}$$

Rearranging this equation gives

$$RC\frac{dT}{dt} + T = T_L$$

This equation, a first-order differential equation, describes how the temperature indicated by the thermometer T will vary with time when the thermometer is inserted into a hot liquid.

In the above thermal system the parameters have been considered to be lumped. This means, for example, that there has been assumed to be just one temperature for the thermometer and just one for the liquid, i.e. the temperatures are only functions of time and not position within a body.

To illustrate the above consider figure 8.24 which shows a thermal system consisting of an electric fire in a room. The fire emits heat at the rate q_1 and the room loses heat at the rate q_2. Assuming that the air in the room is at a uniform temperature T and that there is no heat storage in the walls of the room, derive an equation describing how the room temperature will change with time.

If the air in the room has a thermal capacity C then

$$q_1 - q_2 = C\frac{dT}{dt}$$

If the temperature inside the room is T and that outside the room T_0 then

$$q_2 = \frac{T - T_0}{R}$$

where R is the resistivity of the walls. Substituting for q_2 gives

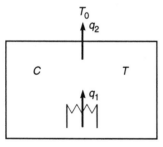

Fig. 8.24 Example

$$q_1 - \frac{T - T_0}{R} = C\frac{dT}{dt}$$

Hence

$$RC\frac{dT}{dt} + T = Rq_1 + T_0$$

Problems

1 Derive an equation relating the input, force F, with the output, displacement x, for the systems described by figure 8.25.

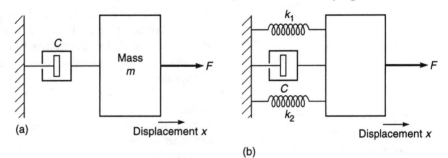

(a)   Displacement x

Fig. 8.25 Problem 1

(b)   Displacement x

2 Propose a model for a stepped shaft (i.e. a shaft where there is a change in diameter) used to rotate a mass and derive an equation relating the input torque and the angular rotation. You may neglect damping.

3 Derive the relationship between the output, the potential difference across the resistor R of v_R, and the input v for the circuit shown in figure 8.26 which has a resistor in series with a capacitor.

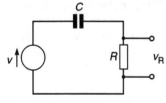

Fig. 8.26 Problem 3

4 Derive the relationship between the output, the potential difference across the resistor R of v_R, and the input v for the series LCR circuit shown in figure 8.27.

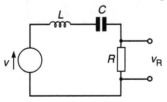

Fig. 8.27 Problem 4

5 Derive the relationship between the output, the potential difference across the capacitor C of v_C, and the input v for the circuit shown in figure 8.28.

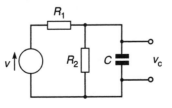

Fig. 8.28 Problem 5

6 Derive the relationship between the height h_2 and time for the hydraulic system shown in figure 8.29. Neglect inertance.

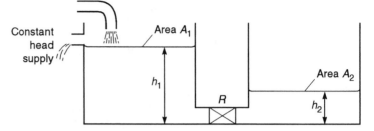

Fig. 8.29 Problem 6

7 A hot object, capacitance C, and temperature T cools in a large room at temperature T_r. If the thermal system has a resistance R derive an equation describing how the temperature of the hot object changes with time and give an electrical analogue of the system.

8 Figure 8.30 shows a thermal system involving two compartments, with one containing a heater. If the temperature of the compartment containing the heater is T_1, the temperature of the other compartment T_2 and the temperature surrounding the compartments T_3, develop equations describing how the temperatures T_1 and T_2 will vary with time. All the walls of the containers have the same resistance and negligible capacity. The two containers have the same capacity C.

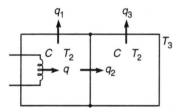

Fig. 8.30 Problem 8

9 Derive the differential equation relating the pressure input p to a diaphragm actuator (as in figure 5.18) to the displacement x of the stem.

10 Derive the differential equation for a motor driving a load through a gear system (figure 8.31) which relates the angular displacement of the load with time.

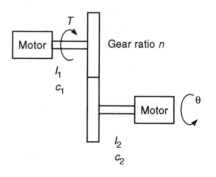

Fig. 8.31 Problem 10

9 System models

9.1 Engineering systems

In chapter 8 the basic building blocks of translational mechanical, rotational mechanical, electrical, fluid and thermal systems were separately considered. However, many systems encountered in engineering involve aspects of more than one of these disciplines. For example, an electric motor involves both electrical and mechanical elements. This chapter looks at how single discipline building blocks can be combined to give models for such multi-discipline systems.

In combining blocks we are assuming that the relationship for each block is linear. The following is a discussion of linearity and how, because many real engineering items are non-linear, we can make a linear approximation for a non-linear item.

9.1.1 Linearity

The relationship between the force F and the extension x produced for an ideal spring is linear, being given by $F = kx$. This means that if force F_1 produces an extension x_1 and force F_2 produces an extension x_2, a force equal to $(F_1 + F_2)$ will produce an extension $(x_1 + x_2)$. This is called the *principle of superposition* and is a necessary condition for a system that can be termed a *linear system*. Another condition for a linear system is that if an input F_1 produces an extension x_1 then an input cF_1 will produce an output cx_1, where c is a constant multiplier. A graph of the force F plotted against the extension x is a straight line passing through the origin when the relationship is linear (figure 9.1(a)).

Real springs, like any other real components, are not perfectly linear (figure 9.1(b)). However, there is often a range of operation for which linearity can be assumed. Thus for the spring giving the graph in figure 9.1(b), linearity can be assumed provided the spring is only used over the central part of its graph. For many system components, linearity can be assumed for operations within a range of values of the variable about some operating point.

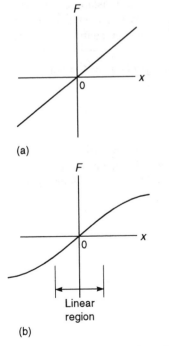

(a)

(b)

Fig. 9.1 Springs: (a) ideal, (b) real

155

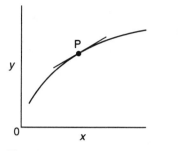

Fig. 9.2 A non-linear relationship

For some system components (figure 9.2) the relationship is non-linear. For such components the best that can be done to obtain a linear relationship is to just work with the straight line which is the slope of the graph at the operating point. Thus for the relationship between y and x in figure 9.2, at the operating point P where the slope has the value m

$$\Delta y = m\, \Delta x$$

where Δy and Δx are small changes in input and output signals at the operating point.

Thus, for example, the rate of flow of liquid q through an orifice is given by

$$q = c_d A \sqrt{\frac{2(p_1 - p_2)}{\rho}}$$

where c_d is a constant called the discharge coefficient, A the cross-sectional area of the orifice, ρ the fluid density and $(p_1 - p_2)$ the pressure difference. For a constant cross-sectional area and density the equation can be written as

$$q = C \sqrt{p_1 - p_2}$$

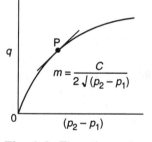

Fig. 9.3 Flow through an orifice

where C is a constant. This is a non-linear relationship between the rate of flow and the pressure difference. We can obtain a linear relationship by considering the straight line representing the slope of the rate of flow/pressure difference graph (figure 9.3) at the operating point. The slope m is $dq/d(p_1 - p_2)$ and has the value

$$m = \frac{dq}{d(p_1 - p_2)} = \frac{C}{2\sqrt{p_{o1} - p_{o2}}}$$

where $(p_{o1} - p_{o2})$ is the value at the operating point. For small changes about the operating point we will assume that we can replace the non-linear graph by the straight line of slope m and therefore can write $m = \Delta q / \Delta(p_1 - p_2)$ and hence

$$\Delta q = m\, \Delta(p_1 - p_2)$$

Hence, if we had $C = 2$ m³/s per kPa, i.e. $q = 2(p_1 - p_2)$, then for an operating point of $(p_1 - p_2) = 4$ kPa with $m = 2/(2\sqrt{4}) = 0.5$, the linearised version of the equation would be

$$\Delta q = 0.5\, \Delta(p_1 - p_2)$$

In the above discussion it was assumed that the flow was through an orifice of constant cross-sectional area. If the orifice is a flow control valve then this is not the case, the cross-sectional area being adjusted to vary the flow rate. In such a situation,

$$q = CA \sqrt{p_1 - p_2}$$

Since both A and $(p_1 - p_2)$ can change, then we have to obtain the linearised equation when either or both these variables can change. Because of the principle of superposition we can consider each of these variables changing independently and then add the two results to obtain the equation for when both change. Thus for changes about the operating point the slopes of a graph of q against A would be

$$m_1 = \frac{dq}{dA} = C \sqrt{p_{o1} - p_{o2}}$$

and thus $\Delta q = m_1 \, \Delta A$. The subscript o is used to indicate values at the operating point. For a graph of q against $(p_1 - p_2)$

$$m_2 = \frac{dq}{d(p_1 - p_2)} = \frac{CA_o}{2\sqrt{p_{o1} - p_{o2}}}$$

and thus $\Delta q = m_2 \, \Delta(p_1 - p_2)$. The linearised version when both variables can change is thus

$$\Delta q = m_1 \, \Delta A + m_2 \, \Delta(p_1 - p_2)$$

with m_1 and m_2 having the values indicated above.

Linearised mathematical models are used because most of the techniques of control systems are based on there being linear relationships for the elements of such systems. Also, because most control systems are maintaining an output equal to some reference value, the variations from this value tend to be rather small and so the linearised model is perfectly appropriate.

To illustrate the above, consider a thermistor being used for temperature measurement in a control system. The relationship between the resistance R of the thermistor and its temperature T is given by

$$R = k e^{-cT}$$

We can linearise this equation about an operating point of T_o. The slope m of a graph of R against T at the operating point T_o is given by dR/dT. Thus

$$m = \frac{dR}{dT} = -kc \, e^{-cT_o}$$

Hence

$$\Delta R = m \, \Delta T = (-kc \, e^{-cT_0})\Delta T$$

9.2 Rotational–translational systems

There are many mechanisms which involve the conversion of rotational motion to translational motion or vice versa. For example, there are rack and pinion, shafts with lead screws, pulley and cable systems, etc.

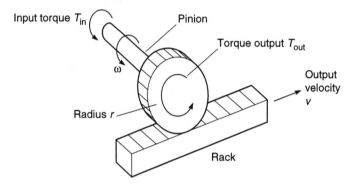

Fig. 9.4 Rack and pinion

To illustrate how such systems can be analysed, consider a rack and pinion system (figure 9.4). The rotational motion of the pinion is transformed into translational motion of the rack. Consider first the pinion element. The net torque acting on it is $(T_{in} - T_{out})$. Thus, considering the moment of inertia element, and assuming negligible damping,

$$T_{in} - T_{out} = I \frac{d\omega}{dt}$$

where I is the moment of inertia of the pinion and ω its angular velocity. The rotation of the pinion will result in a translational velocity v of the rack. If the pinion has a radius r, then $v = r\omega$. Hence we can write

$$T_{in} - T_{out} = \frac{I}{r} \frac{dv}{dt}$$

Now consider the rack element. There will be a force of T/r acting on it due to the movement of the pinion. If there is a frictional force of cv then the net force is

$$\frac{T_{out}}{r} - cv = m \frac{dv}{dt}$$

Eliminating T_{out} from the two equation gives

$$T_{in} - rcv = \left(\frac{I}{r} + mr\right)\frac{dv}{dt}$$

and so

$$\frac{dv}{dt} = \left(\frac{r}{I + mr^2}\right)(T_{in} - rcv)$$

The result is a first-order differential equation describing how the output is related to the input.

9.3 Electromechanical systems

Electromechanical devices, such as potentiometers, motors and generators, transform electrical signals to rotational motion or vice versa. This section is a discussion of how we can derive models for such systems. A potentiometer has an input of a rotation and an output of a potential difference. An electric motor has an input of a potential difference and an output of rotation of a shaft. A generator has an input of rotation of a shaft and an output of a potential difference.

9.3.1 Potentiometer

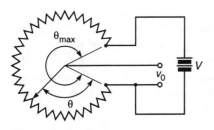

For the *potentiometer* (figure 9.5), which is a potential divider,

$$\frac{v_o}{V} = \frac{\theta}{\theta_{max}}$$

Fig. 9.5 Potentiometer

where V is the potential difference across the full length of the potentiometer track and θ_{max} is the total angle swept out by the slider in being rotated from one end of the track to the other. The output is v_o for the input θ.

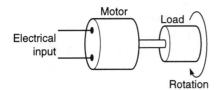

9.3.2 D.C. motor

The d.c. motor is used to convert an electrical input signal into a mechanical output signal (figure 9.6(a)). The motor basically consists of a coil, the armature coil, which is free to rotate. This coil is located in the magnetic field provided by, usually, a current through field coils. When a current i_a flows through the armature coil then, because it is in a magnetic field, forces act on the coil and cause it to rotate (figure 9.6(b)). The force F acting on a wire carrying a current i_a and of length L in a magnetic field of flux density B at right angles to the wire is given by

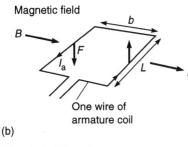

$$F = Bi_a L$$

With N wires

Fig. 9.6 The d.c. motor:
(a) driving a load, (b) basic motor principle

$$F = NBi_a L$$

The forces on the armature coil wires result in a torque T, where $T = Fb$, with b being the breadth of the coil. Thus

$$T = NBi_a Lb$$

The resulting torque is thus proportional to (Bi_a), the other factors all being constants. Hence we can write

$$T = k_1 Bi_a$$

Since the armature is a coil rotating in a magnetic field, a voltage will be induced in it as a consequence of electromagnetic induction. This voltage will be in such a direction as to oppose the change producing it and is thus called the back e.m.f. This back e.m.f. v_b is proportional to the rate or rotation of the armature and the flux linked by the coil, hence the flux density B. Thus

$$v_b = k_2 B\omega$$

where ω is the shaft angular velocity and k_2 a constant.

Consider a d.c. motor which has the armature and field coils separately excited (see figure 7.11(d) and associated discussion). With a so-called *armature-controlled motor* the field current i_f is held constant and the motor controlled by adjusting the armature voltage v_a. A constant field current means a constant magnetic flux density B for the armature coil. Thus

$$v_b = k_2 B\omega = k_3 \omega$$

where k_3 is a constant. The armature circuit can be considered to be a resistance R_a in series with an inductance L_a (figure 9.7). If v_a is the voltage applied to the armature circuit then, since there is a back e.m.f. of v_b, we have

$$v_a - v_b = L_a \frac{di_a}{dt} + R_a i_a$$

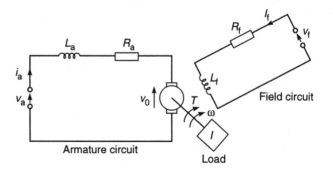

Fig. 9.7 D.C. motor circuits

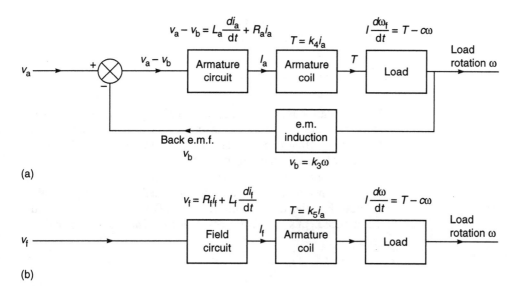

Fig. 9.8 D.C. motors:
(a) armature-controlled,
(b) field-controlled

We can think of this equation in terms of the block diagram shown in figure 9.8(a). The input to the motor part of the system is v_a and this is summed with the feedback signal of the back e.m.f. v_b to give an error signal which is the input to the armature circuit. The above equation thus describes the relationship between the input of the error signal to the armature coil and the output of the armature current i_a. Substituting for v_b

$$v_a - k_3\omega = L_a\frac{di_a}{dt} + R_a i_a$$

The current i_a in the armature generates a torque T. Since, for the armature-controlled motor, B is constant we have

$$T = k_1 B i_a = k_4 i_a$$

where k_4 is a constant. This torque then becomes the input to the load system. The net torque acting on the load will be

Net torque = T – damping torque

The damping torque is $c\omega$, where c is a constant. Hence, if any effects due to the torsional springiness of the shaft are neglected,

Net torque = $k_4 i_a - c\omega$

This will cause an angular acceleration of dω/dt, hence

$$I\frac{d\omega}{dt} = k_4 i_a - c\omega$$

We thus have two equations that describe the conditions occurring for an armature-controlled motor, namely

$$v_a - k_3\omega = L_a\frac{di_a}{dt} + R_ai_a \quad \text{and} \quad I\frac{d\omega}{dt} = k_4i_a - c\omega$$

We can thus obtain the equation relating the output ω with the input v_a to the system by eliminating i_a. See the brief discussion of the Laplace transform in chapter 10, or that in Appendix A, for details of how this might be done.

With a so-called *field-controlled motor* the armature current is held constant and the motor controlled by varying the field voltage. For the field circuit (figure 9.7) there is essentially just inductance L_f in series with a resistance R_f. Thus for that circuit

$$v_f = R_fi_f + L_f\frac{di_f}{dt}$$

We can think of the field-controlled motor in terms of the block diagram shown in figure 9.8(b). The input to the system is v_f. The field circuit converts this into a current i_f, the relationship between v_f and i_f being the above equation. This current leads to the production of a magnetic field and hence a torque acting on the armature coil, as given by $T = k_1Bi_a$. But the flux density B is proportional to the field current i_f and i_a is constant, hence

$$T = k_1Bi_a = k_5i_f$$

where k_5 is a constant. This torque output is then converted by the load system into an angular velocity ω. As earlier, the net torque acting on the load will be

Net torque = T − damping torque

The damping torque is $c\omega$, where c is a constant. Hence, if any effects due to the torsional springiness of the shaft are neglected,

Net torque = $k_5i_f - c\omega$

This will cause an angular acceleration of $d\omega/dt$, hence

$$I\frac{d\omega}{dt} = k_5i_f - c\omega$$

The conditions occurring for a field-controlled motor are thus described by the equations

$$v_f = R_fi_f + L_f\frac{di_f}{dt} \quad \text{and} \quad I\frac{d\omega}{dt} = k_5i_f - c\omega$$

We can thus obtain the equation relating the output ω with the input v_f to the system by eliminating i_f. See the brief discussion of the Laplace transform in chapter 10, or that in Appendix A, for details of how this might be done.

9.4 Hydraulic–mechanical systems

Hydraulic–mechanical converters involve the transformation of hydraulic signals to translational or rotational motion, or vice versa. Thus, for example, the movement of a piston in a cylinder as a result of hydraulic pressure involves the transformation of a hydraulic pressure input to the system to a translational motion output.

Figure 9.9 shows a hydraulic system in which an input of displacement x_i is, after passing through the system, transformed into a displacement x_o of a load. The system consists of a *spool valve* and a *cylinder*. The input displacement x_i to the left results in the hydraulic fluid supply pressure p_s causing fluid to flow into the left hand side of the cylinder. This pushes the piston in the cylinder to the right and expels the fluid in the right hand side of the chamber through the exit port at the right hand end of the spool valve. The rate of flow of fluid to and from the chamber depends on the extent to which the input motion has uncovered the ports allowing the fluid to enter or leave the spool valve. When the input displacement x_i is to the right the spool valve allows fluid to move to the right hand end of the cylinder and so results in a movement of the piston in the cylinder to the left.

The rate of flow of fluid q through an orifice, which is what the ports in the spool valve are, is a non-linear relationship (see section 9.1.1) depending on the pressure difference between the two sides of the orifice and its cross-sectional area A. However, a linearised version of the equation can be used (see section 9.1.1 for its derivation)

$$\Delta q = m_1 \, \Delta A + m_2 \, \Delta(\text{pressure difference})$$

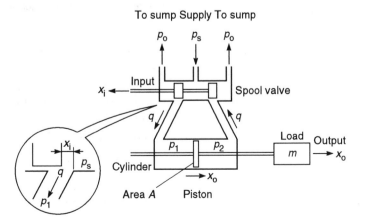

Fig. 9.9 Hydraulic system and load

where m_1 and m_2 are constants at the operating point. For the fluid entering the chamber the pressure difference is $(p_s - p_1)$ and for the exit $(p_2 - p_o)$. If the operating point about which the equation is linearised is taken to be the point at which the spool valve is central and the ports connecting it to the cylinder are both closed, then for this condition q is zero, and so $\Delta q = q$, A is proportional to x_s if x_s is measured from this central position, and the change in pressure on the inlet side of the piston is $-\Delta p_1$ relative to p_s and on the exit side Δp_2 relative to p_o. Thus, for the inlet port the equation can be written as

$$q = m_1 x_i + m_2(-\Delta p_1)$$

and for the exit port

$$q = m_1 x_i + m_2 \Delta p_2$$

Adding the two equations gives

$$2q = 2m_1 x_i - m_2(\Delta p_1 - \Delta p_2)$$

$$q = m_1 x_i - m_3(\Delta p_1 - \Delta p_2)$$

where $m_3 = m_2/2$.

For the cylinder the change in the volume of fluid entering the left hand side of the chamber, or leaving the right hand side, when the piston moves a distance x_o is $A x_o$, where A is the cross-sectional area of the piston. Thus the rate at which the volume is changing is $A(dx_o/dt)$. The rate at which fluid is entering the left hand side of the cylinder is q. However, since there is some leakage flow of fluid from one side of the piston to the other

$$q = A\frac{dx_o}{dt} + q_L$$

where q_L is the rate of leakage. Substituting for q, gives

$$m_1 x_i - m_3(\Delta p_1 - \Delta p_2) = A\frac{dx_o}{dt} + q_L$$

The rate of leakage flow q_L is a flow through an orifice, the gap between the piston and the cylinder. This is of constant cross-section and has a pressure difference $(\Delta p_1 - \Delta p_2)$. Hence, using the linearised equation for such a flow,

$$q_L = m_4(\Delta p_1 - \Delta p_2)$$

Thus, using this equation to substitute for q_L

$$m_1x_i - m_3(\Delta p_1 - \Delta p_2) = A\frac{dx_o}{dt} + m_4(\Delta p_1 - \Delta p_2)$$

$$m_1x_i - (m_3 + m_4)(\Delta p_1 - \Delta p_2) = A\frac{dx_o}{dt}$$

The pressure difference across the piston results in a force being exerted on the load, the force exerted being $(\Delta p_1 - \Delta p_2)A$. There is however some damping of motion, i.e. friction, of the mass. This is proportional to the velocity of the mass, i.e. (dx_o/dt). Hence the net force acting on the load is

$$\text{net force} = (\Delta p_1 - \Delta p_2)A - c\frac{dx_o}{dt}$$

This net force causes the mass to accelerate, the acceleration being (d^2x_o/dt^2). Hence

$$m\frac{d^2x_o}{dt^2} = (\Delta p_1 - \Delta p_2)A - c\frac{dx_o}{dt}$$

Rearranging this equation gives

$$\Delta p_1 - \Delta p_2 = \frac{m}{A}\frac{d^2x_o}{dt^2} + \frac{c}{A}\frac{dx_o}{dt}$$

Using this equation to substitute for the pressure difference in the earlier equation,

$$m_1x_i - (m_3 + m_4)\left(\frac{m}{A}\frac{d^2x_o}{dt^2} + \frac{c}{A}\frac{dx_o}{dt}\right) = A\frac{dx_o}{dt}$$

Rearranging gives

$$\frac{(m_3 + m_4)m}{A}\frac{d^2x_o}{dt^2} + \left(A + \frac{c(m_3 + m_4)}{A}\right)\frac{dx_o}{dt} = m_1x_i$$

and rearranging this equation leads to

$$\frac{(m_3 + m_4)m}{A^2 + c(m_3 + m_4)}\frac{d^2x_o}{dt^2} + \frac{dx_o}{dt} = \frac{Am_1}{A^2 + c)m_3 + m_4)}x_i$$

This equation can be simplified by introducing two constants k and τ, the latter constant being called the time constant (see chapter 10). Hence

$$\tau\frac{d^2x_o}{dt^2} + \frac{dx_o}{dt} = kx_i$$

Thus the relationship between input and output is described by a second-order differential equation.

Problems

1 The relationship between the force F used to stretch a spring and its extension x is given by $F = kx^2$, where k is a constant. Linearise this equation for an operating point of x_o.

2 The relationship between the e.m.f. E generated by a thermo-couple and the temperature T is given by

$$E = aT + bT^2$$

where a and b are constants. Linearise this equation for an operating point of temperature T_o.

3 The relationship between the torque T applied to a simple pendulum and the angular deflection (figure 9.10) is given by

$$T = mgL \sin \theta$$

where m is the mass of the pendulum bob, L the length of the pendulum and g the acceleration due to gravity. Linearise this equation for the equilibrium angle θ of $0°$.

4 Derive a differential equation relating the input voltage to a d.c. servo motor and the output angular velocity, assuming that motor is armature controlled and the equivalent circuit for the motor has an armature with just resistance, its inductance being neglected.

5 Derive differential equations for a d.c. generator. The generator may be assumed to have a constant magnetic field. The armature circuit has the armature coil, having both resistance and inductance, in series with the load. Assume that the load has both resistance and inductance.

6 Derive differential equations for a permanent magnet d.c. motor.

7 Consider a solenoid actuator in which a current passing through the solenoid results in movement of a rod actuator into or out off the solenoid. Propose models for the arrange-ment which could then be used to develop a differential equation relating the input of current to the output of displacement.

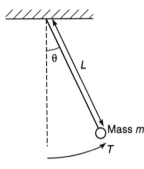

Fig. 9.10 Problem 3

10 Dynamic responses of systems

10.1 Modelling dynamic systems

The most important function of a model devised for measurement or control systems is to be able to predict what the output will be for a particular input. We are not just concerned with a static situation, i.e. that after some time when the steady state has been reached an output of x corresponds to an input of y. We have to consider how the output will change with time when there is a change of input or when the input changes with time. For example, how will the temperature of a temperature-controlled system change with time when the thermostat is set to a new temperature? For a control system, how will the output of the system change with time when the set value is set to a new value or perhaps increased at a steady rate?

Chapters 8 and 9 were concerned with models of systems when the inputs varied with time, with the results being expressed in terms of differential equations. This chapter is about how we can use such models to make predictions about how outputs will change with time when the input changes with time.

10.1.1 Transient and steady-state responses

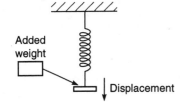

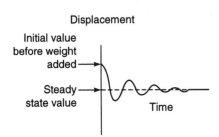

Fig. 10.1 Transient and steady-state responses of a spring system

The total response of a control system, or element of a system, can be considered to be made up of two aspects, the steady-state response and the transient response. The *transient response* is that part of a system response which occurs when there is a change in input and which dies away after a short interval of time. The *steady-state response* is the response that remains after all transient responses have died down.

To give a simple illustration of this, consider a vertically suspended spring (figure 10.1) and what happens when a weight is suddenly suspended from it. The deflection of the spring abruptly increases and then may well oscillate until after some time it settles down to a steady value. The steady value is the steady-state response of the spring system, the oscillation that occurs prior to

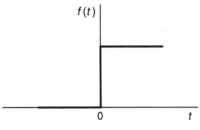

Fig. 10.2 Step function at $t = 0$

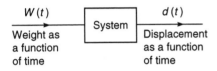

Fig. 10.3 The spring system

this steady state is the transient response. The input to the spring system, the weight, is a quantity which varies with time. Up to some particular time there is no added weight, i.e. no input, then after that time there is an input which remains constant for the rest of time. This type of input is known as a *step input* and has a graph of the form shown in figure 10.2.

Both the input and the output are functions of time. One way of indicating this is to write them in the form $f(t)$, where f is the function and (t) indicates that its value depends on time t. Thus for the weight W input to the spring system we could write $W(t)$ and for the deflection d output $d(t)$. A block diagram of the system would thus look like figure 10.3.

10.1.2 Differential equations

To describe the relationship between the input to a system and its output we must describe the relationship between inputs and outputs which are both possible functions of time. We thus need a form of equation which will indicate how the system output will vary with time when the input is varying with time. This can be done by the use of a *differential equation*. Such an equation includes derivatives with respect to time and so gives information about how the response of a system varies with time. A derivative dx/dt describes the rate at which x varies with time, the derivative d^2x/dt^2 (or $d(dx/dt)/dt$) states how dx/dt varies with time. Differential equations can be classed as *first-order*, *second-order*, *third-order*, etc. according to the highest order of the derivative in the equation. For a first-order equation the highest order will be dx/dt, with a second-order d^2x/dt^2, with a third-order d^3x/dt^3, with nth-order d^nx/dt^n.

This chapter is about the types of responses we can expect from first-order and second-order systems and the solution of such differential equations in order that the response of the system to different types of input can be obtained. This chapter uses the 'try a solution' approach in order to find a solution. The Laplace transformation method is introduced in chapter 11 to indicate how it can be used in such situations. A more detailed discussion of the transform is given in Appendix A. For a more detailed consideration of differential equations, the reader is referred to *Ordinary Differential Equations* by W. Bolton (Longman 1994) and of the Laplace transform to *Laplace and z-Transforms* by W. Bolton (Longman 1994), both texts appearing in their Mathematics for Engineers series.

10.2 First-order systems

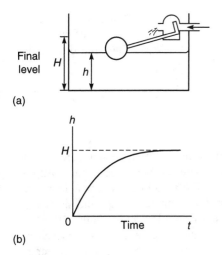

Final level

(a)

(b)

Fig. 10.4 Float-controlled water tank

An example of a first-order system is a float-controlled water tank (figure 10.4(a)). For such a system the rate at which water enters the tank, and hence the rate at which the height of the water in the tank changes with time, depends on the difference in the height h of the water in the tank from the height H at which the float completely switches off the water, i.e.

Rate of change of height is proportional to $(H - h)$

Hence

$$\frac{dh}{dt} = k(H - h)$$

where dh/dt is the rate of change of height and k a constant. The more the water level rises in the tank the smaller becomes the value of $(H - h)$ and so the smaller becomes the rate of change of height with time (dh/dt). A graph of water height against time will look like that in figure 10.4(b). The equation describing this graph is

$$h = H(1 - e^{-kt})$$

All first-order systems have the characteristic that the rate of change of some variable is proportional to the difference between this variable and some set value of the variable (this could however be zero). All have solutions of the form shown above.

As another example, consider a thermometer being placed in a hot liquid at some temperature θ_H. The rate at which the reading of the thermometer θ changes with time is proportional to the difference between θ and θ_H. What will be (a) the form of the differential equation describing how the thermometer temperature changes with time and (b) the equation of a graph of θ with time?

With this system the rate of change of θ is proportional to $(\theta_H - \theta)$. The differential equation is thus of the form

$$\frac{d\theta}{dt} = k(\theta - \theta_H)$$

where k is some constant. This is a relationship typical of a first-order system.

For a first-order system with an input θ_H and an output θ we will have the type of equation typical of a first-order system, namely

$$\theta = \theta_H(1 - e^{-kt})$$

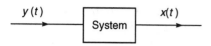

Fig. 10.5 The system

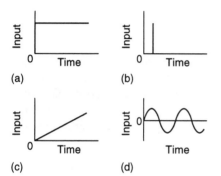

Fig. 10.6 Input signals: (a) step, (b) impulse, (c) ramp, and (d) sinusoidal

10.2.1 The first-order differential equation

Consider a first-order system (figure 10.5) with $y(t)$ as the input to the system and $x(t)$ the output and for which the rate of change of output signal is proportional to $(b_0y - a_0x)$. Then the differential equation is of the form

$$a_1\frac{dx}{dt} + a_0x = b_0y$$

where a_1, a_0 and b_0 are constants.

The input signal $y(t)$ to systems can take many forms. A common form is a step input. This is when the input abruptly changes in value, as in figure 10.6(a). Other common forms of input that are often encountered are impulse, ramp, and sinusoidal signals. An impulse is a very short duration input (figure 10.6(b)). A ramp is a steadily increasing input (figure 10.6(c)) and can be described by an equation of the form $y = kt$, where k is a constant. A sinusoidal input (figure 10.6(d)) can be described by an equation of the form $y = k \sin \omega t$, with ω being the so-called angular frequency and equal to $2\pi f$, with f being the frequency.

10.2.2 Solving a first-order differential equation

A method that we can use to solve a first-order equation and give an equation which directly indicates how the output varies with time involves the recognition of the type of solution that would fit the differential equation and then establishing that such a solution is valid.

Consider a first-order system which has an input of $y(t)$ and an output of $x(t)$ and a differential equation of the form

$$a_1\frac{dx}{dt} + a_0x = b_0y$$

The input $y(t)$ can take many forms. Consider first the situation when the input is 0. Because there is no input to the system we have no signal forcing the system to respond in any way other than its natural response with no input. The differential equation is then

$$a_1\frac{dx}{dt} + a_0x = 0$$

Let us try a solution of the form $x = A\,e^{st}$, where A and s are constants. We then have $dx/dt = sA\,e^{st}$ and so when these values are substituted in the differential equation we obtain

$$a_1sA\,e^{st} + a_0A\,e^{st} = 0$$

and so

$$a_1 s + a_0 = 0$$

with $s = -a_0/a_1$. Thus the solution is

$$x = A \, e^{-a_0 t/a_1}$$

This is termed the *natural response* since there is no forcing function.

Now consider the differential equation when there is a *forcing function*, i.e.

$$a_1 \frac{dx}{dt} + a_0 x = b_0 y$$

Consider the solution to this equation to be made up of two parts, i.e. $x = u + v$. One part represents the transient part of the solution and the other the steady-state part. Substituting this into the differential equation gives

$$a_1 \frac{d(u + v)}{dt} + a_0 (u + v) = b_0 y$$

Rearranging this gives

$$\left(a_1 \frac{du}{dt} + a_0 u \right) + \left(a_1 \frac{dv}{dt} + a_0 v \right) = b_0 y$$

If we let

$$a_1 \frac{dv}{dt} + a_0 v = b_0 y$$

then we must have

$$a_1 \frac{du}{dt} + a_0 u = 0$$

and so two differential equations, one of which contains a forcing function and one which is just the natural response equation. This last equation is just the natural equation which we solved earlier in this section and so will have a solution of the form

$$u = A \, e^{-a_0 t/a_1}$$

The other differential equation contains the forcing function y. For this differential equation the form of solution we try depends on the form of the input signal y. For a step input when y

is constant for all times greater than 0, i.e. $y = k$, we can try a solution $v = A$, where A is a constant. If we have an input signal of the form $y = a + bt + ct^2 + \ldots$, where a, b and c are constants which can be zero, then we can try a solution which is of the form $v = A + Bt + Ct^2 + \ldots$. For a sinusoidal signal we can try a solution of the form $v = A \cos \omega t + B \sin \omega t$.

To illustrate this, assume there is a step input at a time of $t = 0$ with the size of the step being k. Then we try a solution of the form $v = A$. Differentiating a constant gives 0; thus when this solution is substituted into the differential equation we obtain

$$a_0 A = b_0 k$$

Thus $v = (b_0/a_0)k$

The full solution will be given by $y = u + v$ and so be

$$y = A\,e^{-a_0 t/a_1} + \frac{b_0}{a_0}k$$

We can determine the value of the constant A given some initial (boundary) conditions. Thus if the output $y = 0$ when $t = 0$ then

$$0 = A + \frac{b_0}{a_0}k$$

Thus $A = -(b_0/a_0)k$. The solution then becomes

$$x = \frac{b_0}{a_0}k(1 - e^{-a_0 t/a_1})$$

When $t \to \infty$ the exponential term tends to 0. The exponential term thus gives that part of the response which is the transient solution. The steady-state response is the value of x when $t \to \infty$ and so is $(b_0/a_0)k$. Thus the equation can be written as

$$x = \text{steady-state value} \times (1 - e^{-a_0 t/a_1})$$

Figure 10.7 shows a graph of how the output x varies with time for a step input.

To illustrate the above, consider an electrical transducer system which consists of a resistance in series with a capacitor. When subject to a step input of size V it is found to give an output of a potential difference across the capacitor v which is given by the differential equation

$$RC\frac{dv}{dt} + v = V$$

What is the solution of the differential equation, i.e. what is the

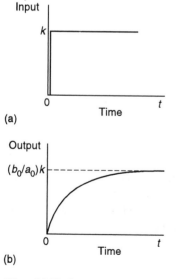

Input

k

0 Time t

(a)

Output

$(b_0/a_0)k$

0 Time t

(b)

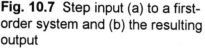

Fig. 10.7 Step input (a) to a first-order system and (b) the resulting output

response of the system and how does v vary with time?

Comparing the differential equation with the equation solved earlier: $a_1 = RC$, $a_0 = 1$, and $b_0 = 1$. Then the solution is of the form

$$v = V(1 - e^{-t/RC})$$

Consider another example but this time with a ramp input. An electrical circuit consists of a 1 MΩ resistance in series with a 2 μF capacitance. At a time $t = 0$ the circuit is subject to a ramp voltage of $4t$ V, i.e. the voltage increases at the rate of 4 V every 1 s. Determine how the voltage across the capacitor will vary with time.

The differential equation will be of a similar form to that given in the previous example but with the step voltage V of that example replaced by the ramp voltage $4t$, i.e.

$$RC\frac{dv}{dt} + v = 4t$$

Thus, using the values given in the question,

$$2\frac{dv}{dt} + v = 4t$$

Taking $v = v_n + v_f$, i.e. the sum of the natural and forced responses, we have

$$2\frac{dv_n}{dt} + v_n = 0$$

and

$$2\frac{dv_f}{dt} + v_f = 4t$$

For the natural response differential equation we can try a solution of the form $v_n = A\,e^{st}$. Hence, using this value

$$2As\,e^{st} + A\,e^{st} = 0$$

Thus $s = -\tfrac{1}{2}$ and so $v_n = A\,e^{-t/2}$. For the forced response differential equation, since the right hand side of the equation is $4t$ we can try a solution of the form $v_f = A + Bt$. Using this value gives

$$2B + A + Bt = 4t$$

Thus we must have $B = 4$ and $A = -2B = -8$. Hence the solution is $v_f = -8 + 4t$. Thus the full solution is

$$v = v_n + v_f = A\,e^{-t/2} - 8 + 4t$$

Since $v = 0$ when $t = 0$ we must have $A = 8$. Hence

$$v = 8\,e^{-t/2} - 8 + 4t$$

As another example, consider a motor when the relationship between the output angular velocity ω and the input voltage v for a motor is given by

$$\frac{IR}{k_1 k_2}\frac{d\omega}{dt} + \omega = \frac{1}{k_1}v$$

What will be the steady-state value of the angular velocity when the input is a step of size 1 V?

Comparing the differential equation with the equation solved earlier, then $a_1 = IR/k_1 k_2$, $a_0 = 1$ and $b_0 = 1/k_1$. The steady-state value for a step input is then given by $(b_0/a_0)k = 1/k_1$.

10.2.3 The time constant

For a first-order system subject to a step input of size k we have an output y which varies with time t according to

$$x = \frac{b_0}{a_0}k(1 - e^{-a_0 t/a_1})$$

or

$$x = \text{steady-state value} \times (1 - e^{-a_0 t/a_1})$$

When the time $t = (a_1/a_0)$ then the exponential term has the value $e^{-1} = 0.37$ and

$$x = \text{steady-state value} \times (1 - 0.37)$$

In this time the output has risen to 0.63 of its steady-state value. This time is called the *time constant* τ. In a time of $2(a_1/a_0) = 2\tau$, the exponential term becomes $e^{-2} = 0.14$ and so

$$x = \text{steady-state value} \times (1 - 0.14)$$

In this time the output has risen to 0.86 of its steady state value. In a similar way, values can be calculated for the output after 3τ, 4τ, 5τ, etc. Table 10.1 shows the results of such calculations and figure 10.8 the graph of how the output varies with time for a unit

Table 10.1 Response of a first-order system to a step input

Time t	Fraction of steady-state output
0	0
1τ	0.63
2τ	0.86
3τ	0.95
4τ	0.98
5τ	0.99
∞	1

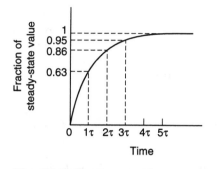

Fig. 10.8 Response of a first-order system to a step input

step input. In terms of the time constant τ, we can write the equation as

$$x = \text{steady-state value} \times (1 - e^{-t/\tau})$$

The time constant τ is (a_1/a_0), thus we can write our general form of the first-order differential equation

$$a_1 \frac{dx}{dt} + a_0 x = b_0 y$$

as

$$\tau \frac{dx}{dt} + x = \frac{b_0}{a_0} y$$

But b_0/a_0 is the factor by which the input y is multiplied to give the steady-state value. We can term it the *steady-state gain* since it is the factor stating by how much bigger the output is than the input under steady-state conditions. Thus if we denote this by G_{ss} then the differential equation can be written in the form

$$\tau \frac{dx}{dt} + x = G_{ss} y$$

To illustrate this consider figure 10.9 which shows how the output v_o of a first-order system varies with time when subject to a step input of 5 V. The time constant is the time taken for a first-order system output to change from 0 to 0.63 of its final steady-state value. In this case this time is about 3 s. We can check this value, and that the system is first order, by finding the value at 2, i.e. 6 s. With a first-order system it should be 0.86 of the steady-state value. In this case it is. The steady-state output is 10 V. Thus the steady-state gain G_{ss} is (steady-state output/input) = 10/5 = 2. The differential equation for a first-order system can be written as

$$\tau \frac{dy}{dt} + y = G_{ss} x$$

Thus, for this system, we have

$$3 \frac{dv_o}{dt} + v_o = 2v_i$$

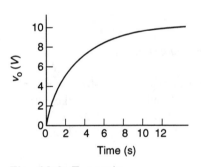

Fig. 10.9 Example

10.3 Second-order systems

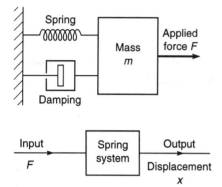

Many second-order systems can be considered to be essentially just a stretched spring with a mass and some means of providing damping. Figure 10.10 shows the basis of such a system. Such a system was analysed in section 8.1.2. The equation describing the relationship between the input of force F and the output of a displacement x is

$$m\frac{d^2x}{dt^2} + c\frac{dx}{dt} + kx = F$$

where m is the mass, c the damping constant and k the spring constant.

This is a second-order differential equation of a system which has had an abrupt application of the force F, i.e. a step input. The way in which the resulting displacement x will vary with time will depend on the amount of damping in the system. If there was no damping at all then the mass would freely oscillate on the spring and the oscillations would continue indefinitely. No damping means $c\,dx/dt = 0$. However, damping will cause the oscillations to die away until a steady displacement of the mass is obtained. If the damping is high enough there will be no oscillations and the displacement of the mass will just slowly increase with time and gradually the mass will move towards its steady displacement position (figure 10.11).

Fig. 10.10 A second-order spring system

10.3.1 The second-order differential equation

Consider a mass on the end of a spring. In the absence of any damping and left to freely oscillate without being forced the output of the second-order system is a continuous oscillation (simple harmonic motion). Thus, suppose we describe this oscillation by the equation

$$x = A \sin \omega_n t$$

where x is the displacement at a time t, A the amplitude of the oscillation and ω_n the angular frequency of the free undamped oscillations. Differentiating this gives

$$\frac{dy}{dt} = \omega_n A \cos \omega_n t$$

Differentiating a second time gives

$$\frac{d^2y}{dt^2} = -\omega_n^2 A \sin \omega_n t = -\omega_n^2 y$$

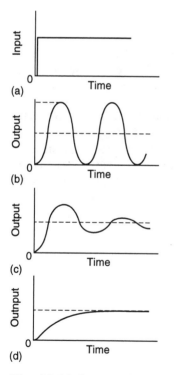

Fig. 10.11 Outputs for a step input (a): with (b) no damping, (c) some damping, (d) high damping

This can be reorganised to give the differential equation

$$\frac{d^2y}{dt^2} + \omega_n^2 y = 0$$

But for a mass m on a spring of stiffness k we have a restoring force of

$$m\frac{d^2x}{dt^2} = -kx$$

and thus

$$m\frac{d^2x}{dt^2} + kx = 0$$

$$\frac{d^2x}{dt^2} + \frac{k}{m}x = 0$$

Thus, comparing the two differential equations, we must have

$$\omega_n^2 = \frac{k}{m}$$

and $x = A \sin \omega_n t$ is the solution to the differential equation.

Now consider when we have damping. The motion of the mass on the spring (figure 10.10) when subject to a step input F is then described by

$$m\frac{d^2x}{dt^2} + c\frac{dx}{dt} + kx = F$$

We can solve this second-order differential equation by the same method used earlier for the first-order differential equation and consider the solution to be made up of two elements, a transient response and a forced response, i.e. $x = x_n + x_f$. Substituting for x in the above equation then gives

$$m\frac{d^2(x_n + x_f)}{dt^2} + c\frac{d(x_n + x_f)}{dt} + k(x_n + x_f) = F$$

If we let

$$m\frac{d^2x_n}{dt^2} + c\frac{dx_n}{dt} + kx_n = 0$$

then we must have

$$m\frac{d^2x_f}{dt^2} + c\frac{dx_f}{dt} + kx_f = F$$

To solve the transient equation we can try a solution of the form $x_n = A\,e^{st}$. With such a solution $dx_n/dt = As\,e^{st}$ and $d^2x_n/dt^2 = As^2\,e^{st}$. Thus, substituting these values in the differential equation gives

$$mAs^2\,e^{st} + cAs\,e^{st} + kA\,e^{st} = 0$$

$$ms^2 + cs + k = 0$$

Thus $x_n = A\,e^{st}$ can only be a solution provided the above equation equals 0. This equation is called the *auxiliary equation*. The roots of the equation can be obtained by factoring or using the formula for the roots of a quadratic equation. Thus

$$s = \frac{-c \pm \sqrt{c^2 - 4mk}}{2m} = -\frac{c}{2m} \pm \sqrt{\left(\frac{c}{2m}\right)^2 - \frac{k}{m}}$$

$$= -\frac{c}{2m} \pm \sqrt{\frac{k}{m}\left(\frac{c^2}{4mk}\right) - \frac{k}{m}}$$

But $\omega_n^2 = k/m$ and so, if we let $\zeta^2 = c^2/4mk$ we can write the above equation as

$$s = -\zeta\omega_n \pm \omega_n\sqrt{\zeta^2 - 1}$$

ζ is termed the *damping factor*.

The value of s obtained from the above equation depends very much on the value of the square root term. Thus when ζ^2 is greater than 1 the square root term gives a square root of a positive number, and when ζ^2 is less than 1 we have the square root of a negative number. The damping factor in determining whether the square root term is that of a positive or negative number is a crucial factor in determining the form of the output from the system.

With $\zeta > 1$ there are two different real roots s_1 and s_2, where

$$s_1 = -\zeta\omega_n + \omega_n\sqrt{\zeta^2 - 1}$$

$$s_2 = -\zeta\omega_n - \omega_n\sqrt{\zeta^2 - 1}$$

and so the general solution for x_n is

$$x_n = A\,e^{s_1t} + B\,e^{s_2t}$$

For such conditions the system is said to be *over-damped*.

When $\zeta = 1$ there are two equal roots with $s_1 = s_2 = -\omega_n$. For this condition, which is called *critically damped*,

$$x_n = (At + B)\,e^{-\omega_n t}$$

It may seem that the solution for this case should be $x_n = A\,e^{st}$, but two constants are required and so the solution is of this form (see *Ordinary Differential Equations* by W. Bolton (Longman 1994) for a discussion of this).

With $\zeta < 1$ there are two complex roots since the roots both involve the square root of (-1).

$$s = -\zeta\omega_n \pm \omega_n \sqrt{\zeta^2 - 1} = -\zeta\omega_n \pm \omega_n \sqrt{-1}\,\sqrt{1 - \zeta^2}$$

and so writing j for $\sqrt{-1}$,

$$s = -\zeta\omega_n \pm j\omega_n \sqrt{1 - \zeta^2}$$

If we let

$$\omega = \omega_n \sqrt{1 - \zeta^2}$$

then we can write

$$s = -\zeta\omega_n \pm j\omega$$

and so the two roots are

$$s_1 = -\zeta\omega_n + j\omega$$

$$s_2 = -\zeta\omega_n - j\omega$$

The term ω is the angular frequency of the motion when it is in the damped condition specified by ζ. The solution under these conditions is thus

$$x_r = A\,e^{(-\zeta\omega_n + j\omega)t} + B\,e^{(-\zeta\omega_n - j\omega)t}$$

$$= e^{-\zeta\omega_n t}(A\,e^{j\omega t} + B\,e^{-j\omega t})$$

But $e^{j\omega t} = \cos \omega t + j \sin \omega t$ and $e^{-j\omega t} = \cos \omega t - j \sin \omega t$. Hence

$$x_n = e^{-\zeta\omega_n t}(A \cos \omega t + jA \sin \omega t + B \cos \omega t - jB \sin \omega t)$$

$$= e^{-\zeta\omega_n t}[(A + B) \cos \omega t + j(A - B) \sin \omega t]$$

If we substitute constants P and Q for $(A + B)$ and $j(A - B)$, then

$$x_n = e^{-\zeta\omega_n t}(P \cos \omega t + Q \sin \omega t)$$

For such conditions the system is said to be *under-damped*.

The above has thus given the solutions for the natural part of the solution. To solve the forcing equation,

$$m\frac{d^2 x_f}{dt^2} + c\frac{dx_f}{dt} + kx_f = F$$

we need to consider a particular form of input signal and then try a solution. Thus for a step input of size F at time $t = 0$ we can try a solution $x_f = A$, where A is a constant (see the discussion of the solution of first-order differential equations for a discussion of the choice of solutions). Then $dx_f/dt = 0$ and $d^2 x_f/dt^2 = 0$. Thus, when these are substituted in the differential equation

$$0 + 0 + kA = F$$

Thus $A = F/k$ and so $x_f = F/k$. The complete solution, which is the sum of natural and forced solutions, is thus for the over-damped system

$$x = A e^{s_1 t} + B e^{s_2 t} + \frac{F}{k}$$

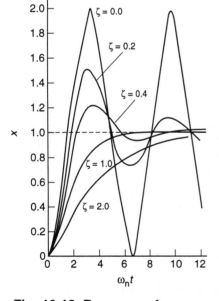

for the critically damped system

$$x = (At + B) e^{-\omega_n t} + \frac{F}{k}$$

and for the under-damped system

$$x = e^{-\zeta\omega_n t}(P \cos \omega t + Q \sin \omega t) + \frac{F}{k}$$

When $t \to \infty$ the above three equations all lead to the solution $x = F/k$. This is the steady-state condition.

Figure 10.12 shows graphs of the output as a function of time for different degrees of damping, i.e. different values of the damping factor ζ. The time axis is $\omega_n t$. This has been used so that the graphs fit second-order systems regardless of their value of ω_n. As a consequence, the $x = 0$ values for the undamped oscillation occur for $\omega_n t = 0, 2, 4$, etc.

In general we can write a second-order differential equation in the form

$$a_2\frac{d^2 x}{dt^2} + a_1\frac{dx}{dt} + a_0 x = b_0 y$$

Fig. 10.12 Response of a second-order system to a unit step input

with

$$\omega_n^2 = \frac{a_0}{a_2}$$

and

$$\zeta^2 = \frac{a_1^2}{4a_2a_0}$$

The following examples are designed to illustrate the points made above.

Consider a series RLC circuit (figure 10.13) with $R = 100\ \Omega$, $L = 2.0$ H and $C = 20\ \mu$F. The current i in the circuit is given by

$$\frac{d^2i}{dt^2} + \frac{R}{L}\frac{di}{dt} + \frac{1}{LC}i = \frac{V}{LC}$$

when there is a step input V. If we compare the equation with the general second-order differential equation given above, the natural angular frequency is given by

$$\omega_n^2 = \frac{1}{LC} = \frac{1}{2.0 \times 20 \times 10^{-6}}$$

and so $\omega_n = 158$ Hz. Comparison with the general second-order equation also gives

$$\zeta^2 = \frac{(R/L)^2}{4 \times (1/LC)} = \frac{R^2C}{4L} = \frac{100^2 \times 20 \times 10^{-6}}{4 \times 2.0}$$

Thus $\zeta = 0.16$. Since ζ is less than 1 the system is under-damped. The damped oscillation frequency ω is given by

$$\omega = \omega_n\sqrt{1-\zeta^2} = 158\sqrt{1-0.16^2} = 156\text{ Hz}$$

Because the system is under-damped the solution will be of the same form as

$$x = e^{-\zeta\omega_n t}(P\cos\omega t + Q\sin\omega t) + \frac{F}{k}$$

and so

$$i = e^{-0.16\times158t}(P\cos156t + Q\sin156t) + V$$

Since $i = 0$ when $t = 0$, then

$$0 = 1(P + 0) + V$$

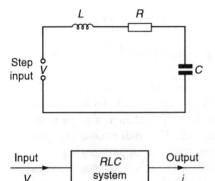

Fig. 10.13 Series RLC circuit

Thus $P = -V$. Since $di/dt = 0$ when $t = 0$ then differentiating the above equation and equating it to zero gives

$$\frac{di}{dt} = e^{-\zeta\omega_n t}(\omega P \sin \omega t - \omega Q \cos \omega t)$$
$$- \zeta\omega_n e^{-\zeta\omega_n t}(P \cos \omega t + Q \cos \omega t)$$

$$0 = 1(0 - \omega Q) - \zeta\omega_n(P + 0)$$

$$Q = \frac{\zeta\omega_n P}{\omega} = -\frac{\zeta\omega_n V}{\omega} = -\frac{0.16 \times 158V}{156} \approx -0.16V$$

Thus the solution of the differential equation is

$$i = V - Ve^{-25.3t}(\cos 156t + 0.16 \sin 156t)$$

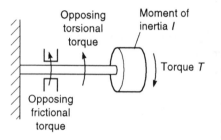

Fig. 10.14 Torsional system

As another example, consider the system shown in figure 10.14. The input, a torque T, is applied to a disc with a moment of inertia I about the axis of the shaft. The shaft is free to rotate at the disc end but fixed at its far end. The shaft rotation is opposed by the torsional stiffness of the shaft, an opposing torque of $k\theta_o$ occurring for an input rotation of θ_o. k is a constant. Frictional forces damp the rotation of the shaft and provide an opposing torque of $c\, d\theta_o/dt$, where c is a constant. What is the condition for this system to be critically damped?

We first need to obtain the differential equation for the system. The net torque is

$$\text{net torque} = T - c\frac{d\theta_o}{dt} - k\theta_o$$

The net torque is $I\, d^2\theta_o/dt$, hence

$$I\frac{d^2\theta_o}{dt^2} = T - c\frac{d\theta_o}{dt} - k\theta_o$$

$$I\frac{d^2\theta_o}{dt^2} + c\frac{d\theta_o}{dt} + k\theta_o = T$$

The condition for critical damping is given when the damping ratio ζ equals 1. Comparing the above differential equation with the general form of the second-order differential equation, then

$$\zeta^2 = \frac{a_1^2}{4a_2 a_0} = \frac{c^2}{4Ik}$$

Thus for critical damping we must have $c = \sqrt{(Ik)}$.

10.4 Performance measures for second-order systems

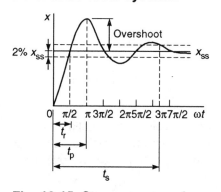

Fig. 10.15 Step response of an under-damped system

Figure 10.15 shows the typical form of the response of an under-damped second-order system to a step input. Certain terms are used to specify such a performance.

The *rise time* t_r is the time taken for the response x to rise from 0 to the steady-state value x_{ss} and is a measure of how fast a system responds to the input. This is the time for the oscillating response to complete a quarter of a cycle, i.e. $\frac{1}{2}\pi$. Thus

$$\omega t_r = \frac{1}{2}\pi$$

The rise time is sometimes specified as the time taken for the response to rise from some specified percentage of the steady-state value, e.g. 10%, to another specified percentage, e.g. 90%.

The *peak time* t_p is the time taken for the response to rise from 0 to the first peak value. This is the time for the oscillating response to complete one half cycle, i.e. π. Thus

$$\omega t_p = \pi$$

The *overshoot* is the maximum amount by which the response overshoots the steady-state value. It is thus the amplitude of the first peak. The overshoot is often written as a percentage of the steady-state value. For the under-damped oscillations of a system we can write

$$x = e^{-\zeta\omega_n t}(P \cos \omega t + Q \sin \omega t) + \text{steady-state value}$$

Since $x = 0$ when $t = 0$ then

$$0 = 1(P + 0) + x_{ss}$$

and so $P = -x_{ss}$. The overshoot occurs at $\omega t = \pi$ and thus

$$x = e^{-\zeta\omega_n \pi/\omega}(P + 0) + x_{ss}$$

The overshoot is the difference between the output at that time and the steady-state value. Hence

$$\text{Overshoot} = x_{ss}\, e^{-\zeta\omega_n \pi/\omega}$$

Since $\omega = \omega_n\sqrt{(1 - \zeta^2)}$ then we can write

$$\text{Overshoot} = x_{ss} \exp\left(\frac{-\zeta\omega_n \pi}{\omega_n\sqrt{1-\zeta^2}}\right)$$

$$= x_{ss} \exp\left(\frac{-\zeta\pi}{\sqrt{1-\zeta^2}}\right)$$

Expressed as a percentage of x_{ss},

$$\text{Percentage overshoot} = \exp\left(\frac{-\zeta\pi}{\sqrt{1-\zeta^2}}\right) \times 100\%$$

Table 10.2 Percentage peak overshoot

Damping ratio	Percentage overshoot
0.2	52.7
0.4	25.4
0.6	9.5
0.8	1.5

Table 10.2 gives values of the percentage overshoot for particular damping ratios.

An indication of how fast oscillations decay is provided by the *subsidence ratio* or *decrement*. This is the amplitude of the second overshoot divided by that of the first overshoot. The first overshoot occurs when we have $\omega t = \pi$, the second overshoot when $\omega t = 2\pi$. Thus,

$$\text{First overshoot} = x_{ss}\exp\left(\frac{-\zeta\pi}{\sqrt{1-\zeta^2}}\right)$$

and

$$\text{Second overshoot} = x_{ss}\exp\left(\frac{-2\zeta\pi}{\sqrt{1-\zeta^2}}\right)$$

Thus,

$$\text{Subsidence ratio} = \frac{\text{second overshoot}}{\text{first overshoot}}$$

$$= \exp\left(\frac{-\zeta\pi}{\sqrt{1-\zeta^2}}\right)$$

The *settling time* t_s is used as a measure of the time taken for the oscillations to die away. It is the time taken for the response to fall within and remain within some specified percentage, e.g. 2%, of the steady-state value (see figure 10.14). This means that the amplitude of the oscillation should be less than 2% of x_{ss}. We have

$$x = e^{-\zeta\omega_n t}(P\cos\omega t + Q\sin\omega t) + \text{steady-state value}$$

and, as derived earlier, $P = -x_{ss}$. The amplitude of the oscillation is $(x - x_{ss})$ when x is a maximum value. The maximum values occur when ωt is some multiple of π and thus we have $\cos\omega t = 1$ and $\sin\omega t = 0$. For the 2% settling time, the settling time t_s is when the maximum amplitude is 2% of x_{ss}, i.e. $0.02x_{ss}$. Thus

$$0.02 x_{SS} = e^{-\zeta \omega_n t_s}(x_{SS} \times 1 + 0)$$

$$0.02 = e^{-\zeta \omega_n t_s}$$

Taking logarithms

$$\ln 0.02 = -\zeta \omega_n t_s$$

$\ln 0.02 = -3.9$ or approximately 4. Thus

$$t_s = \frac{4}{\zeta \omega_n}$$

The above is the value of the settling time if the specified percentage is 2%. If the percentage is 5% the equation becomes

$$t_s = \frac{3}{\zeta \omega_n}$$

Since the time taken to complete one cycle, i.e. the periodic time, is $1/f$, where f is the frequency, and since $\omega = 2\pi f$ then the time to complete one cycle is

$$\text{Periodic time} = \frac{2\pi}{\omega}$$

Hence, in a settling time of t_s the number of oscillations that occur is

$$\text{Number of oscillations} = \frac{\text{settling time}}{\text{periodic time}}$$

and thus for a settling time defined for 2% of the steady-state value,

$$\text{Number of oscillations} = \frac{4/\zeta \omega_n}{2\pi/\omega}$$

Since $\omega = \omega_n \sqrt{(1 - \zeta^2)}$, then

$$\text{Number of oscillations} = \frac{2\omega_n \sqrt{1 - \zeta^2}}{\pi \zeta \omega_n}$$

$$= \frac{2}{\pi} \sqrt{\frac{1}{\zeta^2} - 1}$$

To illustrate the above, consider a second-order system which has a natural angular frequency of 2.0 Hz and a damped frequency of 1.8 Hz. Since $\omega = \omega_n \sqrt{(1 - \zeta^2)}$, then the damping factor is given by

$$1.8 = 2.0\sqrt{1 - \zeta^2}$$

and so $\zeta = 0.44$. Since $\omega t_r = \frac{1}{2}\pi$, then the 100% rise time is given by

$$t_r = \frac{\pi}{2 \times 1.8} = 0.87 \text{ s}$$

The percentage overshoot is given by

$$\text{Percentage overshoot} = \exp\left(\frac{-\zeta\pi}{\sqrt{1 - \zeta^2}}\right) \times 100\%$$

$$= \exp\left(\frac{-0.44\pi}{\sqrt{1 - 0.44^2}}\right) \times 100\%$$

$$= 21\%$$

The 2% settling time is given by

$$t_s = \frac{4}{\zeta\omega_n} = \frac{4}{0.44 \times 2.0} = 4.5 \text{ s}$$

The number of oscillations occurring within the 2% settling time is given by

$$\text{Number of oscillations} = \frac{2}{\pi}\sqrt{\frac{1}{\zeta^2} - 1} = \frac{2}{\pi}\sqrt{\frac{1}{0.44^2} - 1} = 1.3$$

Problems

1   A first-order system has a time constant of 4 s and a steady-state transfer function of 6. What is the form of the differential equation for this system?

2   A mercury-in-glass thermometer has a time constant of 10 s. If it is suddenly taken from being at 20°C and plunged into hot water at 80°C, what will be the temperature indicated by the thermometer after (a) 10 s, (b) 20 s?

3   A circuit consists of a resistor R in series with an inductor L. When subject to a step input voltage V at time $t = 0$ the differential equation for the system is

$$\frac{di}{dt} + \frac{R}{L}i = \frac{V}{L}$$

What is (a) the solution for this differential equation, (b) the time constant, (c) the steady-state current i?

4 Describe the form of the output variation with time for a step input to a second-order system with a damping factor of (a) 0, (b) 0.5, (c) 1.0, (d) 1.5.

5 A *RLC* circuit has a current i which varies with time t when subject to a step input of V according to the following differential equation

$$\frac{d^2 i}{dt^2} + 10\frac{di}{dt} + 16i = 16V$$

What is (a) the undamped frequency, (b) the damping ratio, (c) the solution to the equation if $i = 0$ when $t = 0$ and $di/dt = 0$ when $t = 0$?

6 A system has an output x which varies with time t when subject to a step input of y according to the following differential equation

$$\frac{d^2 x}{dt^2} + 10\frac{dx}{dt} + 25x = 50y$$

What is (a) the undamped frequency, (b) the damping ratio, (c) the solution to the equation if $x = 0$ when $t = 0$ and $dx/dt = -2$ when $t = 0$ and there is a step input of size 3 units?

7 An accelerometer (an instrument for measuring acceleration) has an undamped angular frequency of 100 Hz and a damping factor of 0.6. What will be (a) the maximum percentage overshoot and (b) the rise time when there is a sudden change in acceleration?

8 What will be (a) the undamped angular frequency, (b) the damping factor, (c) the damped angular frequency, (d) the rise time, (e) the percentage maximum overshoot and (f) the 0.2 % settling time for a system which gave the following differential equation for a step input y?

$$\frac{d^2 x}{dt^2} + 5\frac{dx}{dt} + 16x = 16y$$

9 When a voltage of 10 V is suddenly applied to a moving coil voltmeter it is observed that the pointer of the instrument rises to 11 V before eventually settling down to read 10 V. What is (a) the damping factor and (b) the number of oscillations the pointer will make before it is within 0.2% of its steady-state value?

11 System transfer functions

11.1 The transfer function

For an amplifier system it is customary to talk of the *gain* of the amplifier. This states how much bigger the output signal will be when compared with the input signal. It enables the output to be determined for specific inputs. Thus, for example, an amplifier with a voltage gain of 10 will give, for an input voltage of 2 mV, an output of 20 mV; or if the input is 1 V an output of 10 V. The gain states the mathematical relationship between the output and the input for the block.

$$\text{Gain} = \frac{\text{output}}{\text{input}}$$

However, for many systems the relationship between the output and the input is in the form of a differential equation and so a statement of the function as just a simple number like the gain of 10 is not possible. We cannot just divide the output by the input because the relationship is a differential equation and not a simple algebraic equation. We can however transform a differential equation into an algebraic equation by using what is termed the *Laplace transform*. Differential equations describe how systems behave with time and are transformed by means of the Laplace transform into simple algebraic equations, not involving time, where we can carry out normal algebraic manipulations of the quantities. We talk of behaviour in the *time domain* being transformed to the *s-domain*. Then we can define the relationship between output and input in terms of a *transfer function*. The transfer function states the relationship between the Laplace transform of the output and the Laplace transform of the input, i.e.

$$\text{Transfer function} = \frac{\text{Laplace transform of output}}{\text{Laplace transform of input}}$$

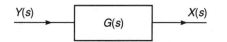

Fig. 11.1 Block diagram

We can indicate when a signal is in the time domain, i.e. is a function of time, by writing it as $f(t)$. When in the s-domain a function is written, since it is a function of s, as $F(s)$. It is usual to use a capital letter F for the Laplace transform and a lower-case letter f for the time-varying function $f(t)$.

Suppose that the input to a linear system has a Laplace transform of $Y(s)$ and the Laplace transform of the output is $X(s)$. The *transfer function* $G(s)$ of the system is then defined as

$$G(s) = \frac{X(s)}{Y(s)}$$

with all the initial conditions being zero, i.e. we assume zero output when zero input, zero rate of change of output with time when zero rate of change of input with time. Thus the output transform $X(s) = G(s)Y(s)$ and is thus the product of the input transform and the transfer function. If we represent a system by a block diagram (figure 11.1) then $G(s)$ is the function in the box which takes an input of $Y(s)$ and converts it to an output of $X(s)$.

This chapter gives an indication of how Laplace transforms can be used in relation to the transfer functions of systems. For more details the reader is referred to Appendix A or *Laplace and z-Transforms* by W. Bolton (Longman 1994), part of the Mathematics for Engineers series.

11.1.1 Laplace transforms

To obtain the Laplace transform of some differential equation which includes quantities which are functions of time we can use tables coupled with a few basic rules, Appendix A includes such a table and more details of the rules.

The following are a few basic transforms:

1   A unit impulse signal which occurs at time $t = 0$ has a transform of 1.
2   A unit step signal, i.e. a signal which jumps to the constant value of 1 at a time $t = 0$, has a transform of $1/s$.
3   A unit ramp signal which starts at time $t = 0$, i.e. is described by the equation input $= 1t$, has a transform of $1/s^2$.
4   A unit amplitude sine wave signal, i.e. is described by the equation input $= 1 \sin \omega t$, has a transform of $\omega/(s^2 + \omega^2)$.
5   A unit amplitude cosine wave signal, i.e. is described by the equation output $= 1 \cos \omega t$, has a transform of $s/(s^2 + \omega^2)$.

The following are few of the basic rules involved in working with Laplace transforms:

1 If a function of time is multiplied by a constant then the Laplace transform is multiplied by the same constant. For example, the Laplace transform of a step input of 6 V to an electrical system is just 6 times the transform for a unit step and thus $6s$.

2 If an equation includes the sum of, say, two separate quantities which are functions of time, then the transform of the equation is the sum of the two separate Laplace transforms.

3 The Laplace transform of a first derivative of a function is

$$\text{Transform of } \left\{\frac{\text{d}}{\text{d}t}f(t)\right\} = sF(s) - f(0)$$

where $f(0)$ is the value of $f(t)$ when $t = 0$. However, when we are dealing with a transfer function we have all initial conditions zero.

4 The Laplace transform for the second derivative of a function is

$$\text{Transform of } \left\{\frac{\text{d}^2}{\text{d}t^2}f(t)\right\} = s^2F(s) - sf(0) - \frac{\text{d}}{\text{d}t}f(0)$$

where $\text{d}f(0)/\text{d}t$ is the value of the first derivative of $f(t)$ when we have $t = 0$. However, when we are dealing with a transfer function we have all initial conditions zero.

5 The Laplace transform of an integral of a function is

$$\text{Transform of } \left\{\int_0^t f(t)\,\text{d}t\right\} = \frac{1}{s}F(s)$$

When algebraic manipulations have occurred in the s-domain, then the outcome can be transformed back to the time domain by using the table of transforms in the inverse manner, i.e. finding the time domain function which fits the s-domain result. Often the transform has to be rearranged to be put into a form given in the table. The following are some useful such inversions; for more inversions see the table given in Appendix A.

1 $\dfrac{1}{s+a}$ gives e^{-at}

2 $\dfrac{a}{s(s+a)}$ gives $(1 - e^{-at})$

3 $\dfrac{b-a}{(s+a)(s+b)}$ gives $e^{-at} - e^{-bt}$

4 $\dfrac{s}{(s+a)^2}$ gives $(1 - at)\,e^{-at}$

5    $\dfrac{a}{s^2(s+a)}$ gives $t - \dfrac{1 - e^{-at}}{a}$

The following sections illustrate the application of the above to first-order and second-order systems.

11.2 First-order systems

Consider a system where the relationship between the input and the output is in the form of a first-order differential equation. The differential equation of a first-order system is of the form

$$a_1 \frac{dx}{dt} + a_0 x = b_0 y$$

where a_1, a_0, b_0 are constants, y is the input and x the output, both being functions of time. The Laplace transform of this, with all initial conditions zero, is

$$a_1 s X(s) + a_0 X(s) = b_0 Y(s)$$

and so we can write the transfer function $G(s)$ as

$$G(s) = \frac{X(s)}{Y(s)} = \frac{b_0}{a_1 s + a_0}$$

This can be rearranged to give

$$G(s) = \frac{b_0/a_0}{(a_1/a_0)s + 1} = \frac{G}{\tau s + 1}$$

where G is the *gain* of the system when there are steady-state conditions, i.e. there is no dx/dt term. (a_1/a_0) is the time constant τ of the system (see section 10.2.3).

When a first-order system is subject to a unit step input then $Y(s) = 1/s$ and the output transform $X(s)$ is

$$X(s) = G(s)Y(s) = \frac{G}{s(\tau s + 1)}$$

$$= G \frac{(1/\tau)}{s(s + 1/\tau)}$$

Hence, since we have the transform in the form $a/s(s + a)$, using the inverse transformation listed as item 2 in the previous section gives

$$y = G(1 - e^{-t/\tau})$$

The following examples illustrate the above points in the consideration of the transfer function of a first-order system and behaviour when subject to a step input.

Consider a circuit which has a resistance R in series with a capacitance C. The input to the circuit is v and the output is the potential difference v_C across the capacitor. The differential equation relating the input and output is

$$v = RC\frac{dv_C}{dt} + v_C$$

Taking the Laplace transform, with all initial conditions zero, then

$$V(s) = RCsV_C(s) + V_C(s)$$

Hence the transfer function is

$$G(s) = \frac{V_C(s)}{V(s)} = \frac{1}{RCs + 1}$$

Now consider a thermocouple which has a transfer function linking its voltage output V and temperature input of

$$G(s) = \frac{30 \times 10^{-6}}{10s + 1} \text{ V/°C}$$

Since the transform of the output is equal to the product of the transfer function and the transform of the input, then

$$V(s) = G(s) \times \text{input } (s)$$

Suppose there is a step input of size 100°C, i.e. the temperature of the thermocouple is abruptly increased by 100°C. Its transform is $100/s$. Thus

$$V(s) = \frac{30 \times 10^{-6}}{10s + 1} \times \frac{100}{s} = \frac{30 \times 10^{-4}}{10s(s + 0.1)}$$

$$= 30 \times 10^{-4}\frac{0.1}{s(s + 0.1)}$$

The fraction element is of the form $a/s(s + a)$ and so the inverse transform is

$$V = 30 \times 10^{-4} (1 - e^{-0.1t}) \text{ V}$$

The final value, i.e. the steady-state value, is when $t \to \infty$ and so is when the exponential term is zero. The final value is therefore 30×10^{-4} V. Thus the time taken to reach, say, 95% of this is given by

$$0.95 \times 30 \times 10^{-4} = 30 \times 10^{-4} (1 - e^{-0.1t})$$

Thus $0.05 = e^{-0.1t}$ and $\ln 0.05 = -0.1t$. The time is thus 30 s.

Now consider a ramp input to the thermocouple of $5t$ °C/s, i.e. the temperature is being raised by 5°C every second. Its transform is $5/s^2$. Thus

$$V(s) = \frac{30 \times 10^{-6}}{10s + 1} \times \frac{5}{s^2} = 150 \times 10^{-6} \frac{0.1}{s^2(s + 0.1)}$$

The inverse transform can be obtained using item 5 in the list given in the previous section. Thus

$$V = 150 \times 10^{-6}\left(t - \frac{1 - e^{-0.1t}}{0.1}\right)$$

After a time of, say, $t = 12$ s we would have $V = 7.5 \times 10^{-4}$ V. Now consider an impulse input of size 100°C, i.e. the thermocouple is subject to a momentary temperature increase of 100°C. It has a transform of 100. Hence

$$V(s) = \frac{30 \times 10^{-6}}{10s + 1} \times 100 = 3 \times 10^{-4} \frac{1}{s + 0.1}$$

Hence $V = 3 \times 10^{-4} e^{-0.1t}$ V. After, say, $t = 2$ s, the thermocouple voltage $V = 1.8 \times 10^{-4}$ V.

11.3 Second-order systems

For a second-order system, the relationship between the input y and the output x is described by a differential equation of the form

$$a_2\frac{d^2x}{dt^2} + a_1\frac{dx}{dt} + a_0x = b_0y$$

where a_2, a_1, a_0 and b_0 are constants. The Laplace transform of this equation, with all initial conditions zero, is

$$a_2s^2X(s) + a_1sX(s) + a_0X(s) = b_0Y(s)$$

Hence

$$G(s) = \frac{X(s)}{Y(s)} = \frac{b_0}{a_2s^2 + a_1s + a_0}$$

An alternative way of writing the differential equation for a second-order system is

$$\frac{d^2x}{dt^2} + 2\zeta\omega_n\frac{dx}{dt} + \omega_n^2x = b_0\omega_n^2y$$

where ω_n is the natural angular frequency with which the system oscillates and ζ is the damping ratio. The Laplace transform of this equation gives

$$G(s) = \frac{X(s)}{Y(s)} = \frac{b_0\omega_n^2}{s^2 + 2\zeta\omega_n s + \omega_n^2}$$

The above are the general forms taken by the transfer function for a second-order system.

When a second-order system is subject to a unit step input, i.e. $X(s) = 1/s$, then the output transform is

$$X(s) = \frac{b_0\omega_n^2}{s(s^2 + 2\zeta\omega_n s + \omega_n)}$$

This can be rearranged as

$$X(s) = \frac{b_0\omega_n^2}{s(s+p_1)(s+p_2)}$$

where p_1 and p_2 are the roots of the equation

$$s^2 + 2\zeta\omega_n s + \omega_n = 0$$

Hence, using the equation for the roots of a quadratic equation,

$$p = \frac{-2\zeta\omega_n \pm \sqrt{4\zeta^2\omega_n^2 - 4\omega_n^2}}{2}$$

and so

$$p_1 = -\zeta\omega_n + \omega_n\sqrt{\zeta^2 - 1}$$

$$p_2 = -\zeta\omega_n - \omega_n\sqrt{\zeta^2 - 1}$$

With $\zeta > 1$ the square root term is real and the system is over-damped. The inverse transform of the equation is then given by item 14 in the table in Appendix A as

$$x = \frac{b_0\omega_n^2}{p_1 p_2}\left[1 - \frac{p_2}{p_2 - p_1}e^{-p_2 t} + \frac{p_1}{p_2 - p_1}e^{-p_1 t}\right]$$

With $\zeta = 1$ the square root term is zero and so $p_1 = p_2 = -\omega_n$. The system is critically damped. The equation then becomes

$$X(s) = \frac{b_0\omega_n^2}{s(s+\omega_n)^2}$$

This equation can be expanded by means of partial fractions (see Appendix A) to give

$$Y(s) = b_0\omega_n^2 \left[\frac{1}{s} - \frac{1}{s+\omega_n} - \frac{\omega_n}{(s+\omega_n)^2} \right]$$

Hence

$$y = b_0\omega_n^2[1 - e^{-\omega_n t} - \omega_n t e^{-\omega_n t}]$$

With $\zeta < 1$ then the inverse transform, using item 28 in the table in Appendix A, is

$$y = b_0 \left[1 - \frac{e^{-\zeta\omega_n t}}{\sqrt{1-\zeta^2}} \sin\left(\omega_n\sqrt{(1-\zeta^2)}\, t + \phi\right) \right]$$

where $\cos\phi = \zeta$. This is an under-damped oscillation.

To illustrate the above, consider what will be the state of damping of a system having the following transfer function and subject to a unit step input.

$$G(s) = \frac{1}{s^2 + 8s + 16}$$

For a unit step input $Y(s) = 1/s$ and so the output transform is

$$X(s) = \frac{1}{s(s^2 + 8s + 16)}$$

This can be simplified to

$$X(s) = \frac{1}{s(s+4)(s+4)}$$

The roots of $s^2 + 8s + 16$ are thus $p_1 = p_2 = -4$. Both the roots are real and the same and so the system is critically damped.

Now consider a second-order system subject to a ramp input. A robot arm having the following transfer function is subject to a unit ramp input.

$$G(s) = \frac{K}{(s+3)^2}$$

The output transform $X(s)$ is given by $X(s) = G(s)Y(s)$ and so is

$$X(s) = \frac{K}{(s+3)^2} \times \frac{1}{s^2}$$

We can rewrite this, using partial fractions (see Appendix A), as

$$X(s) = \frac{K}{9s^2} - \frac{2K}{9(s+3)} + \frac{K}{9(s+3)^2}$$

Hence the inverse transform is

$$x = \tfrac{1}{9}Kt - \tfrac{2}{9}Ke^{-3t} + \tfrac{1}{9}Kte^{-3t}$$

This equation thus describes how the output will vary with time.

11.4 Systems in series

If a system consists of a number of subsystems in series, as in figure 11.2, then the transfer function $G(s)$ of the system is given by

$$G(s) = \frac{X(s)}{Y(s)}$$

$$= \frac{X_1(s)}{Y(s)} \times \frac{X_2(s)}{X_1(s)} \times \frac{X(s)}{X_2(s)}$$

$$= G_1(s) \times G_2(s) \times G_3(s)$$

The transfer function of the system as a whole is the product of the transfer functions of the series elements. It has been assumed that when the subsystems were linked together that no interaction occurs between the blocks which would result in changes in the transfer functions. Thus if the subsystems are electrical circuits there can be problems when the circuits interact and load each other.

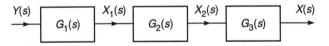

Fig. 11.2 Systems in series

To illustrate the above consider what will be the transfer function for a system consisting of three elements in series, the transfer functions of the elements being 10, 2/s, and 4/(s + 3). Using the equation developed above

$$G(s) = 10 \times \frac{2}{s} \times \frac{4}{s+3} = \frac{80}{s(s+3)}$$

As another illustration, consider a field-controlled d.c. motor. This consists of three subsystems in series, the field circuit, the armature coil and the load. Figure 11.3 illustrates the arrangement and the transfer functions of the subsystems. The overall transfer function is the product of the transfer functions of the series elements. Thus

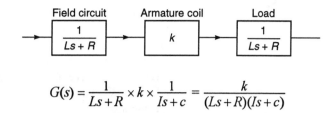

Fig. 11.3 Example

$$G(s) = \frac{1}{Ls+R} \times k \times \frac{1}{Is+c} = \frac{k}{(Ls+R)(Is+c)}$$

11.5 Systems with feedback loops

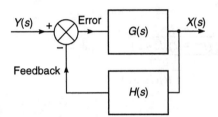

Fig. 11.4 Negative feedback system

Figure 11.4 shows a simple system having negative feedback. With *negative feedback* the system input and the feedback signals are subtracted at the summing point. The term *forward path* is used for the path having the transfer function $G(s)$ in the figure and *feedback path* for the one having $H(s)$. The entire system is referred to as a *closed-loop system*.

For the negative feedback system, the input to the subsystem having the transfer function $G(s)$ is $Y(s)$ minus the feedback signal. The feedback loop has a transfer function of $H(s)$ and has as its input $X(s)$, thus the feedback signal is $H(s)X(s)$. Thus the $G(s)$ element has an input of $Y(s) - H(s)X(s)$ and an output of $X(s)$. Hence

$$G(s) = \frac{X(s)}{Y(s) - H(s)X(s)}$$

This can be rearranged to give

$$[1 + G(s)H(s)]X(s) = G(s)Y(s)$$

$$\frac{X(s)}{Y(s)} = \frac{G(s)}{1 + G(s)H(s)}$$

Hence the overall transfer function for the negative feedback system $T(s)$ is

$$T(s) = \frac{X(s)}{Y(s)} = \frac{G(s)}{1 + G(s)H(s)}$$

To illustrate the above, consider what will be the overall transfer function for a closed-loop system having a forward path transfer function of $2/(s + 1)$ and a negative feedback path transfer function of $5s$. Using the equation developed above

$$T(s) = \frac{G(s)}{1 + G(s)H(s)} = \frac{2/(s + 1)}{1 + [2/(s + 1)]5s}$$

$$= \frac{2}{11s + 1}$$

As another illustration, consider an armature-controlled d.c. motor. This has a forward path consisting of three elements: the armature circuit with a transfer function $1/(Ls + R)$, the armature coil with a transfer function k and the load with a transfer function $1/(Is + c)$. There is a negative feedback path with a transfer function K. The forward path transfer function for the series elements is

$$G(s) = \frac{1}{Ls+R} \times k \times \frac{1}{Is+c} = \frac{k}{(Ls+R)(Is+c)}$$

The feedback path has a transfer function of K. Thus the overall transfer function is

$$T(s) = \frac{G(s)}{1+G(s)H(s)} = \frac{\dfrac{k}{(Ls+R)(Is+c)}}{1 + \dfrac{kK}{(Ls+R)(Is+c)}}$$

$$= \frac{k}{(Ls+R)(Is+c)+kK}$$

As another illustration, consider a position control system which has a negative feedback path with a transfer function of 1 and two sub-systems in its forward path: a controller with a transfer function of K and a motor/drive system with a transfer function of

$$\frac{1}{s(s+1)}$$

What value of K is necessary for the system to be critically damped? The forward path has a transfer function of $K/s(s + 1)$ and the feedback path a transfer function of 1. Thus the overall transfer function of the system is

$$T(s) = \frac{G(s)}{1+G(s)H(s)} = \frac{\dfrac{K}{s(s+1)}}{1 + \dfrac{K}{s(s+1)}} = \frac{K}{s(s+1)+K}$$

The denominator is thus $s^2 + s + K$. This will have the roots

$$s = \frac{-1 \pm \sqrt{1-4K}}{2}$$

To be critically damped we must have $1 - 4K = 0$ and thus $K = \frac{1}{4}$.

Problems

1 What are the transfer functions for systems giving the following input/output relationships?

(a) A hydraulic system has an input q and an output h where

$$q = A\frac{dh}{dt} + \frac{\rho g h}{R}$$

(b) A spring–dashpot–mass system with an input F and an output x, where

$$m\frac{d^2x}{dt^2} + c\frac{dx}{dt} + kx = F$$

(c) An RLC circuit with an input v and output v_C, where

$$v = RC\frac{dv_C}{dt} + LC\frac{d^2v_C}{dt^2} + v_C$$

2 What are the time constants of the systems giving the following transfer functions?

(a) $G(s) = \dfrac{5}{3s+1}$, (b) $G(s) = \dfrac{3}{2s+3}$

3 Determine how the outputs of the following systems vary with time when subject to a unit step input at time $t = 0$.

(a) $G(s) = \dfrac{2}{s+2}$, (b) $G(s) = \dfrac{10}{s+5}$

4 What is the state of the damping for the systems having the following transfer functions?

(a) $G(s) = \dfrac{5}{s^2 - 6s + 16}$, (b) $G(s) = \dfrac{10}{s^2 + s + 100}$,

(c) $G(s) = \dfrac{2s+1}{s^2 + 2s + 1}$, (d) $G(s) = \dfrac{3s+20}{s^2 + 2s + 20}$

5 What is the output of a system with the following transfer function and subject to a unit step input at time $t = 0$?

$$G(s) = \frac{s}{(s+3)^2}$$

6 What is the output of a system having the following transfer function and subject to a unit impulse?

$$G(s) = \frac{2}{(s+3)(s+4)}$$

7 What are the overall transfer functions of the following negative feedback systems?

(a) Forward path $G(s) = \dfrac{4}{s(s+1)}$, feedback path $H(s) = \dfrac{1}{s}$,

(b) Forward path $G(s) = \dfrac{2}{s+1}$, feedback path $H(s) = \dfrac{1}{s+2}$,

(c) Forward path $G(s) = \dfrac{4}{(s+2)(s+3)}$, feedback path $H(s) = 5$,

(d) Forward path with two elements in series $G_1(s) = \dfrac{2}{s+2}$ and $G_2(s) = \dfrac{1}{s}$, feedback path $H(s) = 10$.

8 What is the overall transfer function for a closed-loop system having a forward path transfer function of $5/(s+3)$ and a negative feedback path transfer function of 10?

9 A closed-loop system has a forward path having two series elements with transfer functions 5 and $1/(s+1)$. If the feedback path has a transfer function $2/s$, what is the overall transfer function of the system?

10 A closed-loop system has a forward path having two series elements with transfer functions of 2 and $1/(s+1)$. If the feedback path has a transfer function of s, what is the overall transfer function of the system?

12 Frequency response

12.1 Sinusoidal input

In the previous two chapters, the response of systems to step, impulse and ramp inputs has been considered. This chapter extends this to when there is a sinusoidal input. While for many control systems a sinusoidal input might not be encountered normally, it is a useful testing input since the way the system responds to such an input is a very useful source of information to aid the design and analysis of systems.

Consider a first-order system which is described by the differential equation

$$a_1 \frac{dx}{dt} + a_0 x = b_0 y$$

where y is the input and x the output. Suppose we have the unit amplitude sinusoidal input of $y = \sin \omega t$. What will the output be? Well we know that when we add $a_1\, dx/dt$ and $a_0 x$ we must end up with the sinusoid $b_0 \sin \omega t$. But sinusoids have the property that when differentiated the result is also a sinusoid and with the same frequency (a cosine is a sinusoid, being just $\sin (\omega t + 90°)$). This applies no matter how many times we carry out the differentiation. Thus we should expect that the steady-state response x will also be sinusoidal and with the same frequency. The output may, however, differ in amplitude and phase from the input.

12.2 Phasors

In discussing sinusoidal signals it is convenient to use *phasors*. Consider a sinusoid described by the equation $v = V \sin (\omega t + \phi)$, where V is the amplitude, ω the angular frequency and ϕ the phase angle. The phasor can be represented by a line of length $|V|$ making an angle of ϕ with the phase reference axis. The $| \; |$ lines are used to indicate that we are only concerned with the magnitude or size of the quantity when specifying its length. A phasor quantity in order to be specified always requires a magnitude and angle to

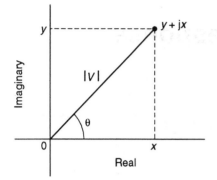

Fig. 12.1 Complex representation of a phasor

be specified. The convention adopted in this book is to write a phasor in bold, non-italic, print, e.g. **V**. When such a symbol is seen it implies a quantity having both a magnitude and an angle.

Such a phasor can be described by means of complex number notation. A complex quantity can be represented by $(x + jy)$, where x is the real part and y the imaginary part of the complex number. On a graph with the imaginary component as the y-axis and the real part as the x-axis, x and y are Cartesian coordinates of the point representing the complex number (figure 12.1). If we take the line joining this point to the graph origin to represent a phasor, then we have the phase angle ϕ of the phasor represented by

$$\tan \phi = \frac{y}{x}$$

and its length by the use of Pythagoras' theorem as

$$\text{Length of phasor } |V| = \sqrt{x^2 + y^2}$$

and

$$x = |V| \cos \phi$$

$$y = |V| \sin \phi$$

Thus we can write

$$\mathbf{V} = x + jy = |V| (\cos \theta + j \sin \theta)$$

Thus a specification of the real and imaginary parts of a complex number enables a phasor to be specified.

Consider a phasor of length 1 and phase angle 0° (figure 12.2(a)). It will have a complex representation of $1 + j0$. Now consider the same length phasor but with a phase angle of 90° (figure 12.2(b)). It will have a complex representation of $0 + j1$. Thus rotation of a phasor anticlockwise by 90° corresponds to multiplication of the phasor by j. If we now rotate this phasor by a further 90° (figure 12.2(c)), then following the same multiplication rule we have the original phasor multiplied by j^2. But the phasor is just the original phasor in the opposite direction, i.e. just multiplied by -1. hence $j^2 = -1$ and so $j = \sqrt{(-1)}$. Rotation of the original

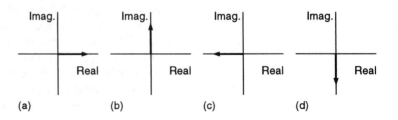

Fig. 12.2 Phasor rotation (a) 0°, (b) 90°, (c) 180°, (d) 270°

phasor through a total of 270°, i.e. $3 \times 90°$, is equivalent to multiplying the original phasor by $j^3 = j(j^2) = -j$.

For further discussion of complex numbers and their application in engineering, the reader is referred to *Complex Numbers* by W. Bolton (Longman 1994), part of their Mathematics for Engineers series.

To illustrate the above, consider a voltage v which varies sinusoidally with time according to the equation

$$v = 10 \sin (\omega t + 30°) \text{ V}$$

When represented by a phasor, what are (a) its length, (b) its angle relative to the reference axis, (c) its real and imaginary parts when represented by a complex number?

(a) The phasor will have a length scaled to represent the amplitude of the sinusoid and so is 10 V.
(b) The angle of the phasor relative to the reference axis is equal to the phase angle and so is 30°.
(c) The real part is given by the equation $x = 10 \cos 30° = 8.7$ V and the imaginary part by $y = 10 \sin 30° = 5.0$ V. Thus the phasor is specified by $8.7 + j5.0$ V.

12.2.1 Phasor equations

Consider a phasor to represent the unit amplitude sinusoid of $x = \sin \omega t$. Differentiation of the sinusoid gives $dx/dt = \omega \cos \omega t$. But we can also write this as $dx/dt = \omega \sin (\omega t + 90°)$. In other words, differentiation has just results in a phasor with a length increased by a factor of ω and which is rotated round by 90° from the original phasor. Thus, in complex notation, we have multiplied the original phasor by $j\omega$, since multiplication by j is equivalent to a rotation through 90°.

Thus the differential equation

$$a_1 \frac{dx}{dt} + a_0 x = b_0 y$$

can be written, in complex notation, as a *phasor equation*

$$j\omega a_1 \mathbf{X} + a_0 \mathbf{X} = b_0 \mathbf{Y}$$

where the bold, non-italic, letters indicate that the data refers to phasors. We can say that the differential equation, which was an equation in the time domain, has been transformed into an equation in the *frequency domain*. The frequency domain equation can be rewritten as

$$(j\omega a_1 + a_0)\mathbf{X} = b_0\mathbf{Y}$$

$$\frac{\mathbf{X}}{\mathbf{Y}} = \frac{b_0}{j\omega a_1 + a_0}$$

But, in section 11.2, when the same differential equation was written in the s-domain, we had

$$G(s) = \frac{X(s)}{Y(s)} = \frac{b_0}{a_1 s + a_0}$$

If we replace s by $j\omega$ we have the same equation. It turns out that we can always do this to convert from the s-domain to the frequency domain. This thus leads to a definition of a *frequency-response function* or *frequency transfer function* $G(j\omega)$, for the steady state, as

$$G(j\omega) = \frac{\text{output phasor}}{\text{input phasor}}$$

To illustrate the above consider the determination of the frequency-response function for a system having a transfer function of

$$G(s) = \frac{1}{s+1}$$

The frequency-response function is obtained by replacing s by $j\omega$. Thus

$$G(j\omega) = \frac{1}{j\omega + 1}$$

12.3 Frequency response

A first-order system has a transfer function which can be written as

$$G(s) = \frac{1}{1 + \tau s}$$

where τ is the time constant of the system (see section 11.2). The frequency-response function $G(j\omega)$ can be obtained by replacing s by $j\omega$. Hence

$$G(j\omega) = \frac{1}{1 + j\omega\tau}$$

We can put this into a more convenient form by multiplying the top and bottom of the expression by $(1 - j\omega\tau)$ to give

$$G(j\omega) = \frac{1}{1+j\omega\tau} \times \frac{1-j\omega\tau}{1-j\omega\tau} = \frac{1-j\omega\tau}{1+j^2\omega^2\tau^2}$$

But $j^2 = -1$, thus

$$G(j\omega) = \frac{1}{1+\omega^2\tau^2} - j\frac{\omega\tau}{1+\omega^2\tau^2}$$

This is of the form $x + jy$ and so, since $G(j\omega)$ is the output phasor divided by the input phasor, we have the size of the output phasor bigger than that of the input phasor by a factor which can be written as $|G(j\omega)|$, with

$$|G(j\omega)| = \sqrt{x^2 + y^2} = \sqrt{\left(\frac{1}{1+\omega^2\tau^2}\right)^2 + \left(\frac{\omega\tau}{1+\omega^2\tau^2}\right)^2}$$

$$= \frac{1}{\sqrt{1+\omega^2\tau^2}}$$

$|G(j\omega)|$ tells us how much bigger the amplitude of the output is than the amplitude of the input. It is generally referred to as the *magnitude* or *gain*. The phase difference ϕ between the output phasor and the input phasor is given by

$$\tan\phi = \frac{y}{x} = -\omega\tau$$

The negative sign indicates that the output phasor lags behind the input phasor by this angle.

To illustrate the above, consider a system (an electrical circuit with a resistor in series with a capacitor across which the output is taken) which has a transfer function of

$$G(s) = \frac{1}{RCs + 1}$$

The system is first-order with a time constant τ of RC. The frequency-response function can be obtained by substituting $j\omega$ of s and so give

$$G(j\omega) = \frac{1}{j\omega RC + 1}$$

We can multiply the top and bottom of the above equation by $1 - j\omega RC$ and then rearrange the result to give

$$G(j\omega) = \frac{1}{1+\omega^2(RC)^2} - j\frac{\omega(RC)}{1+\omega^2(RC)^2}$$

Hence, as before,

$$|G(j\omega)| = \frac{1}{\sqrt{1+\omega^2(RC)^2}}$$

and

$$\tan\phi = -\omega RC$$

Consider another problem involving the determination of the steady-state magnitude and phase of the output from a system when subject to a sinusoidal input of 2 sin (3*t* + 60°) if it has a transfer function of

$$G(s) = \frac{4}{s+1}$$

The frequency-response function is obtained by replacing *s* by jω. Thus

$$G(j\omega) = \frac{4}{j\omega+1}$$

Multiplying top and bottom of the equation by (−jω + 1) gives

$$G(j\omega) = \frac{-j4\omega+4}{\omega^2+1} = \frac{4}{\omega^2+1} - j\frac{4\omega}{\omega^2+1}$$

The magnitude for a complex number $x+jy$ is given by $\sqrt{x^2+y^2}$ and so

$$|G(j\omega)| = \sqrt{\frac{4^2}{(\omega^2+1)^2} + \frac{4^2\omega^2}{(\omega^2+1)^2}} = \frac{4}{\sqrt{\omega^2+1}}$$

and the phase angle is given by $\tan\phi = y/x$ and so

$$\tan\phi = -\omega$$

For the specified input we have $\omega = 3$ rad/s. Thus the magnitude is

$$|G(j\omega)| = \frac{4}{\sqrt{3^2+1}} = 1.3$$

and the phase given by

$$\tan\phi = -3$$

Thus $\phi = -72°$. This is the phase angle between the input and the output. Thus the output is $2.6 \sin(3t - 12°)$.

12.3.1 Frequency response for a second-order system

Consider a second-order system with the transfer function (see section 11.3)

$$G(s) = \frac{\omega_n^2}{s^2 + 2\zeta\omega_n s + \omega_n^2}$$

where ω_n is the natural angular frequency and ζ the damping ratio. The frequency-response function is obtained by replacing s by $j\omega$. Thus

$$G(j\omega) = \frac{\omega_n^2}{-\omega^2 + j2\zeta\omega\omega_n + \omega_n^2} = \frac{\omega_n^2}{(\omega_n^2 - \omega^2) + j2\zeta\omega_n}$$

$$= \frac{1}{\left[1 - \left(\frac{\omega}{\omega_n}\right)^2\right] + j2\zeta\left(\frac{\omega}{\omega_n}\right)}$$

Multiplying the top and bottom of the expression by

$$\left[1 - \left(\frac{\omega}{\omega_n}\right)^2\right] - j2\zeta\left(\frac{\omega}{\omega_n}\right)$$

gives

$$G(j\omega) = \frac{\left[1 - \left(\frac{\omega}{\omega_n}\right)^2\right] - j2\zeta\left(\frac{\omega}{\omega_n}\right)}{\left[1 - \left(\frac{\omega}{\omega_n}\right)^2\right]^2 + \left[2\zeta\left(\frac{\omega}{\omega_n}\right)\right]^2}$$

This is of the form $x + jy$ and so, since $G(j\omega)$ is the output phasor divided by the input phasor, we have the size or magnitude of the output phasor bigger than that of the input phasor by a factor which is given by $\sqrt{(x^2 + y^2)}$ as

$$|G(j\omega)| = \frac{1}{\sqrt{\left[1 - \left(\frac{\omega}{\omega_n}\right)^2\right]^2 + \left[2\zeta\left(\frac{\omega}{\omega_n}\right)\right]^2}}$$

The phase ϕ difference between the input and output is given by $\tan \phi = x/y$ and so

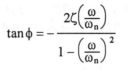

$$\tan \phi = -\frac{2\zeta\left(\dfrac{\omega}{\omega_n}\right)}{1-\left(\dfrac{\omega}{\omega_n}\right)^2}$$

The minus sign indicates that the output phase lags behind the input.

12.4 Bode plots

The frequency response of a system is the set of values of the magnitude $|G(j\omega)|$ and phase angle ϕ that occur when a sinusoidal input signal is varied over a range of frequencies. This can be expressed as two graphs, one of the magnitude $|G(j\omega)|$ plotted against the angular frequency ω and the other of the phase ϕ plotted against ω. The magnitude and angular frequency are plotted using logarithmic scales. Such a pair of graphs is referred to as a *Bode plot*.

The magnitude is expressed in decibel units (dB).

$$|G(j\omega)| \text{ in dB} = 20 \lg_{10} |G(j\omega)|$$

Thus, for example, a magnitude of 20 dB means that

$$20 = 20 \lg_{10} |G(j\omega)|$$

so $1 = \lg_{10} |G(j\omega)|$

and $10^1 = |G(j\omega)|$

Thus a magnitude of 20 dB means the magnitude is 10, therefore the amplitude of the output is 10 times that of the input. A magnitude of 40 dB would mean a magnitude of 100 and so the amplitude of the output is 100 times that of the input.

12.4.1 Examples of Bode plots

Consider the Bode plot for a system having the transfer function $G(s) = K$, where K is a constant. The frequency-response function is thus $G(j\omega) = K$. The magnitude $|G(j\omega)| = K$ and so, in decibels, $|G(j\omega)| = 20 \lg K$. The magnitude plot is thus a line of constant magnitude, changing K merely shifts the magnitude line up or down by a certain number of decibels. The phase is zero. Figure 12.3 shows the Bode plot.

Consider the Bode plot for a system having a transfer function $G(s) = 1/s$. The frequency-response function $G(j\omega)$ is thus $1/j\omega$. Multiplying this by j/j gives $G(j\omega) = -j/\omega$. The magnitude $|G(j\omega)|$ is thus $1/\omega$. In decibels this is $20 \lg (1/\omega) = -20 \lg \omega$. When $\omega = 1$ rad/s the magnitude is 0. When $\omega = 10$ rad/s it is -20 dB. When $\omega = 100$ rad/s it is -40 dB. For each tenfold increase in

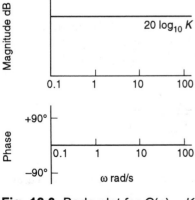

Fig. 12.3 Bode plot for $G(s) = K$

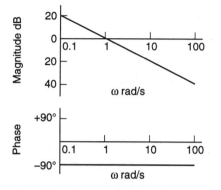

Fig. 12.4 Bode plot for $G(s) = 1/s$

angular frequency the magnitude drops by -20 dB. The magnitude plot is thus a straight line of slope -20 dB per decade of frequency which passes through 0 dB at $\omega = 1$ rad/s. The phase of such a system is given by

$$\tan \phi = \frac{-\dfrac{1}{\omega}}{0} = -\infty$$

Hence $\phi = -90°$ for all frequencies. Figure 12.4 shows the Bode plot.

Consider the Bode plot for a first-order system for which the transfer function is given by

$$G(s) = \frac{1}{\tau s + 1}$$

The frequency-response function is then

$$G(j\omega) = \frac{1}{j\omega\tau + 1}$$

The magnitude (see section 12.2.1) is then

$$|G(j\omega)| = \frac{1}{\sqrt{1 + \omega^2\tau^2}}$$

In decibels this is

$$20 \lg \left(\frac{1}{\sqrt{1 + \omega^2\tau^2}} \right)$$

When $\omega \ll 1/\tau$ then $\omega^2\tau^2$ is negligible compared with 1 and so the magnitude is $20 \lg 1 = 0$ dB. Hence at low frequencies there is a straight line magnitude plot at a constant value of 0 dB. For higher frequencies, when $\omega \gg 1/\tau$, $\omega^2\tau^2$ is much greater than 1 and so the 1 can be neglected. The magnitude is then $20 \lg (1/\omega\tau)$, i.e. $-20 \lg \omega\tau$. This is a straight line of slope -20 dB per decade of frequency which intersects the 0 dB line when $\omega\tau = 1$, i.e. when $\omega = 1/\tau$. Figure 12.5 shows these lines for low and high frequencies with their intersection, or so-called *break point* or *corner frequency*, at $\omega = 1/\tau$. The two straight lines are called the asymptotic approximation to the true plot. The true plot rounds off the intersection of the two lines. The difference between the true plot and the approximation is a maximum of 3 dB at the break point.

The phase for the first-order system (see section 12.2.1) is given by

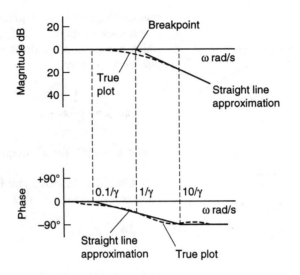

Fig. 12.5 Bode plot for $G(s) = 1/(\tau s + 1)$

$$\tan \phi = -\omega\tau$$

At low frequencies, when ω is less than about $0.1/\tau$, the phase is virtually $0°$. At high frequencies, when ω is more than about $10/\tau$, the phase is virtually $-90°$. Between these two extremes the phase angle can be considered to give a reasonable straight line on the Bode plot (figure 12.5). The maximum error in assuming a straight line is $5\frac{1}{2}°$.

Consider a second-order system with a transfer function of

$$G(s) = \frac{\omega_n^2}{s^2 + 2\zeta\omega_n s + \omega_n^2}$$

The frequency-response function is obtained by replacing s by $j\omega$.

$$G(j\omega) = \frac{\omega_n^2}{-\omega^2 + j2\zeta\omega_n\omega + \omega_n^2}$$

The magnitude is then (see section 12.2.2)

$$|G(j\omega)| = \frac{1}{\sqrt{\left[1 - \left(\frac{\omega}{\omega_n}\right)^2\right]^2 + \left[2\zeta\left(\frac{\omega}{\omega_n}\right)\right]^2}}$$

Thus, in decibels, the magnitude is

$$20 \lg \frac{1}{\sqrt{\left[1 - \left(\frac{\omega}{\omega_n}\right)^2\right]^2 + \left[2\zeta\left(\frac{\omega}{\omega_n}\right)\right]^2}}$$

$$= -20 \lg \sqrt{\left[1 - \left(\frac{\omega}{\omega_n}\right)^2\right]^2 + \left[2\zeta\left(\frac{\omega}{\omega_n}\right)\right]^2}$$

For $(\omega/\omega_n) \ll 1$ the magnitude approximates to $-20 \lg 1$ or 0 dB. For $(\omega/\omega_n) \gg 1$ the magnitude approximates to $-20 \lg (\omega/\omega_n)^2$. Thus when ω increases by a factor of 10 the magnitude increases by a factor of $-20 \lg 100$ or -40 dB. Thus at low frequencies the magnitude plot is a straight line at 0 dB, while at high frequencies it is a straight line of -40 dB per decade of frequency. The intersection of these two lines, i.e. the break point, is at $\omega = \omega_n$. The magnitude plot is thus approximately given by these two asymptotic lines. The true value, however, depends on the damping ratio ζ. Figure 12.6 shows the two asymptotic lines and the true plots for a number of damping ratios.

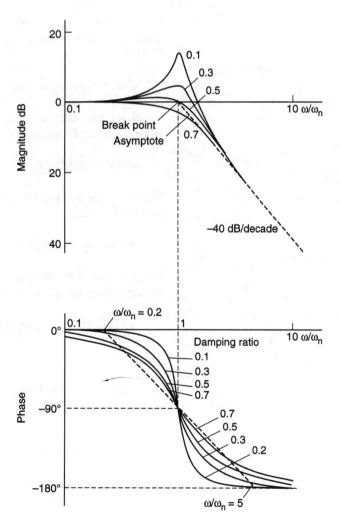

Fig. 12.6 Bode plot for second-order system

The phase is given by (see section 12.2.2)

$$\tan\phi = -\frac{2\zeta\left(\frac{\omega}{\omega_n}\right)}{1-\left(\frac{\omega}{\omega_n}\right)^2}$$

For $(\omega/\omega_n) \ll 1$, e.g. $(\omega/\omega_n) = 0.2$, then $\tan\phi$ is approximately 0 and so $\phi = 0°$. For $(\omega/\omega_n) \gg 1$, e.g. $(\omega/\omega_n) = 5$, $\tan\phi$ is approximately $-(-\infty)$ and so $\phi = -180°$. When $\omega = \omega_n$ then we have $\tan\phi = -\infty$ and so $\phi = -90°$. A reasonable approximation is given by a straight line drawn through $-90°$ at $\omega = \omega_n$ and the points $0°$ at $(\omega/\omega_n) = 0.2$ and $-180°$ at $(\omega/\omega_n) = 5$. Figure 12.6 shows the graph.

12.4.2 Building up Bode plots

Consider a system involving a number of elements in series. The transfer function of the system as a whole is given by (see section 11.4)

$$G(s) = G_1(s)G_2(s)G_3(s) \dots \text{etc.}$$

Hence the frequency-response function for a two-element system, when s is replaced by $j\omega$, is

$$G(j\omega) = G_1(j\omega)G_2(j\omega)$$

We can write the transfer function $G_1(j\omega)$ as a complex number (see section 12.2), i.e.

$$x + jy = |G_1(j\omega)| \,(\cos\phi_1 + j\sin\phi_1)$$

where $|G(j\omega)|$ is the magnitude and ϕ_1 the phase of the frequency-response function. Similarly we can write $G_2(j\omega)$ as

$$|G_2(j\omega)| \,(\cos\phi_2 + j\sin\phi_2)$$

Thus

$$\begin{aligned}
G(j\omega) = {} & |G_1(j\omega)| \,(\cos\phi_1 + j\sin\phi_1) \\
& \times |G_2(j\omega)| \,(\cos\phi_2 + j\sin\phi_2) \\[6pt]
= {} & |G_1(j\omega)| \,|G_2(j\omega)| \,[\cos\phi_1 \cos\phi_2 \\
& + j(\sin\phi_1 \cos\phi_2 + \cos\phi_1 \sin\phi_2) + j^2 \sin\phi_1 \sin\phi_2]
\end{aligned}$$

But $j^2 = -1$ and, since $\cos\phi_1 \cos\phi_2 - \sin\phi_1 \sin\phi_2 = \cos(\phi_1 + \phi_2)$ and $\sin\phi_1 \cos\phi_2 + \cos\phi_1 \sin\phi_2 = \sin(\phi_1 + \phi_2)$, thus

$$G(j\omega) = |G_1(j\omega)|\,|G_2(j\omega)|\,[\cos\,(\phi_1 + \phi_2) + j\,\sin\,(\phi_1 + \phi_2)]$$

The frequency-response function of the system has a magnitude which is the product of the magnitudes of the separate elements and a phase which is the sum of the phases of the separate elements, i.e.

$$|G(j\omega)| = |G_1(j\omega)|\,|G_2(j\omega)|\,|G_3(j\omega)|\,\ldots\text{ etc.}$$

$$\phi = \phi_1 + \phi_2 + \phi_3 + \ldots\text{ etc.}$$

Now, considering the Bode plot where the logarithm of the magnitudes are plotted.

$$\lg|G(j\omega)| = \lg|G_1(j\omega)| + \lg|G_2(j\omega)| + \lg|G_3(j\omega)| + \ldots\text{ etc.}$$

Thus we can obtain the Bode plot of a system by adding together the Bode plots of the magnitudes of the constituent elements. Likewise the phase plot is obtained by adding together the phases of the constituent elements.

By using a number of basic elements, the Bode plot for a wide range of systems can be readily obtained. The basic elements used are:

1   $G(s) = K$. This gives the Bode plot shown in figure 12.3.
2   $G(s) = 1/s$. This gives the Bode plot shown in figure 12.4.
3   $G(s) = s$. This gives a Bode plot which is a mirror image of that in figure 11.4. $|G(j\omega)| = 20$ dB per decade of frequency, passing through 0 dB at $\omega = 1$ rad/s. ϕ is constant at $90°$.
4   $G(s) = 1/(\tau s + 1)$. This gives the Bode plot shown in figure 12.5.
5   $G(s) = \tau s + 1$. This gives a Bode plot which is a mirror image of that in figure 12.5. For the magnitude plot, the break point is at $1/\tau$ with the line prior to it being at 0 dB and after it at a slope of 20 dB per decade of frequency. The phase is zero at $0.1/\tau$ and rises to $+90°$ at $10/\tau$.
6   $G(s) = \omega_n^2/(s^2 + 2\zeta\omega_n s + \omega_n^2)$. This gives the Bode plot shown in figure 12.6.
7   $G(s) = (s^2 + 2\zeta\omega_n s + \omega_n^2)/\omega_n^2$. This gives a Bode plot which is a mirror image of that in figure 12.6.

To illustrate the above, consider the drawing of the asymptotes of the Bode plot for a system having a transfer function of

$$G(s) = \frac{10}{2s + 1}$$

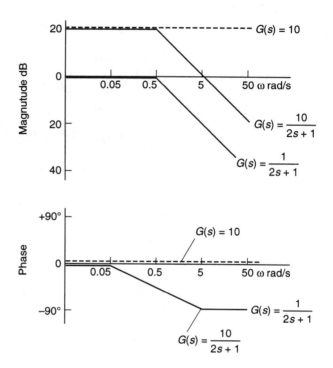

Fig. 12.7 Example

The transfer function is made up of two elements, one with a transfer function of 10 and one with transfer function $1/(2s + 1)$. The Bode plots can be drawn for each of these and then added together to give the required plot. The Bode plot for transfer function 10 will be of the form given in figure 12.3 with $K = 10$ and that for $1/(2s + 1)$ like that given in figure 12.5 with $\tau = 2$. The result is shown in figure 12.7.

As another example, consider the drawing of the asymptotes of the Bode plot for a system having a transfer function of

$$G(s) = \frac{2.5}{s(s^2 + 3s + 25)}$$

The transfer function is made up of three components, one with a transfer function of 0.1, one with transfer function $1/s$ and one with transfer function $25/(s^2 + 3s + 25)$. The transfer function of 0.1 will give a Bode plot like that of figure 12.3 with $K = 0.1$. The transfer function of $1/s$ will give a Bode plot like that of figure 12.4. The transfer function of $25/(s^2 + 3s + 25)$ can be represented as $\omega_n^2/(s^2 + 2\zeta\omega_n + \omega_n^2)$ with $\omega_n = 5$ rad/s and $\zeta = 0.3$. The break point will be when $\omega = \omega_n = 5$ rad/s. The asymptote for the phase passes through $-90°$ at the break point, and is $0°$ when we have $(\omega/\omega_n) = 0.2$ and $-180°$ when $(\omega/\omega_n) = 5$. Figure 12.8 shows the resulting Bode plot.

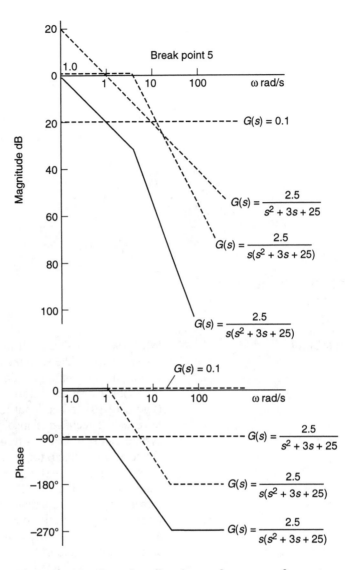

Fig. 12.8 Example

12.5 Performance specifications

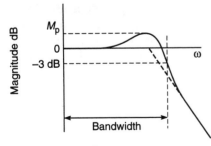

Fig. 12.9 Performance specifications

The terms used to describe the performance of a system when subject to a sinusoidal input are peak resonance and bandwidth. The *peak resonance* M_p is defined as being the maximum value of the magnitude (figure 12.9). A large value of the peak resonance corresponds to a large value of the maximum overshoot of a system. For a second-order system it can be directly related to the damping ratio by comparison of the response with the Bode plot of figure 12.6, a low damping ratio corresponding to a high peak resonance. The *bandwidth* is defined as the frequency band between which the magnitude does not fall below −3 dB. For the system giving the Bode plot in figure 12.9, the bandwidth is the spread between zero frequency and the frequency at which the magnitude drops below −3 dB.

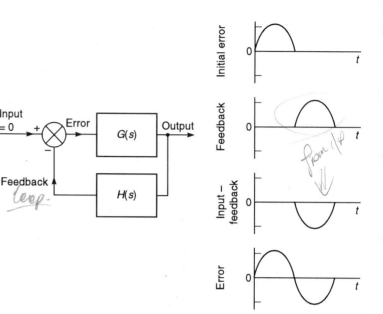

Fig. 12.10 Self-sustaining
oscillations

12.6 Stability

When there is a sinusoidal input to a system, the output from that
system is sinusoidal with the same angular frequency but can have
an output with an amplitude and phase which differ from that of
the input. Consider a closed-loop system with negative feedback
(figure 12.10) and no input to the system. Suppose, somehow, we
have a half-rectified sinusoidal pulse as the error signal in the
system and that it passes through to the output and is fed back to
arrive at the comparator element with amplitude unchanged but a
phase change of −180°, i.e. as shown in the figure. When this
signal is subtracted from the input signal we have a resulting error
signal which just continues the initial half-rectified pulse. This
pulse then goes back round the feedback loop and once again
arrives just in time to continue the signal. Thus we have a self-
sustaining oscillation.

For self-sustained oscillations to occur we must have a
system which has a frequency-response function with a magnitude
of 1 and a phase of −180°. The system through which the signal
passes is $G(s)$ in series with $H(s)$. If the magnitude is less than 1
then each succeeding half wave pulse is smaller in size and so the
oscillation dies away. If the magnitude is greater than 1 then each
succeeding pulse is larger than the previous one and so the wave
builds up and the system is unstable.

1   A control system will oscillate with a constant amplitude if the
    magnitude resulting from the system $G(s)$ in series with $H(s)$ is
    1 and the phase is −180°.
2   A control system will oscillate with a diminishing amplitude if
    the magnitude resulting from the system $G(s)$ in series with
    $H(s)$ is less than 1 and the phase is −180°.

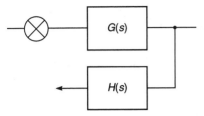

Fig. 12.11 Open-loop transfer function

3 A control system will oscillate with an increasing amplitude, and so is unstable, if the magnitude resulting from the system $G(s)$ in series with $H(s)$ is greater than 1 and the phase is $-180°$.

The transfer function for $G(s)$ in series with $H(s)$ is called the *open-loop transfer function*, since effectively we are considering the closed-loop situation shown in figure 12.10 with a break in the feedback loop at the comparator, as illustrated in figure 12.11. The open-loop transfer function is thus the product $G(s)H(s)$.

A good, stable control system usually requires that the magnitude of the open-loop transfer function, i.e. $|G(s)H(s)|$, should be significantly less than 1. Typically a value between 0.4 and 0.5 is used. In addition, the phase angle should be between about $-115°$ and $-125°$. Such values produces a slightly underdamped control system which gives, with a step input, about a 20 to 30% overshoot with a subsidence ratio of about 3 to 1 (see section 10.4 for an explanation of these terms).

A measure of the stability of a control system is given by the Bode plot for the open-loop transfer function. The term *phase crossover* is used for the frequency in the phase plot at which the phase angle first reaches $-180°$. The term *gain margin* is used for the factor by which the magnitude ratio must be multiplied at phase crossover to make it have the value 1 (figure 12.12). The term *gain crossover* is used for the frequency in the magnitude plot at which the open-loop magnitude first reaches the value 1. The term *phase margin* is used for the number of degrees by which the phase angle is numerically smaller than $-180°$ at gain crossover (figure 12.12). The rules considered above for a good, stable control system mean a gain margin of between 2 and 2.5 and a phase margin between $45°$ and $65°$.

To illustrate the above, consider the Bode plot in figure 12.13 for the open-loop transfer function of a control system. The gain margin is the magnitude value when the phase is $-180°$ and so is about 8 dB. This means

$$8 = 20 \lg (\text{magnitude})$$

and so the magnitude is $10^{8/20} = 2.5$. The phase margin is the phase difference from $-180°$ when the magnitude is zero. It is thus about $40°$. The system can thus be expected to have good stability.

As another example, consider the determination of the value of K for a system with the following open-loop transfer function

$$\frac{K}{s(2s+1)(s+1)}$$

which will give a system with a gain margin of 3 dB, i.e. about 2. The open-loop frequency-response function is

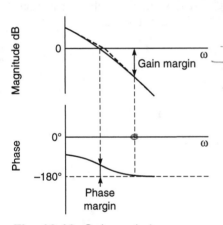

Fig. 12.12 Gain and phase margins

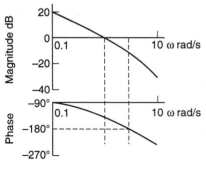

Fig. 12.13 Example

$$G_o(j\omega) = \frac{K}{j\omega(j2\omega + 1)(j\omega + 1)} = \frac{K}{-3\omega^2 + j\omega(1 - 2\omega^2)}$$

Multiplying the top and bottom by $-3\omega^2 - j\omega(1 - 2\omega^2)$ gives

$$G_o(j\omega) = \frac{-3K\omega^2 + jK\omega(1 - 2\omega^2)}{9\omega^4 + \omega^2(1 - 2\omega^2)^2}$$

The magnitude of this open-loop frequency-response function is thus

$$|G_o(j\omega)| = \frac{K}{\sqrt{9\omega^4 + \omega^2(1 - 2\omega^2)^2}}$$

and the phase is

$$\tan \phi = \frac{1 - 2\omega^2}{3\omega}$$

For the system to have a gain margin of 3 dB then this must be the magnitude value at $\phi = -180°$. At this angle the above equation gives $1 - 2\omega^2 = 0$ and so $\omega = 1/\sqrt{2}$. Thus, at this angle,

$$\text{Gain margin} = -20 \lg |G_o(j\omega)|$$

and so

$$3 = -20 \lg \left(\frac{K}{\sqrt{9\omega^4 + \omega^2(1 - 2\omega^2)^2}} \right)$$

$$3 = -20 \lg \left(\frac{K}{\sqrt{9/4 + 0}} \right)$$

Thus $K/(3/2) = 10^{-3/20}$ and $K = 1.06$.

Problems

1   What are the magnitudes and phases of the systems having the following transfer functions?

(a) $\dfrac{5}{s+2}$, (b) $\dfrac{2}{s(s+1)}$, (c) $\dfrac{1}{(2s+1)(s^2+s+1)}$

2   What will be the steady-state response of a system with a transfer function

$$\frac{1}{s+2}$$

when subject to the input 3 sin $(5t + 30°)$?

3 What will be the steady-state response of a system with a transfer function

$$\frac{5}{s^2 + 3s + 10}$$

when subject to the input 2 sin $(2t + 70°)$?

4 For systems with the following transfer functions, determine the values of the magnitudes and phase at angular frequencies of (i) 0 rad/s, (ii) 1 rad/s, (iii) 2 rad/s, (iv) ∞ rad/s.

(a) $G(s) = \dfrac{1}{s(2s + 1)}$, (b) $G(s) = \dfrac{1}{3s + 1}$

5 Draw the asymptotes of the Bode plots for the systems having the transfer functions:

(a) $G(s) = \dfrac{10}{s(0.1s + 1)}$, (b) $G(s) = \dfrac{1}{(2s + 1)(0.5s + 1)}$,

(c) $G(s) = \dfrac{5}{(2s + 1)(s^2 + 3s + 25)}$

6 For the Bode plot given in figure 12.14, estimate the gain margin and the phase margin.

7 Determine the value of K for a system with the following open-loop transfer function

$$G_o(s) = \frac{K}{s(2s + 1)(3s + 1)}$$

which will give (a) a marginally stable system, (b) a gain margin of 2 dB.

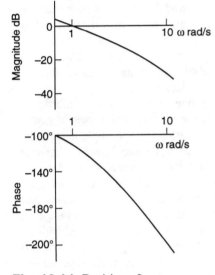

Fig. 12.14 Problem 6

13 Closed-loop controllers

13.1 Continuous and discrete processes

Process controllers can be considered in terms of those for the control of continuous processes and those for the control of discrete processes. In this section the requirements for both types are discussed but the remaining part of the chapter, and chapter 14, is concerned with the control of continuous processes, discrete process control being discussed in chapter 17.

Continuous processes are ones which have uninterrupted inputs and outputs. The controller in a *closed-loop control system* with continuous processes detects the error and implements some method of control, converting the error, i.e. the difference between the required and actual conditions, into a control action designed to reduce the error. The error might arise as a result of some change in the conditions being controlled or because the set value is changed, e.g. there is a step input to the system to change the set value to a new value. In this chapter we are concerned with the ways in which controllers can react to error signals, i.e. the *control modes* as they are termed, which occur with continuous processes. Such controllers might, for example, be pneumatic systems or operational amplifier systems. However, computer systems are rapidly replacing many of these. The term *direct digital control* is used when the computer is in the feedback loop and exercising control in this way.

Discrete processes are ones for which the control involves the sequencing of operations. A domestic washing machine (see section 1.5) where a number of actions have to be carried out in a predetermined sequence is an example. Another example is the manufacture of a product which involves the assembly of a number of discrete parts in a specific sequence by some controlled system. The sequence of operations might be *clock-based* or *event-based* or a combination of the two. With a clock-based system the actions are carried out at specific times. With some systems the actions are not carried out at particular times but in response to some event. Such systems are said to be *event-based*. The time at

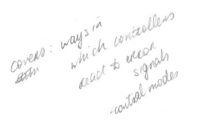

which the response occurs is thus determined by the process and not the control system clock. There are some systems which are termed *interactive*. With such systems, the relationships are more loosely defined than in clock-based or event-based systems. Such systems are not tightly synchronised to an external process. The system gives a response to a signal from the process but at a time determined by the control system. A simple example of such a system is a bank cash machine. The input to the system is the instructions tapped out on the keyboard for a certain amount of cash. There is not an immediate response but a response within an acceptable time interval, at a time convenient to the bank's computer. The response time will depend on how busy the communication lines to the central computer and the computer are. The term *real time* is used for a computer control system in which the lag from input time to output time is sufficiently small for the response to be accepted as effectively directly following from the input. Real-time systems often contain a mixture of activities, some of which can be classified as clock based, some event based and some interactive. See chapter 16 for a discussion of the control of digital processes.

The term *progammable logic controller* (PLC) is used for a simple controller based on a microprocessor. Such a controller operates by examining the input signals from sensors and carrying out logic instructions which have been programmed into its memory. The output after such processing are signals which feed into correcting/actuator units. Thus it can carry out sequences of operations. The main difference between a PLC and a computer is that programming is predominantly concerned with logic and switching operations and the interfacing for input and output devices is inside the controller. Such controllers are discussed in more detail in chapter 17.

In many processes there can be a mixture of continuous and discrete control. For example, in the domestic washing machine there will be sequence control for the various parts of the washing cycle with feedback loop control of the temperature of the hot water and the level of the water.

13.2 Control modes

There are a number of ways by which a control unit for a continuous process can react to an error signal.

1 The *two-step mode* in which the controller is just a switch activated by the error signal, i.e. the control action is just on–off.
2 The *proportional mode* (P) which produces a control action that is proportional to the error.
3 The *derivative mode* (D) which produces a control action that is proportional to the rate at which the error is changing.

4 The *integral mode* (I) which produces a control action that continues to increase as long the error persists.
5 Combinations of modes: proportional plus derivative modes (PD), proportional plus integral modes (PI), proportional plus integral plus derivative modes (PID).

These five modes of control are discussed in the following sections of the chapter. The controller can function by means of pneumatic circuits, analogue electronic circuits involving operational amplifiers or digital electronic circuits using microprocessors.

13.2.1 Lag

In any control system there are lags. Thus, for example, a change in the condition being controlled does not immediately produce a correcting response from the control system. This is because time is required for the system to make the necessary responses. For example, in the control of the temperature in a room by means of a central heating system, a lag will occur between the room temperature falling below the required temperature and the control system responding and switching on the heater. This is not the only lag. Even when the control system has responded there is a lag in the room temperature responding due to the time taken for the heat to transfer from the heater to the air in the room.

13.2.2 Steady-state error

We might get an error occurring as a result of the controlled variable changing or a change in the set value input. The controller then has an error input and will produce a response attempting to reduce the error. The term *steady-state error* is used for the error input to the controller that has to exist in steady-state conditions, i.e. after all transients have died away. It thus indicates the error occurring between the set value input and the controlled output. The steady-state error is thus a measure of the accuracy of the control system in coping with a change in the controlled variable or tracking the set value input.

13.3 Two-step mode

An example of the *two-step mode* of control is the bimetallic thermostat (see figure 2.23) that might be used with a domestic central heating system. This is just a switch which is switched on or off according to the temperature. If the room temperature is above the required temperature then the bimetallic strip is in an off position and the heater is off. If the room temperature falls below the required temperature then the bimetallic strip moves into an on position and the heater is switched fully on. The controller in this case can be in only two positions, on or off, as indicated by figure 13.1.

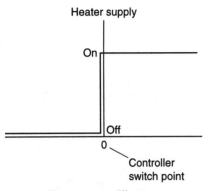

Fig. 13.1 Two-step control

With the two-step mode the control action is discontinuous. A consequence of this is that oscillations occur about the required condition. This is because of lags in the time the control system and the process take to respond. For example, in the case of the domestic central heating system, when the room temperature drops below the required level the time that elapses before the control system responds and switches the heater on might be very small in comparison with the time that elapses before the heater begins to have an effect on the room temperature. In the meantime the temperature has fallen even more. The reverse situation occurs when the temperature has risen to the required temperature. Since time elapses before the control system reacts and switches the heater off, and yet more time while the heater cools and stops heating the room, the room temperature goes beyond the required value. The result is that the room temperature oscillates above and below the required temperature (figure 13.2).

With the simple two-step system described above there is the problem that when the room temperature is hovering about the set value the thermostat might be almost continually switching on or off, reacting to very slight changes in temperature. This can be avoided if, instead of just a single temperature value at which the controller switches the heater on or off, two values are used and the heater is switched on at a lower temperature than the one at which it is switched off (figure 13.3).

Two-step control action tends to be used where changes are taking place very slowly, i.e. with a process with a large capacitance. Thus, in the case of heating a room, the effect of switching on or off the heater on the room temperature is only a slow

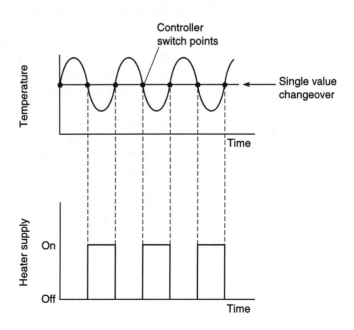

Fig. 13.2 Oscillations with two-step control

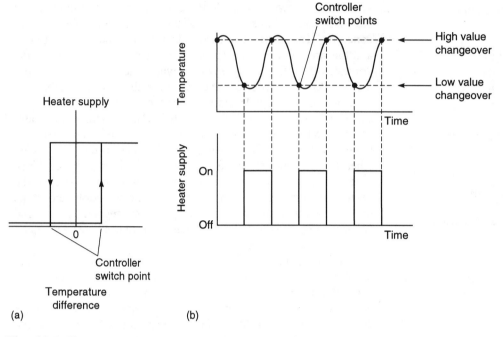

Fig. 13.3 Two controller switch points

change. The result of this is an oscillation with a long periodic time. Two-step control is thus not very precise, but it does involve simple devices and is thus fairly cheap.

13.4 Proportional control

With the two-step method of control, the controller output is either an on or an off signal, regardless of the magnitude of the error. With *proportional control* the size of the controller output is proportional to the size of the error. This means the correction element of the control system, e.g. a valve, will receive a signal which is proportional to the size of the correction required.

Figure 13.4 shows how the output of such a controller varies with the size and sign of the error. The linear relationship between controller output and error tends to exist only over a certain range of errors, this range being called the *proportional band*. For the relationship shown in the figure this corresponds to a controller output which varies from 4 mA to 20 mA. This output can then correspond to, say, a correction valve changing from fully closed to fully open. Within the proportional band the equation of the straight line can be represented by

$$\text{Change in controller output from set point} = K_\mathrm{p}e$$

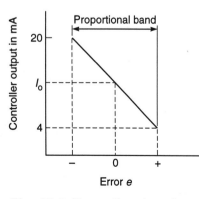

Fig. 13.4 Proportional mode

where e is the error and K_p a constant. K_p is thus the gradient of the straight line in figure 13.4. However, the output of a controller is generally expressed as a percentage of the full range of possible values. Thus, for figure 13.4, the 4 mA output corresponds to 0% and the 20 mA to 100%. Similarly, the error is expressed as a percentage of the full-range value, i.e. the error range corresponding to the 0 to 100% controller output. Thus

% change in controller output from set point
$= K_P \times$ % change in error

Hence, since 100% controller output corresponds to an error percentage equal to the proportional band

$$K_P = \frac{100}{\text{proportional band}}$$

We can rewrite the equation as

$$I_{\text{out}} - I_0 = K_P e$$

$$I_{\text{out}} = K_P e + I_0$$

where I_0 is the controller output percentage at zero error, I_{out} the output percentage at percentage error e.

K_P is, within the proportional band, the transfer function of the controller since, taking Laplace transforms,

Change in output $(s) = K_P E(s)$

$$\text{Transfer function} = \frac{\text{change in output } (s)}{E(s)} = K_P$$

Generally a 50% controller output is chosen to be the output when the error is zero. Thus, in the case of the controller being used to control a valve which allows water into a tank, when the error is zero the valve will be half (50%) open. This will give the normal flow rate. Any error will then increase or reduce the flow rate at a value which depends on the size of the error. The result will be to return the error to its zero value and the controller to a 50% output. Suppose the process has the flow of liquid into a tank being controlled and for some reason a new set point is required for the flow rate. We can talk of this change in terms of there being a step input to the control system. This new set value could require the correcting valve to be kept open at a higher percentage, say 60%. This cannot be achieved by the zero error setting but requires a permanent error setting called an *offset* (figure 13.5). There is thus a steady-state error. The size of this offset is directly proportional to the size of the load changes and inversely proportional to the K_P, a higher value of K_P giving a steeper gradient in figure 13.5 and so a smaller error change needed to accommodate a load change.

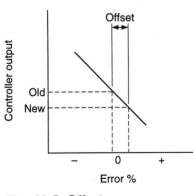

Fig. 13.5 Offset

The proportional mode of control tends to be used in processes where the transfer function K_p can be made large enough to reduce the offset to an acceptable level. However, the larger the transfer function the greater the chance of the system oscillating and becoming unstable.

To illustrate the above discussion of proportional control, consider a proportional controller which is to be used to control the height of water in a tank where the water level can vary from zero to 9.0 m. What proportional band and transfer function will be required if the required height of water is 5.0 m and the controller is to fully close a valve when the water rises to 5.5 m and fully open it when the water falls to 4.5 m? When the error is −0.5 m the controller output must be 100% open and when +0.5 m it must be 0% open. The proportional band must therefore extend from a height error of −0.5 m to one of +0.5 m. Expressed as a percentage, the proportional band extends from

$$-(0.5/9.0) \times 100 = -5.6\% \text{ to } + (0.5/9.0) \times 100 = +5.6\%$$

The proportional band is thus 11.2%. Note that if we work in percentages for the controller, we must work in percentages for the error. This value of proportional band will thus mean a transfer function K_p of $(100\%)/(11.2\%) = 8.9$.

As an illustration of offset error, consider a proportional controller which has a transfer function of 15 and a set point of 50% output. It outputs to a valve which at the set point allows a flow of 200 m³/s. The valve changes its output by 4 m³/s for each percent change in controller output. What will be the controller output and the offset error when the flow has to be changed to 240 m³/s? The new controller setting, as a percentage, for a flow of 240 m³/s is 240/4 = 60%. Hence

$$K_P = 15 = \frac{60 - 50}{e}$$

Thus the offset is $e = 0.67\%$.

13.4.1 Electronic proportional controller

A summing operational amplifier with an inverter can be used as a proportional controller (figure 13.6). For a summing amplifier we have (see section 3.2.3)

$$V_{out} = -R_f\left(\frac{V_0}{R_2} + \frac{V_e}{R_1}\right)$$

The input through R_1 is the zero error voltage value V_0, and the

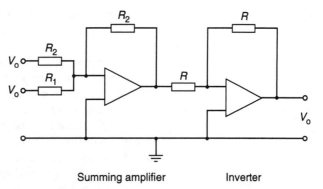

Fig. 13.6 Electronic proportional controller

Summing amplifier   Inverter

input through R_2 is the error signal V_e. But when the feedback resistor $R_f = R_2$, then the equation becomes

$$V_{out} = -\frac{R_2}{R_1}V_e - V_0$$

If this output from the summing amplifier is then passed through an inverter, i.e. an operational amplifier with a feedback resistance equal to the input resistance, then

$$V_{out} = -\frac{R_2}{R_1}V_e - V_0$$

$$V_{out} = K_P V_e + V_0$$

where K_P is the proportionality constant. The result is a proportional controller.

13.5 Derivative control

With *derivative control* the change in controller output from the set point value is proportional to the rate of change with time of the error signal. This can be represented by the equation

$$I_{out} - I_0 = K_D \frac{de}{dt}$$

where I_0 is the set point output value, I_{out} the output value that will occur when the error e is changing at the rate de/dt. It is usual to express these controller outputs as a percentage of the full range of output and the error as a percentage of full range. K_D is the constant of proportionality and is commonly referred to as the *derivative time* since it has units of time.

The transfer function is obtained by taking Laplace transforms, thus

$$(I_{out} - I_0)(s) = K_D sE(s)$$

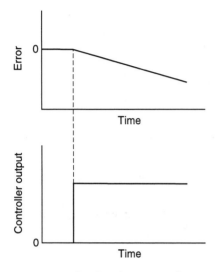

Fig. 13.7 Derivative control

Hence the transfer function is $K_D s$.

With derivative control, as soon as the error signal begins to change there can be quite a large controller output since it is proportional to the rate of change of the error signal and not its value. Rapid initial responses to error signals thus occurs. Figure 13.7 shows the controller output that results when there is a constant rate of change of error signal with time. The controller output is constant because the rate of change is constant and occurs immediately the deviation occurs. Derivative controllers do not, however, respond to steady-state error signals, since with a steady error the rate of change of error with time is zero. Because of this derivative control is often combined with proportional control. Figure 13.8 shows the form of an electronic derivative controller circuit, the circuit involving an operational amplifier connected as a differentiator circuit followed by another operational amplifier connected as an inverter. The derivative time K_D is $R_2 C$.

To illustrate the above, consider a derivative controller which has a set point of 50% and derivative constant K_D of 0.4 s. What will be the controller output when the error (a) changes at 1%/s, (b) is constant at 4%? Using the equation given above

$$I_{out} = K_D \frac{de}{dt} + I_0 = 0.4 \times 1 + 50 = 50.4\ \%$$

With de/dt zero, then I_{out} equals I_0, i.e. 50%. The output only differs from the set point value when the error is changing.

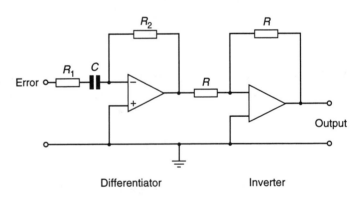

Fig. 13.8 Electronic derivative controller

13.5.1 Proportional plus derivative control

With proportional plus derivative control the change in controller output from the set point value is given by

$$\text{Change in controller output from set point} = K_P\left(e + K_D \frac{de}{dt}\right)$$

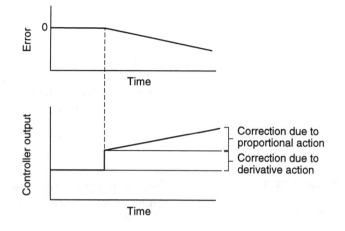

Fig. 13.9 Proportional plus derivative control

Hence

$$I_{out} = K_P\left(e + K_D \frac{de}{dt}\right) + I_0$$

where I_0 is the output at the set point, I_{out} the output when the error is e, K_p is the proportionality constant and K_D the derivative constant, de/dt is the rate of change of error. The system has a transfer function given by

$$(I_{out} - I_0)(s) = K_P E(s) + K_P K_D s E(s)$$

and hence

Transfer function $= K_P(1 + K_D s)$

Figure 13.9 shows how the controller output can vary when there is a constantly changing error. There is an initial quick change in controller output because of the derivative action followed by the gradual change due to proportional action. This form of control can thus deal with fast process changes, however a change in set value will require an offset error (see earlier discussion of proportional control).

To illustrate the above, consider what the controller output will be for a proportional plus derivative controller (a) initially and (b) 2 s after the error begins to change from the zero error at the rate of 1.2%/s. The controller has a set point of 50%, $K_p = 4$ and $K_D = 0.4$ s. Using the equation given above

$$I_{out} = K_P\left(e + K_D \frac{de}{dt}\right) + I_0$$

Initially the error e is zero. Hence, initially when the error begins to change

$$I_{out} = 4(0 + 0.4 \times 1.2) + 50 = 51.9\%$$

Because the rate of change is constant, after 2 s the error will have become 2.4%. Hence, then

$$I_{out} = 4(2.4 + 0.4 \times 1.2) + 50 = 61.52\%$$

13.6 Integral control

Integral control is one where the rate of change of the control output I is proportional to the input error signal e.

$$\frac{dI}{dt} = K_I e$$

K_I is the constant of proportionality and, when the controller output is expressed as a percentage and the error as a percentage, has units of s^{-1}. The reciprocal of K_I is called the *integral time T_I* and is in seconds. Integrating the above equation gives

$$\int_{I_0}^{I_{out}} dI = \int_0^t K_I e \, dt$$

$$I_{out} - I_0 = \int_0^t K_I e \, dt$$

I_0 is the controller output at zero time, I_{out} is the output at time t.

The transfer function is obtained by taking the Laplace transform. Thus

$$(I_{out} - I_0)(s) = \frac{1}{s} K_I E(s)$$

and so

$$\text{Transfer function} = \frac{1}{s} K_I$$

Figure 13.10 illustrates the action of an integral controller when there is a constant error input to the controller. We can consider the graphs in two ways. When the controller output is constant, i.e. dP/dt is zero, the error is zero. When the controller output varies at a constant rate, i.e. dP/dt is constant, the error must have a constant value. The alternative way of considering the graphs is in terms of the area under the error graph.

$$\int_0^t e \, dt = \text{area under the error graph between } t = 0 \text{ and } t$$

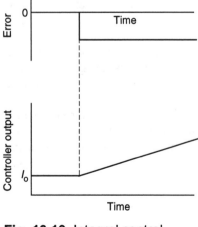

Fig. 13.10 Integral control

Thus up to the time when the error occurs the value of the integral is zero. Hence $I_{out} = I_0$. When the error occurs it maintains a constant value. Thus the area under the graph is increasing as the time increases. Since the area increases at a constant rate the controller output increases at a constant rate.

To illustrate the above, consider an integral controller with a value of K_I of 0.10 s^{-1} and an output of 40% at the set point. What will be the output after times of (a) 1 s, (b) 2 s, if there is a sudden change to a constant error of 20 %? Using the equation developed above

$$I_{out} - I_0 = \int_0^t K_I e\, dt$$

When the error does not vary with time the equation becomes

$$I_{out} = K_I et + I_0$$

Thus for (a) when $t = 1$ s,

$$I_{out} = 0.10 \times 20 \times 1 + 40 = 42\%$$

For (b) when $t = 2$ s,

$$I_{out} = 0.10 \times 20 \times 2 + 40 = 44\%$$

Figure 13.11 shows the form of the circuit used for an electronic integral controller. It consists of an operational amplifier connected as an integrator and followed by another operational amplifier connected as a summer to add the integrator output to that of the controller output at zero time. K_I is $1/R_1C$.

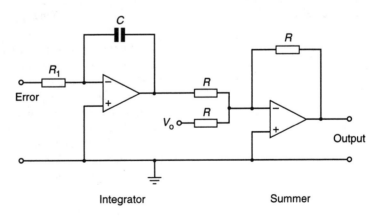

Fig. 13.11 Electronic integral controller

Integrator   Summer

13.6.1 Proportional plus integral control

The integral mode of control is not usually used alone but is frequently used in conjunction with the proportional mode. When integral action is added to a proportional control system the controller output I_{out} is given by

$$I_{out} = K_P(e + K_I \int e \, dt) + I_0$$

where K_P is the proportional control constant, K_I the integral control constant, I_{out} the output when there is an error e and I_0 the output at the set point when the error is zero. Figure 13.12 shows how the system reacts when there is an abrupt change to a constant error. The error gives rise to a proportional controller output which remains constant since the error does not change. There is then superimposed on this a steadily increasing controller output due to the integral action.

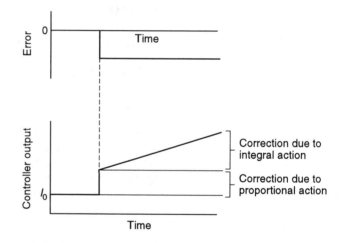

Fig. 13.12 Integral plus proportional control

Suppose there is a change in the controller set point from, say, 50% to 60%. With just a proportional controller this can only be done by having an offset error, i.e. an error value other than zero for the set point value. However, with the combination of integral and proportional control this is not the case. The integral part of the control can provide a change in controller output without any offset error. The controller can be said to reset its set point. Figure 13.13 shows the effects of the proportional action and the integral action if we create an error signal which is increased from the zero value and then decreased back to it again. With proportional action alone the controller mirrors the change and ends up back at its original set point value. With the integral action the controller output increases in proportion to the way the area under the error–time graph increases and since, even when the error has reverted back to zero, there is still a value for the

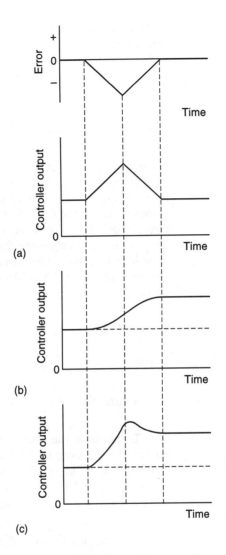

(a)

(b)

(c)

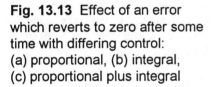

Fig. 13.13 Effect of an error which reverts to zero after some time with differing control: (a) proportional, (b) integral, (c) proportional plus integral

area there is a change in controller output which continues after the error has ceased. The result of combining the proportional and integral actions, i.e. adding the two separate graphs, is thus a change in controller output without an offset error. A step input to the control system can thus give a steady-state value with no error.

Because of the lack of an offset error, this type of controller can be used where there are large changes in the process variable. However because the integration part of the control takes time the changes must be relatively slow to prevent oscillations. Another disadvantage of this form of control is that when the process is started up with controller output at 100% – e.g. with a liquid level control the initial condition may be an empty tank and so the error is so large that the controller has to give a 100% output to fully open a valve – the integral action causes a considerable overshoot of the error before finally settling down.

13.7 PID controller

Combining all three modes of control (proportional, integral and derivative) enables a controller to be produced which has no offset error and reduces the tendency for oscillations. Such a controller is known as a *three-mode controller* or *PID controller*. The equation describing its action is

$$I_{out} = K_P\left(e + K_I\int e\,dt + K_D\frac{de}{dt}\right) + I_0$$

where I_{out} is the output from the controller when there is an error e which is changing with time t, I_0 is the set point output when there is no error, K_P is the proportionality constant, K_I the integral constant and K_D the derivative constant. One way of considering a three-mode controller is as a proportional controller which has integral control to eliminate the offset error and derivative control to reduce time lags. The transfer function of such a controller is obtained by taking the Laplace transform of the above equation. Thus

$$(I_{out} - I_0)(s) = K_P E(s) + \frac{1}{s}K_P K_I E(s) + sK_D E(s)$$

and so

$$\text{transfer function} = K_P\left(1 + \frac{1}{s}K_I + sK_D\right)$$

To illustrate the above, consider what will be the controller output of a three-mode controller having K_P as 4, K_I as 0.6 s^{-1}, K_D as 0.5 s, a set point output of 50% and, subject to the error change shown in figure 13.14, (a) immediately the change starts to occur and (b) 2 s after it starts. Using the equation given above for I_{out}

$$I_{out} = K_P\left(e + K_I\int e\,dt + K_D\frac{de}{dt}\right) + I_0$$

we have for (a) $e = 0$, $de/dt = 1$ s^{-1}, and $\int e\,dt = 0$. Thus

$$I_{out} = 4(0 + 0 + 0.5 \times 1) + 50 = 52.0\%$$

For (b) we have, at 2 s, $e = 1\%$, $\int e\,dt = 1.5$ s since the integral is the area under the error–time graph up to 2 s, and $de/dt = 0$. Thus

$$I_{out} = 4(1 + 0.6 \times 1.5 + 0) + 50 = 57.6\%$$

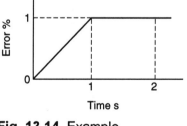

Fig. 13.14 Example

13.8 Direct digital control

Figure 13.15 shows the basis of a direct digital control system that can be used with a continuous process. Microprocessor-based controllers require inputs which are digital, process the inform-ation in digital form and give an output in digital form. Since many

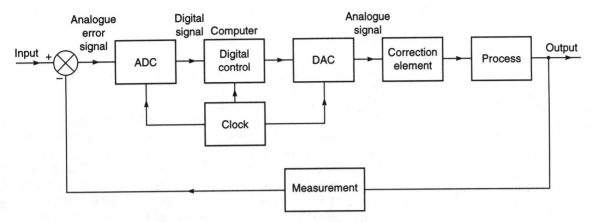

Fig. 13.15 Digital control system

control systems have analogue measurements an analogue-to-digital converter (ADC) is used for the inputs. A clock supplies a pulse at regular time intervals and dictates when samples are taken by the ADC of the controlled variable. These samples are then converted to digital signals which are compared by the micro-processor with the set point value to give the error signal. The microprocessor can then initiate a control mode to process the error signal and give a digital output. The control mode used by the microprocessor is determined by the program of instructions used by the microprocessor for processing the digital signals, i.e. the *software*. The digital output, generally after processing by a digital-to-analogue converter since correcting elements generally require analogue signals, can be used to initiate the correcting action.

Microprocessors as controllers have the advantage over analogue controllers that the form of the controlling action, e.g. proportional or three mode, can be altered by purely a change in the computer software. No change in hardware or electrical wiring is required. Indeed the control strategy can be altered by the computer program during the control action in response to the developing situation.

They also have other advantages. With analogue control separate controllers are required for each process being controlled. With a microprocessor many separate processes can be controlled by sampling processes with a multiplexer (see chapter 3). Digital control gives better accuracy than analogue control because the amplifiers and other components used with analogue systems change their characteristics with time and temperature and so show drift while digital control, because it operates on signals in only the on–off mode, does not suffer from drift in the same way.

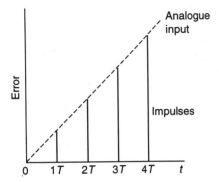

Fig. 13.16 Sampling an analogue error signal

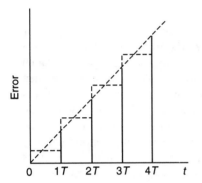

Fig. 13.17 Digital integration

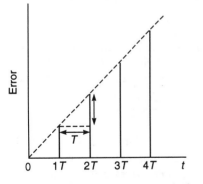

Fig. 13.18 Digital differentiation

13.8.1 Implementing control modes

With digital control the analogue-to-digital converter samples analogue signals at regular time intervals, say every T seconds. Thus the output for an analogue input is a number of impulses, (figure 13.16). The following is an indication of how the various control modes could be realised.

The proportional mode can be realised by just scaling the size of the impulses. With the integral mode, the integral term represents the area under a graph of error against time between time $t = 0$ and t. It can be given approximately by dividing the area into rectangular strips and then summing the area of the strips (figure 13.17). Suppose we consider the width of each strip to be T, the sampling interval. Then the area of a strip is approximately the average of the sizes of the error pulses occurring at the beginning and end of the interval. In this way we can obtain an output with digital pulses which represents the integral mode. The derivative mode can be considered to be the slope of the error–time graph and so approximated by the slope of the line joining two consecutive impulses (figure 13.18). It is thus the difference in the sizes of the pulses divided by the sampling time T.

13.8.2 A computer control system

Typically a computer control system consists of the elements shown in figure 13.15 with set points and control parameters being entered from a keyboard. The software for use with the system will provide the program of instructions needed, for example, for the computer to implement the PID control mode, provide the operator display, recognise and process the instructions inputted by the operator, provide information about the system, provide start-up and shut-down instructions, and supply clock/calendar information. An operator display is likely to show such information as the set point value, the actual measured value, the sampling interval, the error, the controller settings and the state of the correction element. The display is likely to be updated every few seconds.

The computer control system illustrated in figure 13.15 is termed *direct digital control*, the computer being in the control loop. A single computer can, however, be used to control a large number of loops. However, a single loop direct digital control system can now, with the advent of cheap microprocessors, often be cheaper than an analogue system based on operational amplifiers.

For a more detailed discussion of computer control systems the reader is referred to specialist texts, e.g. *Real-time Computer Control* by S. Bennett (Prentice-Hall 1994).

13.9 Control system performance

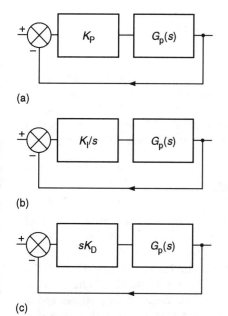

(a)

(b)

(c)

Fig. 13.19 Control systems with differing control: (a) proportional, (b) integral, (c) derivative

The transfer function of a control system is affected by the mode chosen for the controller. Hence the response of the system to, say, a step input is affected. Consider the simple systems shown in figure 13.19. With proportional control the transfer function of the foward path is $K_PG(s)$ and so the transfer function of the feedback system is (see section 11.5)

$$G(s) = \frac{K_P G_p(s)}{1 + K_P G_p(s)}$$

With integral control we have a forward path transfer function of $K_I G_p(s)/s$ and so the system transfer function is

$$G(s) = \frac{K_I G_p(s)}{s + K_I G_p(s)}$$

With derivative control the forward path transfer function is $sK_D G(s)$ and so the system transfer function is

$$G(s) = \frac{sK_D G_p(s)}{1 + K_D G_p(s)}$$

Suppose we have a process which is first order and has a transfer function of $1/(\tau s + 1)$. With proportional control, and unit feedback, the transfer function of the control system becomes

$$G(s) = \frac{K_P/(\tau s + 1)}{1 + K_P/(\tau s + 1)} = \frac{K_P}{\tau s + 1 + K_P}$$

The control system remains a first-order system. The proportional control has had the effect of just changing the form of the first-order response of the process. Without the controller the response to a unit step input was (see section 11.2)

$$y = 1 - e^{-t/\tau}$$

Now it is

$$y = K_P(1 - e^{-t/(\tau/1 + K_P)})$$

The time constant has effectively been reduced. With integral control the transfer function is

$$G(s) = \frac{K_I/s(\tau s + 1)}{1 + K_I/s(\tau s + 1)} = \frac{K_I}{s(\tau s + 1) + K_I} = \frac{K_I}{\tau s^2 + s + K_I}$$

The control system is now a second-order system. Now, with a step input, the system will give a second-order response instead of a first-order response. With derivative control the transfer function is

$$G(s) = \frac{sK_D/(\tau s + 1)}{1 + sK_D/(\tau s + 1)} = \frac{sK_D}{\tau s + 1 + sK_D}$$

13.10 Controller tuning

The term *tuning* is used to describe the process of selecting the best controller settings. With a proportional controller this means selecting the value of K_P, with a PID controller the three constants K_P, K_I and K_D have to be selected. There are a number of methods of doing this; here just two methods will be discussed, both by *Ziegler and Nichols*.

13.10.1 Process reaction method

The procedure with this method is to open the process control loop so that no control action occurs. Generally the break is made between the controller and the correction unit. A test input signal is then applied to the correction unit and the response of the controlled variable determined. The test signal should be as small as possible. Figure 13.20 shows the form of test signal and a typical response. The test signal is a step signal with a step size expressed as the percentage change P in the correction unit. The resulting graph of the measured variable plotted against time is called the *process reaction curve*. The measured variable is expressed as the percentage of the full-scale range.

A tangent is drawn to give the maximum gradient of the graph. For figure 13.20 the maximum gradient R is M/T. The time between the start of the test signal and the point at which this tangent intersects the graph time axis is termed the lag L. Table 13.1 gives the criteria recommended by Ziegler and Nichols for control settings based on the values of P, R and L.

Consider the following example. Determine the settings of K_P, K_I and K_D required for a three-mode controller which gave the process reaction curve shown in figure 13.21 when the test signal was a 6% change in the control valve position. Drawing a tangent to the maximum gradient part of the graph gives a lag L of 150 s and a gradient R of 5/300 = 0.017 s^{-1}. Hence

Table 13.1 Process reaction curve criteria

Control mode	K_P	K_I	K_D
P	P/RL		
PI	$0.9P/RL$	$1/3.33L$	
PID	$1.2P/RL$	$1/2L$	$0.5L$

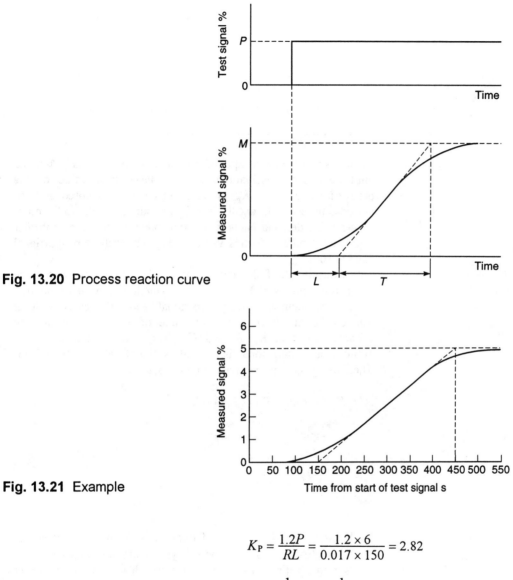

Fig. 13.20 Process reaction curve

Fig. 13.21 Example

$$K_P = \frac{1.2P}{RL} = \frac{1.2 \times 6}{0.017 \times 150} = 2.82$$

$$K_I = \frac{1}{1.2L} = \frac{1}{1.2 \times 150} = 0.0056 \text{ s}$$

$$K_D = 0.5L = 0.5 \times 150 = 75 \text{ s}$$

13.10.2 Ultimate cycle method

With this method, integral and derivative actions are first reduced to their minimum values. The proportional constant K_P is set low and then gradually increased. This is the same as saying that the proportional band is gradually made narrower. While doing this

Table 13.2 Ultimate cycle criteria

Control mode	K_P	K_I	K_D
P	$0.5K_{Pc}$		
PI	$0.45K_{Pc}$	$1.2/T_c$	
PID	$0.6K_{Pc}$	$2.0/T_c$	$T_c/8$

small disturbances are applied to the system. This is continued until continuous oscillations occur. The critical value of the proportional constant K_{Pc} at which this occurs is noted and the periodic time of the oscillations T_c measured. Table 13.2 shows how the Ziegler and Nichols recommended criteria for controller settings are related to this value of K_{Pc}. The critical proportional band is $100/K_{Pc}$.

Consider the following example. When tuning a three-mode control system by the ultimate cycle method it was found that oscillations begin when the proportional band is decreased to 30%. The oscillations have a periodic time of 500 s. What are the suitable values of K_P, K_I and K_D? The critical value of K_{Pc} is 100/critical proportional band and is therefore 100/30 = 3.33. Then, using the criteria given in table 13.2,

$$K_P = 0.6K_{Pc} = 0.6 \times 3.33 = 2.0$$

$$K_I = \frac{2}{T_c} = \frac{2}{500} = 0.004 \text{ s}^{-1}$$

$$K_D = \frac{T_c}{8} = \frac{500}{8} = 62.5 \text{ s}$$

Problems

1 What are the limitations of two-step (on–off) control and in what situation is such a control system commonly used?

2 A two-position mode controller switches on a room heater when the temperature falls to 20°C and off when it reaches 24°C. When the heater is on the air in the room increases in temperature at the rate of 0.5°C per minute; when the heater is off it cools at 0.2°C per minute. If the time lags in the control system are negligible, what will be the times taken for (a) the heater switching on to off, (b) the heater switching off to on?

3 A two-position mode controller is used to control the water level in a tank by opening or closing a valve which in the open position allows water at the rate of 0.4 m³/s to enter the tank. The tank has a cross-sectional area of 12 m² and water leaves it at the constant rate of 0.2 m³/s. The valve opens when the water level reaches 4.0 m and closes at 4.4 m. What will be the

times taken for (a) the valve opening to closing, (b) the valve closing to opening?

4   A proportional controller is used to control the height of water in a tank where the water level can vary from zero to 4.0 m. The required height of water is 3.5 m and the controller is to fully close a valve when the water rises to 3.9 m and fully open it when the water falls to 3.1 m. What proportional band and transfer function will be required?

5   A proportional controller has K_P of 20 and a set point of 50 % output. It output is to have a valve which at the set point allows a flow of 2.0 m³/s. The valve changes its output in direct proportion to the controller output. What will be the controller output and the offset error when the flow has to be changed to 2.5 m³/s?

6   A derivative controller has a set point of 50% and derivative constant K_D of 0.5 s. The error starts at zero and then changes at 2%/s for 3 s before becoming constant for 2 s, after which it decreases at 1%/s to zero. What will be the controller out- put at (a) 0 s, (b) 1 s, (c) 4 s, (d) 6 s?

7   An integral controller has a set point of 50% and a value of K_I of 0.10 s⁻¹. The error starts at zero and changes at 4%/s for 2 s before becoming constant for 3 s. What will be the output after times of (a) 1 s, (b) 3 s?

8   A three mode controller has K_P as 2, K_I as 0.1 s⁻¹, K_D as 1.0 s, and a set point output of 50%. The error starts at zero and changes at 5%/s for 2 s before becoming constant for 3 s. It then decreases at 2%/s to zero and remains at zero. What will be the controller output at (a) 0 s, (b) 3 s, (c) 7 s, (d) 11 s?

9   Describe and compare the characteristics of (a) proportional control, (b) proportional plus integral control, (c) proportional plus integral plus derivative control.

10   Determine the settings of K_P, K_I and K_D required for a three mode controller which gave a process reaction curve with a lag L of 200 s and a gradient R of 0.010%/s when the test signal was a 5% change in the control valve position.

11   When tuning a three-mode control system by the ultimate cycle method it was found that oscillations began when the proportional band was decreased to 20%. The oscillations had a periodic time of 200 s. What are the suitable values of K_P, K_I and K_D?

14 Adaptive control

14.1 Adaptive control

An *adaptive control system* is a control system that automatically is able to achieve a desired response in the presence of extreme changes in the controlled system's parameters and major external disturbances. A conventional control system could have problems dealing with the extreme changes that the adaptive system can cope with. The adaptive control system is based on the use of a microprocessor as the controller. Such a device enables the control mode and the control parameters used to be adapted to fit the circumstances, modifying them as the circumstances change.

An adaptive control system can be considered to have three stages of operation:

1 Starts to operate with controller conditions set on the basis of an assumed condition.
2 The desired performance is continuously compared with the actual system performance.
3 The control system mode and parameters are automatically and continuously adjusted in order to minimise the difference between the desired and actual system performance.

For example, the proportional constant K_p may be automatically adjusted to fit the circumstances, changing as they do.

This chapter is a brief discussion of commonly used forms of adaptive control systems. For more details, the reader is referred to specialist texts such as *Adaptive Control* by K.J. Åstrom and B. Wittenmark (Addison-Wesley 1989).

14.2 Types of adaptive control systems

Adaptive control systems can take a number of forms. Three commonly used forms are:

1 Gain-scheduled control
2 Self-tuning
3 Model-reference adaptive systems

14.2.1 Gain-scheduled control

With *gain-scheduled control* or, as it is sometimes referred to, *pre-programmed adaptive control*, preset changes in the parameters of the controller are made on the basis of some auxiliary measurement of some process variable. Figure 14.1 illustrates this method. The term *gain-scheduled control* was used because the only parameter originally adjusted was the gain, i.e. the proportionality constant K_p.

A disadvantage of this system is that the control parameters have to be determined for many operating conditions so that the controller can select the one to fit the prevailing conditions. An advantage, however, is that the changes in the parameters can be made quickly when the conditions change.

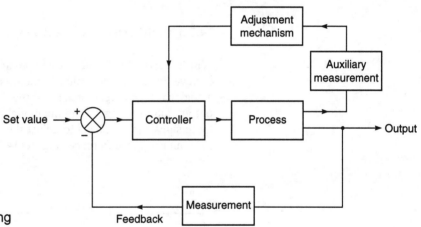

Fig. 14.1 Gain scheduling

14.2.2 Self-tuning

With *self-tuning control* the system continuously tunes its own parameters based on monitoring the variable that the system is controlling and the output from the controller. Figure 14.2 illustrates the features of this system.

Self-tuning is often found in commercial PID controllers, it generally then being referred to as *auto-tuning*. When the operator presses a button, the controller injects a small disturbance into the system and measures the response. This response is compared to the desired response and the control parameters adjusted, by a modified Ziegler–Nichols rule (see section 13.7), to bring the actual response closer to the desired response.

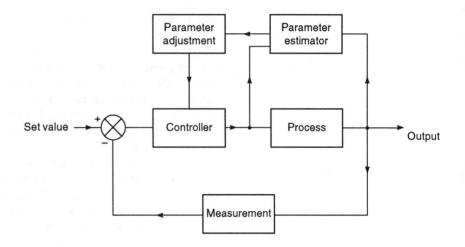

Fig. 14.2 Self-tuning

14.2.3 Model-reference adaptive systems

With the *model-reference* system an accurate model of the system is developed. The set value is then used as an input to both the actual and the model systems and the difference between the actual output and the output from the model compared. The difference in these signals is then used to adjust the parameters of the controller to minimise the difference. Figure 14.3 illustrates the features of the system.

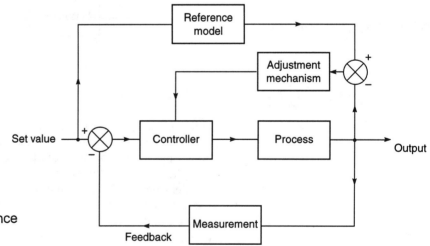

Fig. 14.3 Model-reference adaptive system

Problems

1 Explain what is meant by adaptive control.
2 Explain the basis on which the following forms of adaptive control systems function: (a) gain-scheduled, (b) self-tuning, (c) model-reference.

15 Microprocessors

15.1 Basic structure

Computers have three sections: a *central processing unit* (CPU) to recognise and carry out program instructions, *input and output circuitry/interfaces* to handle communications between the computer and the outside world, and *memory* to hold the program instructions and data. Digital signals move from one section to another along paths called *buses*. A bus, in the physical sense, is just a number of conductors along which electrical signals can be carried. It might be tracks on a printed circuit board or wires in a ribbon cable. The data associated with the processing function of the CPU is carried by the *data bus*, the information for the address of a specific memory location for the accessing of stored data is carried by the *address bus* and the signals relating to control actions are carried by the *control bus*. Figure 15.1 illustrates the general arrangement. In some cases a microprocessor chip constitutes just the CPU while in other cases it might have all the components necessary for the complete computer on one chip. Microprocessors which have memory and various input/output arrangements all on the same chip are called *microcontrollers*. They are effectively a microcomputer on a single chip.

This section is a brief overview of the structure of microprocessors with the following sections discussing programming. Chapter 16 discusses interfacing and chapter 17 programmed logic controllers based on the the use of microprocessors. For more details the reader is referred to the manuals issued by the microprocessor manufacturers, or specialist texts such as *Microprocessors* by M. Rafiquzzaman (Prentice-Hall 1992), or *Microprocessor Fundamentals* by R.L. Tokheim (Schaum Outline Series, McGraw-Hill 1990).

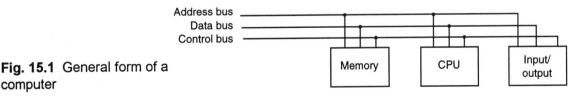

Fig. 15.1 General form of a computer

15.1.1 Buses

The *data bus* is used to transport a word (see section 3.6) between the memory, or the input/output interfaces, to or from the CPU. Word lengths used may be 4, 8, 16 or 32. Each wire in the bus carries a binary signal, i.e. a 0 or a 1. Thus with a four-wire bus we might have the word 1010 being carried, each bit being carried by a separate wire in the bus, as

Word	Bus wire
0 (least significant bit)	First data bus wire
1	Second data bus wire
0	Third data bus wire
1 (most significant bit)	Fourth data bus wire

The more wires the data bus has the longer the word length that can be used. The range of values which a single item of data can have is restricted to that which can be represented by the word length. Thus with a word of length 4 bits the number of values is $2^4 = 16$. Thus if the data is to represent, say, a temperature, then the range of possible temperatures must be divided into 16 segments if we are to represent that range by a 4-bit word. The earliest microprocessors were 4-bit (word length) devices and such 4-bit microprocessors are still widely used in such devices as toys, washing machines and domestic central heating controllers. They were followed by 8-bit microprocessors, e.g. the Rockwell 6502 and the Zilog Z80. 16-bit, 32-bit and 64-bit microprocessors are now available, however, 8-bit microprocessors are still widely used for controllers.

The *address bus* carries signals which indicate where data is to be found and so the selection of certain memory locations or input or output ports. When a particular address is selected by its address being placed on the address bus, only that location is open to the communications from the CPU. The CPU is thus able to communicate with just one location at a time. A computer with an 8-bit data bus has typically a 16-bit wide address bus, i.e. 16 wires. This size of address bus enables 2^{16} locations to be addressed. 2^{16} is 65 536 locations and is usually written as 64K, where K is equal to 1024. The more memory that can be addressed the greater the volume of data that can be stored and the larger and more sophisticated the programs that can be used.

The *control bus* is the means by which signals are sent to synchronise the separate elements. The system clock signals are carried by the control bus. These signals generate time intervals during which system operations can take place. The CPU sends some control signals to other elements to indicate the type of operation being performed, e.g. whether it needs to READ (receive) a signal or WRITE (send) a signal.

15.1.2 CPU

The CPU is the section of the processor which processes the data, fetching instructions from memory, decoding them and executing them. It can be considered to consist of a control unit, an arithmetic and logic unit (ALU) and registers (figure 15.2).

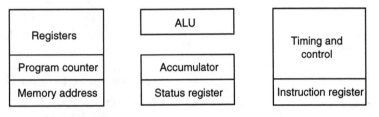

Fig. 15.2 Features of a CPU

The *control unit* determines the timing and sequence of operations. It generates the timing signals used to fetch a program instruction from memory and execute it. The Motorola 6800 uses a clock with a maximum frequency of 1 MHz, i.e. a clock period of 1 μs, and instructions require between two and twelve clock cycles. Operations involving the microprocessor are reckoned in terms of the number of cycles they take.

The *arithmetic and logic unit* is responsible for performing the actual data manipulation. It carries out arithmetic operations such as add and subtract and logic operations such as AND, OR and EXCLUSIVE-OR (see section 1.6.1).

Internal data that the CPU is currently using is temporarily held in a group of registers while instructions are being executed. *Registers* provide means of temporarily storing data in the CPU. There are a number of types of register, the number, size and types of registers varying from one microprocessor to another. The following are common types of registers.

1 The *accumulator register* (A) is where data for an input to the arithmetic and logic unit is temporarily stored. In order for the CPU to be able to access, i.e. read, instructions or data in the memory it has to supply the address of the required memory word using the address bus. When this has been done, the required instructions or data can be read into the CPU using the data bus. Since only one memory location can be addressed at once, temporary storage has to be used when, for example, numbers are combined. For example, in the addition of two numbers, one of the numbers is fetched from one address and placed in the accumulator register while the CPU fetches the other number from the other memory address. Then the two numbers can be processed by the arithmetic and logic section

of the CPU. The result is then transferred back into the accumulator register. The accumulator register is thus a temporary holding register for data to be operated on by the arithmetic and logic unit and also, after the operation, the register for holding the results. It is thus involved in all data transfers associated with the execution of arithmetic and logic operations.

2 The *status register*, or *condition code register* or *flag register*, contains information concerning the result of the latest process carried out in the arithmetic and logic unit. It contains individual bits with each bit having special significance. The bits are called *flags*. The status of the latest operation is indicated by each flag with each flag being set or reset to indicate a specific status. For example, they can be used to indicate whether the last operation resulted in a negative result, a zero result, a carry output occurs (e.g. the sum of two binary numbers such as 101 and 110 has resulted in a result (1)011 which is bigger than the microprocessor's word size and carries a 1 overflow), an overflow occurs or the program is to be allowed to be interrupted to allow an external event to occur. The following are common flags.

Flag	Set, i.e. 1	Reset, i.e. 0
Z	result is zero	result is not zero
N	result is negative	result is not negative
C	carry is generated	carry is not generated
V	overflow occurs	overflow does not occur
I	interrupt is ignored	interrupt is processed normally

3 The *program counter register* (PC), or *instruction pointer* (IP), is the register used to allow the CPU to keep track of its position in a program. This register contains the address of the memory location that contains the next program instruction. As each instruction is executed the program counter register is updated so that it contains the address of the memory location where the next instruction to be executed is stored. The program counter is incremented each time so that the CPU executes instructions sequentially unless an instruction, such as JUMP or BRANCH, changes the program counter out of that sequence.

4 The *memory address register* (MAR) contains the address of data. Thus, for example, in the summing of two numbers the memory address register is loaded with the address of the first number. The data at the address is then moved to the accumulator. The memory address of the second number is then loaded into the memory address register. The data at this

address is then added to the data in the accumulator. The result is then stored in a memory location addressed by the memory address register.

5 The *instruction register* (IR) stores an instruction. After fetching an instruction from the memory, the CPU stores it in the instruction register. It can then be decoded and used to execute an operation.

6 *General-purpose registers* may serve as temporary storage for data or addresses and be used in operations involving transfers between various other registers.

7 The *stack pointer register* (SP) contents form an address which defines the top of the stack in RAM (see section 15.5). The stack is a special area of the memory in which program counter values can be stored when a subroutine part of a program is being used.

The number and form of the registers depends on the microprocessor concerned. For example, the Rockwell 6502 microprocessor has one accumulator register, one status register, two index registers, a stack pointer register and a program counter. The status register has flag bits to show negative, zero, carry, overflow, decimal, interrupt and break. The index registers are used when executing instructions in the indexed addressing mode (see section 15.3.1). The Motorola 6802 microprocessor has two accumulator registers, a status register, an index register, a stack pointer register and a program counter register. The status register has flag bits to show negative, zero, carry, overflow, half-carry and interrupt.

15.1.3 Memory

The memory unit stores binary data and takes the form of one or more integrated circuits. The data may be program instruction codes or numbers being operated on. The size of the memory is determined by the number of wires in the address bus. For data that is stored permanently a memory device called a *read-only memory* (ROM) is used. ROMs are programmed with the required contents during the manufacture of the integrated circuit. No data can then be written into this memory while the memory chip is in the computer. The data can only be read and is used for fixed programs.

The term *erasable and programmable ROM* (EPROM) is used for ROMs that can be programmed and their contents altered. A typical EPROM chip contains a series of small electronic circuits, cells, which can store charge. The program is stored by applying voltages to the integrated circuit connection pins and producing a pattern of charged and uncharged cells. The pattern remains permanently in the chip until erased by shining ultraviolet light through a quartz window on the top of the device. This causes all the cells to become discharged. The chip can then be reprogrammed.

Temporary data, i.e. data currently being operated on, is stored in read/write memory referred to as a *random-access memory* (RAM). Such a memory can be read or written to.

When ROM is used for program storage, then the program is available and ready for use when the system is switched on. Programs stored in ROM are termed *firmware*. Some firmware must always be present. When RAM is used for program storage then such programs are referred to as *software*. When the system is switched on, software may be loaded into RAM from some other peripheral equipment such as a keyboard or hard disk or floppy disk.

15.2 Languages

The inputs that cause a microprocessor to perform a specific action are called *instructions* and the collection of instructions that the microprocessor will recognise is its *instruction set*. The form of the instruction set depends on the microprocessor concerned. The series of instructions that are needed to carry out a particular task is called a *program*.

Microprocessors work in binary code. Instructions written in binary code are referred to as being in *machine code*. Writing a program in such a code is a skilled and very tedious process. It is prone to errors because the program is just a series of 0s and 1s and the instructions are not easily comprehended from just looking at the pattern. An alternative is to use an easily comprehended form of shorthand code for the patterns of 0s and 1s. For example, the operation of adding data to an accumulator might be represented by just ADDA. Such a shorthand code is referred to as a *mnemonic code*, a mnemonic code being a 'memory aiding' code. The term *assembly language* is used for such a code. Writing a program using mnemonics is easier because they are an abbreviated version of the operation performed by the instruction. Also, because the instructions describe the program operations and can easily be comprehended, they are less likely to be used in error than the binary patterns of machine code programming. The assembler program has still, however, to be converted into machine code since this is all the microprocessor will recognise. This conversion can be done by hand using the manufacturer's data sheets which list the binary code for each mnemonic. However, computer programs are available to do the conversion, such programs being referred to as *assembler programs*.

High-level languages are available which provide a type of programming language which is even closer to describing in easily comprehended language the type of operations required. Examples of such languages are BASIC, C, FORTRAN and PASCAL. Such languages have still, however, to be converted into machine code, by a computer program, for the microprocessor to be able to use. Programs written in high-level languages usually require more memory to store them when they have been converted into machine code and thus tend to take longer to run than programs written in assembly language.

15.2.1 Data representation

Assembly programming uses mnemonics for the instructions and can use a shorthand way of writing binary number data. *Hexadecimal* representation of binary numbers is commonly used. The hexadecimal system is to base 16 instead of the base 2 of binary numbers. In writing numbers in the hexadecimal system, the characters 0 to 9 are used together with the first six letters of the alphabet. Table 15.1 shows the decimal and binary equivalents of hexadecimal numbers.

To convert a binary number to a hexadecimal number, group the binary digits in fours. This is because $2^4 = 16$ and thus each

Table 15.1 Hexadecimal system

Hexadecimal	Decimal	Binary
0	0	0
1	1	1
2	2	10
3	3	11
4	4	100
5	5	101
6	6	110
7	7	111
8	8	1000
9	9	1001
A	10	1010
B	11	1011
C	12	1100
D	13	1101
E	14	1110
F	15	1111

block of four binary numbers can be represented by a single hexa-decimal character. For example, consider the binary number

1011100100011110

Grouped into fours we have

1011 1001 0001 1110
 B   9   1   E

Thus the hexadecimal number is B91E.

15.3 Instruction sets

The instructions that may be given to a microprocessor are termed the *instruction set*. In general, instructions can be classified as:

1 Data transfer
2 Arithmetic
3 Logical
4 Program control

The instruction set differs from one microprocessor to another. Certain instructions are, however, reasonably common to most microprocessors. These are:

Data transfer

1 *Load* This instruction reads the contents of a specified memory location and copies it to a specified register location in the CPU. For example, we might have

Before instruction	After instruction
Data in memory location 0010	Data in memory location 0010
	Data from 0010 in accumulator

2 *Store* This instruction copies the current contents of a specified register into a specified memory location. For example, we might have

Before instruction	After instruction
Data in accumulator	Data in accumulator
	Data copied to memory location 0011

Arithmetic

3   *Add* This instruction adds the contents of a specified memory location to the data in some register. For example, we might have

Before instruction	After instruction
Accumulator with data 0001	Accumulator with data 0011
Memory location with data 0010	

4   *Decrement* This instruction subtracts 1 from the contents of a specified location. For example, we might have the accumulator as the specified location and so

Before instruction	After instruction
Accumulator with data 0011	Accumulator with data 0010

5   *Compare* This instruction indicates whether the contents of a register are greater than, less than or the same as the contents of a specified memory location. The result appears in the status register as a flag.

Logical

6   *AND* This instruction carries out the logical AND operation with the contents of a specified memory location and the data in some register. Numbers are ANDed bit by bit. For example, we might have:

Before instruction	After instruction
Accumulator with data 0011	Accumulator with data 0001
Memory location with data 0001	

Only in the least significant bit in the above data have we a 1 in both sets of data and the AND operation only gives a 1 in the least significant bit of the result.

7   *EXCLUSIVE OR* This instruction carries out the logical EXCLUSIVE OR operation with the contents of a specified memory location and the data in some register. The operation is performed bit by bit.

8 *Logical shift (left or right)* Logical shift instructions involve moving the pattern of bits in the register one place to the left or right by moving a 0 into the end of the number. For example, for logical shift right a 0 is shifted into the most significant bit and the least significant bit is moved to the carry flag in the status register.

Before instruction	After instruction
Accumulator with data 0011	Accumulator with data 0001
	Status register indicates Carry 1

9 *Arithmetic shift (left or right)* Arithmetic shift instructions involve moving the pattern of bits in the register one place to the left or right but with copying of the end number into the vacancy created by the shift. For example, for an arithmetic shift right,

Before instruction	After instruction
Accumulator with data 1011	Accumulator with data 1001
	Status register indicates Carry 1

10 *Rotate (left or right)* Rotate instructions involve moving the pattern of bits in the register one place to left or right and the bit that spills out is written back into the other end. For example, for a rotate right,

Before instruction	After instruction
Accumulator with data 0011	Accumulator with data 1001

Program control

11 *Jump* This instruction changes the sequence in which the program steps are carried out. Normally the program counter causes the program to be carried out sequentially in strict numerical sequence. However, the jump instruction causes the program counter to jump to some other specified location in the program. For example, a program might require the following sequence of instructions:

Decrement the accumulator
Jump if the accumulator is not zero to instruction ...

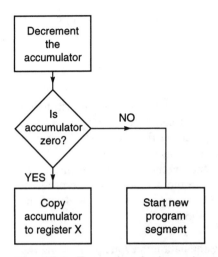

Fig. 15.3 Example of a branch

12 *Branch* This is a conditional instruction which might be *branch if zero* or *branch if plus*. This branch instruction is followed if the right conditions occur. For example, a program might require the sequence of instructions shown in the flow chart in figure 15.3. The diamond-shaped box is used to indicate that a decision is required, ordinary instructions being shown in rectangular boxes.

13 *Halt* This instruction stops all further microprocessor activity.

To illustrate the types of codes used for operations in assembly language, table 15.2 shows them for a number of opera-

Table 15.2 Examples of mnemonics

Instruction	Mnemonic
Rockwell 6502	
Load accumulator	LDA
Load X register	LDX
Add with carry	ADC
Logic shift right by one bit	LSR
Branch if carry clear	BCC
Jump to address	JMP
Motorola 6809	
Load accumulator A	LDAA
Load index register X	LDX
Add with carry to accumulator A	ADCA
Logic shift right memory	LSR
Branch if carry clear	BCC
Jump to address	JMP
Intel 8085A	
Load accumulator	LDA
Move from register 2 to register 1	MOV r1 r2
Add with carry to register r	ADC r
Rotate accumulator left	RLC
Unconditional jump	JMP
Conditional jump if cc is true	Jcc

tions when different microprocessors are involved. Appendix B gives the full range of codes used with the Motorola 6800.

15.3.1 Addressing

When a mnemonic, such as LDA, is used to specify an instruction it will be followed by additional information to specify the sources and destinations of the data required by the instruction. The data following the instruction is referred to as the *operand*.

There are several different methods that are used for specifying data locations and hence the way in which the program causes the microprocessor to obtain its instructions or data. The following are commonly used methods:

1 With *immediate addressing* the data immediately following the instruction is the value to be operated on. For example, LDA B #$FF means load the number FF into accumulator B. The # signifies immediate mode and the $ that the FF refers to a number in hexadecimal notation. The $ sign is not always used in writing instructions, since generally it is obvious which are numbers. Another example is LDX #$C540. This means load the index register with the number C540. This type of operation involves the loading of a predetermined value into a register or memory location.

2 With *direct addressing*, or *zero-page addressing*, it is assumed that the memory location in which the data to be operated on resides is in the lowest 256 bytes of memory and so only eight bits are needed to specify it, i.e. the upper eight bits are all 0s. Thus this mode can only be used for address in locations from $0000 to $00FF. For example, ADD $25 means add the contents of memory location 0025 to the original contents of the accumulator.

3 With *absolute addressing*, or *extended addressing*, the full address of the data to be acted on follows the instruction. For example, LDA A $20F0 means load accumulator A from address $20F0.

4 With *implied addressing*, or *inherent addressing*, the address is implied in the instruction. For example, CLR A means clear accumulator A.

5 With *indexed addressing*, the first byte of the instruction contains the op code and the second byte contains the offset. The offset is added to the contents of the index register to determine the address of the operand. An instruction might thus appear as LDA A $FF,X which means load accumulator A with data at the address given by adding the contents of the index register and FF. Another example is STA A $05,X which means store the contents of accumulator A at the address given by the index register plus 05.

6 *Relative addressing* is used with branch instructions. The branch instructions follow the op codes with a byte called the relative address. This indicates the displacement in address that has to be added to the program counter if the branch occurs. For example, BEQ $F1 indicates that if the data is equal to zero then the next address in the program is F1 further on. The relative address of F1 is added to the address of the next instruction.

The following illustrates the above with regard to a number of instructions:

Address mode	Instruction	
Immediate	LDA A #$F0	Load accumulator A with data F0
Direct	LDA A $50	Load accumulator A with data at address 0050
Extended	LDA A$0F01	Load accumulator A with data at address 0F01
Indexed	LDA A $CF,X	Load accumulator with data at the address given by the index register plus CF
Inherent	CLR A	Clear accumulator A
Extended	CR $2020	Clear address 2020, i.e. store all 0s at address 2020
Indexed	CLR $10,X	Clear the address given by the index register plus 10, i.e. store all 0s at that address
Immediate	ADD A #$FF	Add FF to the data in accumulator A
Direct	ADD A $FF	Add to the data in accumulator A the data stored at address 00FF
Extended	ADD A $00FF	Add to the data in accumulator A the data stored at address 00FF
Indexed	ADD A $FF,X	Add to the data in accumulator A the data stored at the address given by the value in the index register plus FF

15.3.2 Data movement

The following is an example of the type of information that will be found in a manufacturer's (6800) instruction set sheet:

		Addressing modes					
		IMMED			DIRECT		
Operation	Mnemonic	OP	~	#	OP	~	#
Add	ADDA	8B	2	2	9B	3	2

Notes: OP is the operation code (hexadecimal), ~ is the number of microprocessor cycles required, # is the number of program bytes required.

This means that when using the immediate mode of addressing with this processor the Add operation is represented by the mnemonic ADDA. When the immediate form of addressing is used the machine code for it is 8B and it will take two cycles to be fully expressed. The operation will require two bytes in the program. The *op-code* or *operation code* is the term used for the instruction that the microprocessor will act on. For convenience, these are expressed in hexadecimal form. A byte is a group of eight binary digits recognised by the microprocessor as a word. Thus two words are required. With the direct mode of addressing the machine code is 9B and takes three cycles and two program bytes.

To illustrate how information passes between memory and microprocessor, consider the following tasks. The addresses in RAM where a new program may be placed is just a matter of convenience. For the following examples, the addresses starting at 0010 have been used. For direct addressing to be used, the addresses must be on the zero page, i.e. addresses between 0000 and 00FF. The examples are based on the use of the instruction set for the M6800/6802 microprocessors. See Appendix B for details of these microprocessors and their instruction set.

Task: Enter all zeros in accumulator A.

Memory address	Op-code	
0010	8F	CLR A

The next memory address that can be used is 0011 because CLR A only occupies one program byte. This is the inherent mode of addressing.

Task: Add to the contents of accumulator A the data 20.

Memory address	Op-code	
0010	8B 20	ADD A #$20

This uses the immediate form of addressing. The next memory address that can be used is 0012 because, in this form of addressing, ADD A occupies two program bytes.

Task: Load accumulator A with the data contained in memory address 00AF.

Memory address	Op-code	
0010	B6 00AF	LDA A $00AF

This uses the absolute form of addressing. The next memory address that can be used is 0013 because, in this form of addressing, LDA A occupies three program bytes.

Task: Rotate left the data contained in memory location 00AF.

Memory address	Op-code	
0010	79 00AF	ROL $00AF

This uses the absolute form of addressing. The next memory address that can be used is 0013 since ROL, in this mode, occupies three program bytes.

Task: Store the data contained in accumulator A into memory location 0021.

Memory address	Op-code	
0010	D7 21	STA A $21

This uses the direct mode of addressing. The next memory address that can be used is 0012 since STA A, in this mode, occupies two program bytes.

Task: Branch forward four places if the result of the previous instruction is zero.

Memory address	Op-code	
0010	27 04	BEQ $04

This uses the relative mode of addressing. The next memory address if the result is not zero is 0012 since BEQ, in this mode, occupies two program bytes. If the result is zero then the next address is 0012 + 4 = 0016.

15.4 Programming

A commonly used method for the development of programs follows the following step:

1 Define the problem, stating quite clearly what function the program is to perform, the inputs and outputs required, any constraints regarding speed of operation, accuracy, memory size, etc.
2 Define the algorithm to be used. An *algorithm* is the sequence of steps which define a method of solving the problem.
3 Represent the algorithm by means of a *flow chart*. Figure 15.4 shows the standard symbols used in the preparation of flow charts. Each step of an algorithm is represented by one or more of these symbols and linked together by lines to represent the program flow.

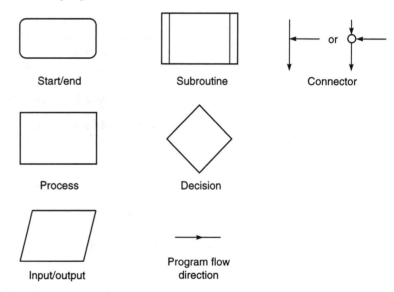

Fig. 15.4 Flow chart symbols

4 Translate the flow chart/algorithm into instructions which the microprocessor can execute. This can be done by writing the instructions in assembly language and then converting these, either manually or by means of a computer program, into machine code.
5 Test and debug the program. Errors in programs are referred to as *bugs* and the process of tracking them down and eliminating them as *debugging*.

15.4.1 Assembly language programs

An assembly language program should be considered as a series of instructions to an assembler which will then produce the machine code program. A program written in assembly language consists of a sequence of statements, one statement per line. A statement contains from one to four sections or *fields*. The fields are:

Label   Op-code   Operand   Comment

A special symbol is used to indicate the beginning or end of a field, the symbols used depending on the microprocessor machine code assembler concerned. With the Motorola 6800 spaces are used. With the Intel 8080 there are a colon after a label, a space after the op-code, commas between entries in the address field and a semicolon before a comment.

The *label* is the name by which a particular entry in the memory is referred to. Labels can consist of letters, numbers and some other characters. With the Motorola 6800, labels are restricted to one to six characters, the first of which must be a letter, and cannot consist of a single letter A, B or X since these are reserved for reference to the accumulator or index register. With the Intel 8080, five characters are permitted with the first character a letter, @, or ?. The label must not use any of the names reserved for registers, instruction codes or pseudo-operations (see later in this section). All labels within a program must be unique. If there is no label then a space must be included in the label field. With the Motorola 6800, an asterisk (*) in the label field indicates that the entire statement is a comment, i.e. a comment included to make the purpose of the program clearer. As such, the comment will be ignored by the assembler during the assembly process for the machine code program.

The *op-code* specifies how data is to be manipulated and is specified by its mnemonic, e.g. LDA A. It is the only field that can never be empty.

In addition, the op-code field may contain directives to the assembler. These are termed *pseudo-operations* since they appear in the op-code field but are not translated into instructions in machine code. They may define symbols, assign programs and data to certain areas of memory, generate fixed tables and data, indicate the end of the program, etc. Common pseudo-operation symbols include:

ORG	Origin: defines the starting memory address of the next part of the program. A program may have several origins.
EQU, SET, DEF	Equates/sets/defines a symbol for a numerical value, another symbol or an expression.
NAM	Name: assigns a name to the program.
RMB, RES	Reserves memory space for future use.
END	Indicates the end of the program.
DATA, DB, DW, FW, FCC, FCB	Enters constant values into memory.

The information included in the *operand* field depends on the mnemonic preceding it and the addressing mode used. It gives the address of the data to be operated on by the process specified by the op-code. It is thus often referred to as the *address field*. This field may be empty if the instructions given by the op-code do not need any data or address. Numerical data in this field may be hexadecimal, decimal, octal, or binary. The assembler assumes that numbers are decimal unless otherwise specified. With the Motorola 6800 a hexadecimal number is preceded by $ or followed by H, an octal number preceded by @ or followed by O or Q, a binary number preceded by % or followed by B. With the Intel 8080, a hexadecimal number is followed by H, an octal number by O or Q, a binary number by B. Hexadecimal numbers must begin with a decimal digit, i.e. 0 to 9, to avoid confusion with names. With the Motorola 6800, the immediate mode of address can be indicated by preceding the operand by #, the indexed mode of address involving the operand being followed by X. No special symbols are used for direct or extended addressing modes. If the address is on the zero page, i.e. FF or less, then the assembler automatically assigns the direct mode. If the address is greater than FF, the assembler assigns the extended mode.

The comment field is optional and is there to allow the programmer to include any comments which may make the program more understandable by the reader. The comment field is ignored by the assembler during the production of the machine code program.

15.4.2 Examples of assembly language programs

The following examples illustrate how some simple programs can be developed. The examples are based on the use of the instruction set for the M6800/6802 microprocessors. See Appendix B for details of these microprocessors and their instruction set.

Problem: The addition of 8-bit numbers located in two memory addresses.

The algorithm is:

1 Start.
2 Load the accumulator with the first number.
3 Add on the second number.
4 Store the sum in a designated memory location.
5 Stop.

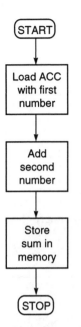

Fig. 15.5 Flow chart for addition of two numbers

Figure 15.5 shows these steps represented as a flow chart. The resulting assembly program is:

Label	Op-code	Operand	Comment
AUGEND	EQU	$0070	
ADDEND	EQU	$0071	
SUM	EQU	$0072	
	ORG	$0010	
	LDA A	AUGEND	First number into accumulator A
	ADD A	ADDEND	Add on second number
	STA	A SUM	Save the sum
	SWI		Program end

The first line in the program specifies the address of the first of the numbers to be added. The second line specifies the address of the number to be added to the first number. The third line specifies where the sum is to be put. The fourth line specifies the memory address at which the program should start. The use of the labels means that the operand involving that data does not have to specify the addresses but merely the labels. When translated into machine code the pseudo-operations indicate the addresses for the items. In machine code the program would become:

```
0010  96  70
0012  9B  71
0014  97  72
0016  3F
```

In many programs there can be a requirement for a task to be carried out a number of times in succession. In such cases the program can be designed so that the operation passes through the same section a number of times. This is termed *looping*, a *loop* being a section of a program that is repeated a number of times. Figure 15.6 shows a flow diagram of a loop. A certain operation has to be performed a number of times before the program proceeds. Only when the number of such operations is completed can the program proceed.

Problem: the addition of numbers located at ten different addresses (these might be, for example, the results of inputs from ten different sensors).

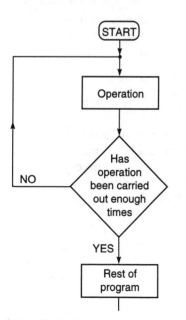

Fig. 15.6 A loop

The algorithm could be:

1 Start.
2 Set the count as 10.
3 Point to location of bottom address number.
4 Add bottom address number.
5 Decrease the count number by 1.
6 Add 1 to the address location pointer.
7 Is count 0 ? If not branch to 4. If yes proceed.
8 Store sum.
9 Stop.

Figure 15.7 shows the flow chart. The resulting assembly program is:

Label	Op-code	Operand	Comment
COUNT	EQU	$0010	
POINT	EQU	$0020	
RESULT	EQU	$0050	
	ORG	$0001	
	LDA B	COUNT	Load count
	LDX	POINT	Initialise index register at start of numbers
SUM	ADD A	X	Add addend
	INX		Add 1 to index register
	DEC B		Subtract 1 from accumulator B
	BNE	SUM	Branch to sum
	STA A	RESULT	Store
	SWI		Stop program

The count number of 10 is loaded into accumulator B. The index register gives the initial address of the data being added. The first summation step is to add the contents of the memory location addressed by the index register to the contents of the accumulator, initially assumed zero (a CLR A instruction could be used to clear it first). The instruction INX adds 1 to the index register so that the next address that will be addressed is 0021. DEC B subtracts 1 from the contents of accumulator B and so indicates that a count of 9 remains. BNE is then the instruction to branch to SUM if not equal to zero, i.e. if the Z flag has a 0 value. The program then loops and repeats the loop until ACC B is zero.

Problem: the determination of the biggest number in a list of

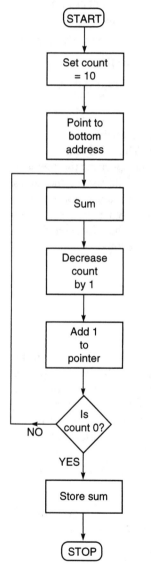

Fig. 15.7 Flow chart for adding ten numbers

numbers (it could be the largest temperature value inputted from a number of temperature sensors).

The algorithm could be:

1 Clear answer address.
2 List starting address.
3 Load the number from the starting address.
4 Compare the number to the number in answer address.
5 Store answer if bigger.
6 Otherwise save number.
7 Increase starting address by 1.
8 Branch to 3 if the address is not the last address.
9 Stop.

Figure 15.8 shows the flow chart. The resulting assembly program is:

Label	Op-code	Operand	Comment
FIRST	EQU	$0030	
LAST	EQU	$0040	
ANSW	EQU	$0041	
	ORG	$0000	
	CLR	ANSW	Clear answer
	LDX	FIRST	Load first address
NUM	LDA A	$30,X	Load number
	CMP A	ANSW	Compare with answer
	BLS	NEXT	Branch to NEXT if lower or same
	STA A	ANSW	Store answer
NEXT	INX		Increment index register
	CPX	LAST	Compare index register with LAST
	BNE	NUM	Branch if not equal to zero
	WAI		Stop program

First the answer address is cleared. The first address is then loaded and the number in that address put into accumulator A. LDA A $30,X means load accumulator A with data at address given by the index register plus 30. Compare the number with the answer, keeping the number if it is greater than the number already in the accumulator, otherwise branch to repeat the loop with the next number.

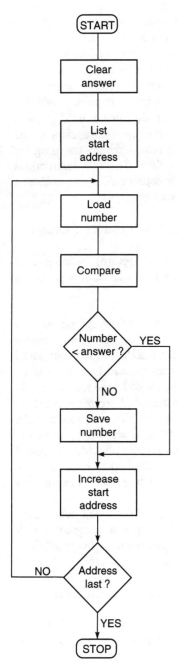

Fig. 15.8 Flow chart for biggest number

15.5 Subroutines

It is often the case that a block of programming, a subroutine, might be required a number of times in a program. For example, a block of programming might be needed to produce a time delay. It would be possible to duplicate the subroutine program a number of times in the main program. This, however, is an inefficient use of memory. Alternatively we could have a single copy of it in memory and branch or jump to it every time the subroutine was required. This, however, presents the problem of knowing, after completion of the subroutine, the point at which to resume in the main program. What is required is a mechanism for getting back to the main program and continuing with it from the point at which it was left to carry out the subroutine. To do this we need to store the contents of the program counter at the time of branching to the subroutine so that this value can be reloaded into the program counter when the subroutine is complete. The two instructions which are provided with most microprocessors to enable a subroutine to be implemented in this way are:

1 JSR (jump to routine), or CALL, which enables a subroutine to be called.
2 RTS (return from subroutine), or RET (return), which is used as the last instruction in a subroutine and returns it to the correct point in the calling program.

Subroutines may be called from many different points in a program. It is thus necessary to store the program counter contents in such a way that we have a last-in-first-out store (LIFO). Such a register is referred to as a *stack*. It is like a stack of plates in that the last plate is always added to the top of the pile of plates and the first plate that is removed from the stack is always the top plate and hence the last plate that was added to the stack. The stack may be a block of registers within a microprocessor or, more commonly, using a section of RAM. A special register within the microprocessor, called the *stack pointer register*, is then used to point to the next free address in the area of RAM being used for the stack.

In addition to the automatic use of the stack when subroutines are used, a programmer can write a program which involves the use of the stack for the temporary storage of data. The two instructions that are likely to be involved are:

1 PUSH, which causes data in specified registers to be saved to the next free location in the stack.
2 PULL, or POP, which causes data to be retrieved from the last used location in the stack and transferred to a specified register.

For example, prior to some subroutine data in some registers may have to be saved and then, after the subroutine, the data restored. The program elements might thus be, with the Motorola 6800:

Label	Op-code	Operand	Comment
SAVE	PSH A		Save accumulator A to stack
	PSH B		Save accumulator B to stack
	TPA		Transfer status register to accumulator A
	PSH A		Save status register to stack
Subroutine			
RESTORE	PUL A		Restore condition code from stack to accumulator A
	TAP		Restore condition code from accumulator A to status register
	PUL B		Restore accumulator B from stack
	PUL A		Restore accumulator A from stack

15.5.1 Delay subroutine

Delay loops are often required when the microprocessor has an input from a device such as an analogue-to-digital converter. The requirement is often to signal to the converter to begin its conversion and then wait a fixed time before reading the data from the converter. This can be done by providing a loop which makes the microprocessor carry out a number of instructions before proceeding with the rest of the program. A simple delay program might be:

Label	Op-code	Operand	Comment
DELAY	LDA A	#$05	Load accumulator A with 05
LOOP	DEC A		Decrement accumulator A by 1
	BNE	LOOP	Branch if not equal to zero
	RTS		Return from subroutine

For each movement through the loop a number of machine cycles are involved. The delay program, when going through the loop five times, thus takes

Instruction	Cycles	Total cycles
LDA A	2	2
DEC A	2	10
BNE	4	20
RTS	1	1

The total delay is thus 33 machine cycles. If each machine cycle takes, say, 1 μs then the total delay is 33 μs. For a longer delay a bigger number can be initially put into accumulator A.

15.6 Trace tables

A program may be checked for possible errors by constructing a *trace table*. This is a table which shows the contents of each register and relevant memory location before and after the execution of each instruction in a program. The following example illustrates this.

In section 15.4.2 a program was developed for the addition of two numbers.

Label	Op-code	Operand	Comment
AUGEND	EQU	$0070	
ADDEND	EQU	$0071	
SUM	EQU	$0072	
	ORG	$0010	
	LDA A	AUGEND	First number into accumulator A
	ADD A	ADDEND	Add on second number
	STA	A SUM	Save the sum
	SWI		Program end

Suppose the numbers to be added are 02H and 06H. The trace table is then

Address bus	Program counter	Acc A	Memory addresses			Status register			
			0070	0071	0072	C	N	V	Z
0010	0011	XX	02	06	XX	X	X	X	X
0012	0013	02	02	06	XX	X	0	0	0
0014	0015	06	02	06	XX	0	0	0	0
0016	0017	06	02	06	08	0	0	0	0

15.7 The automatic camera

To illustrate the uses of microprocessors, consider the automatic camera where the exposure and focusing are automatically determined.

For the exposure, information on light levels might be gathered from a number of sensors, perhaps six, of the light levels at different parts of the image. Thus one sensor might monitor the light intensity at the centre of the image, another might cover a ring surrounding the centre and others the four corner regions. The outputs from these sensors are then fed to a microprocessor, which, on the basis of other inputs regarding the film speed and whether the camera operator wishes to select a particular aperture or a particular speed, then supplies an output to another microprocessor which is used to control the lens. This microprocessor then gives an output which, if the aperture is to be set, drives a stepping motor which rotates a mechanical diaphragm and so sets the aperture.

For the focusing, the arrangement shown in figure 15.9 might be used. When the image is in focus on the plane of the film, the light when split by two lenses gives rise to two images separated by a particular distance. When the image is not in focus the distance between the two images changes, the more out of focus the image the further the images are from the in-focus positions. The positions of the images can be detected by allowing them to fall on an array of very narrow light-sensitive sensors. Thus, which sensors are giving outputs is a measure of the focusing. The outputs from the sensors is fed to a microprocessor which supplies an output to a focusing drive motor.

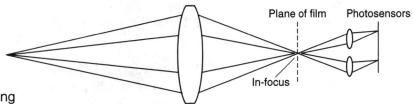

Fig. 15.9 Automatic focusing

Problems

1 Explain the roles of (a) accumulator, (b) status, (c) memory address, and (d) program counter registers.
2 A microprocessor uses eight address lines for accessing memory. What is the maximum number of memory locations that can be addressed?
3 What advantage does assembler language have over machine language?
4 Determine the hexadecimal equivalents of the binary numbers (a) 11010111, (b) 11001111, (c) 10111001.
5 Determine the binary equivalents of the hexadecimal numbers (a) AF, (b) 91, (c) 1B9.

6 Explain the use of (a) the Z flag, (b) the N flag, in a microprocessor status register.

7 Using the following extract from a manufacturer's instruction set (6800), determine the machine codes required for the operation of adding with carry in (a) the immediate address mode, (b) the direct address mode.

| | | Addressing modes | | | | | |
| | | IMMED | | | DIRECT | | |
Operation	Mnemonic	OP	~	#	OP	~	#
Add with carry	ADC A	89	2	2	99	3	2

Notes: OP is the operation code (hexadecimal), ~ is the number of machine cycles required, # is the number of program bytes required.

8 The clear operation with the Motorola 6800 processor instruction set has an entry only in the implied addressing mode column. What is the significance of this?

9 What are the purposes of (a) accumulator, (b) status, (c) program counter registers?

10 What are mnemonics for, say, the Motorola 6800 for (a) clear register A, (b) store accumulator A, (c) load accumulator A, (d) compare accumulators, (e) load index register?

11 Write a line of assembler program for (a) load the accumulator with 20 (hex), (b) decrement the accumulator A, (c) clear the address $0020, (d) ADD to accumulator A the number at address $0020.

12 Explain the operations specified by the following instructions:
(a)   STA B $35, (b) LDA A #$F2, (c) CLC, (d) INC A,
(e) CMP A #$C5, (f) CLR $2000, (g) JMP 05,X

13 Write programs in assembly language to:
(a) Subtract a hexadecimal number in memory address 0050 from a hexadecimal number in memory location 0060 and store the result in location 0070.
(b) Multiply two 8-bit numbers, located at addresses 0020 and 0021 and store the product, an 8-bit number, in location 0022.
(c) Store the hexadecimal numbers 0 to 10 in memory locations starting at 0020.
(d) Move the block of 32 numbers starting at address $2000 to a new start address of $3000.

14 Write, in assembly language, a subroutine that can be used to produce a time delay and which can be set to any value.

15 Write, in assembly language, a routine that can be used so that if the input from a sensor to address 2000 is high the program jumps to one routine starting at address 3000, and if low the program continues.

16 Input/output systems

16.1 Interfacing

When a microprocessor is used to control some system it has to accept input information, respond to it and produce output signals to implement the required control action. Thus there can be inputs from sensors to feed data in and outputs to such external devices as relays and motors. The term *peripheral* is used for a device, such as a sensor, keyboard, actuator, etc. which is connected to a microprocessor. It is, however, not normally possible to connect directly such peripheral devices to a microprocessor bus system due to a lack of compatibility in signal forms and levels. Because of such incompatibility, a circuit known as an *interface* is used between the peripheral items and the microprocessor. Figure 16.1 illustrates the arrangement. The interface is where this incompatibility is resolved.

Fig. 16.1 The interface

This chapter discusses the requirements of such interfaces and the very commonly used Motorola 6820 Peripheral Interface Adapter. For more details the reader is referred to specialist texts such as *Microprocessor Interfacing* by R. Vears (Heinemann 1990).

16.1.1 Interface requirements

The following are some of the actions that are often required of an interface circuit:

1 *Electrical buffering/isolation* Needed when the peripheral operates at a different voltage or current to that on the microprocessor bus system. The output of a tiny signal from a microprocessor might be used to control, via the interface, a much larger out- put power, e.g. that required for an electrical

heater or perhaps a motor. There has therefore often to be isolation between the microprocessor and the higher power system.

2   *Timing control* Needed when the data transfer rates of the peripheral and the microprocessor are different. For example, when interfacing a microprocessor to a slower peripheral a delay program (see section 15.5.1) might have to be used. Another possibility is to use special lines between the microprocessor and the peripheral to control the timing of data transfers. Such lines are referred to as *handshake lines* and the process as *handshaking*. The peripheral sends the data and a DATA READY signal to the input/output section. The CPU then determines that the DATA READY signal is active, a section perhaps holding the signal until the CPU is ready to receive it. The CPU reads the data from the input/output section and sends an INPUT ACKNOWLEDGED signal to the peripheral. This signal indicates that the transfer has been completed and thus the peripheral can send more data. For an output, the peripheral sends an OUTPUT REQUEST or PERIPHERAL READY signal to the input/output section. The CPU determines that the PERIPHERAL READY signal is active. The CPU sends an OUTPUT READY signal to the peripheral and sends the data to the peripheral. The next PERIPHERAL READY signal may be used to inform the CPU that the transfer has been completed.

3   *Code conversion* Needed when the codes used by the peripherals differ from those used by the microprocessor. For example, visual display units and printers require codes which can represent the letters and characters present on a standard keyboard. A common such code is the 7-bit *ASCII code* (American Standard Code for Information Interchange). For example, the number 0 has an ASCII code of 0110000, 1 a code of 0110001, the letter A a code of 1000001 and the letter B a code of 1000010. For use by a microprocessor a code conversion is required, the 0110000 being converted to 0000, the 0110001 to 0001, the 1000001 to 1010 and the 1000010 to 1011.

4   *Changing the number of lines* Microprocessors operate on a fixed word length of 4 bits, 8 bits, or 16 bits. This determines the number of lines in the microprocessor data bus. Peripheral equipment may have a different number of lines, perhaps requiring a longer word than that of the microprocessor (see section 16.1.3).

5   *Serial to parallel, and vice versa, data transfer* Within an 8-bit microprocessor, data is generally manipulated eight bits at a time. To transfer 8 bits simultaneously to a peripheral thus requires eight data paths. Such a form of transfer is termed *parallel data transfer*. It is, however, not always possible to

transfer data in this way. For example, data transfer over the public telephone system can only involve one data path. The data has thus to be transferred sequentially one bit at a time. Such a form of transfer is termed *serial data transfer*. Serial data transfer is a slower method of data transfer than parallel data transfer. Thus, if serial data transfer is used, there will be a need to convert incoming serial data into parallel data for the microprocessor and vice versa for outputs from the micro-processor.

6 *Conversion from analogue to digital and vice versa* The output from sensors is generally analogue and this requires conversion to digital signals for the microprocessor. The output from a microprocessor is digital and this might require conversion to an analogue signal in order to operate some actuator.

7 *Conversion of voltage/current levels* There may be a need to change incoming or outgoing voltage or current levels to more suitable values.

16.2 Memory-mapped system

Microprocessors generally use the same buses for both memory and input/output transfers. With, what is termed, a *memory-mapped* system, the input/output ports function as though they are memory with each port being assigned one, or more, addresses. The instruction STA plus an output port address thus sends data to an output port, LDA plus an input address loads data from an input port. Alternatives to memory mapping are *isolated input/output* in which memory and input/output addresses are separately decoded and *attached input/output* in which the input/output ports are activated by special instructions. The memory-mapped system is probably the most common and can be used with any micro-processor, the isolated input/output approach only being possible with those microprocessors that are specifically designed for this approach and have separate IN and OUT instructions (e.g. Intel).

16.2.1 Multiport transfers

Many peripherals often require several input/output ports. This can be because the data word of the peripheral is longer than that of the CPU. The CPU must then transfer the data in segments. For example, if we require a 16-bit output with an 8-bit CPU the procedure is:

1 The CPU prepares the eight most significant bits of the data.
2 The CPU sends the eight most significant bits of the data to the first port.
3 The CPU prepares the eight least significant bits of the data.
4 The CPU sends the eight least significant bits of the data to the second port.

Thus, after some delay, all the sixteen bits are available to the peripheral.

16.3 Peripheral interface adapters

Interfaces can be specifically designed for particular inputs/outputs; however, programmable input/output interface devices are available which permit various different input and output options to be selected by means of software. Such devices are known as *peripheral interface adapters* (PIA). A commonly used PIA parallel interface is the *Motorola 6820*. The PIA occupies four memory locations and can be directly attached to Motorola 6800 buses. It contains two ports, termed A and B. A is primarily intended as the input port and B as the output port. Each port has:

1   A *peripheral data register*. An output port has to operate in a different way to an input port because the data must be held for the peripheral. Thus for output, a register is used to temporarily store data. The register is said to be *latched*, i.e. connected, when a port is used for output and unlatched when used for input.
2   A *data direction register* that determines whether the input/output lines are inputs or outputs.
3   A *control register* that determines the active logical connections in the peripheral and that also contains the DATA READY or PERIPHERAL READY bits.
4   Two *control lines*, CA1 and CA2 or CB1 and CB2. These are the handshaking lines.

Figure 16.2 shows the basic structure of such an interface and the pin connections. Two microprocessor address lines connect the PIA directly through the two register select lines RS0 and RS1. This gives the PIA four addresses for the six registers. When RS1 is low side A is addressed, when high it is side B. RS0 addresses registers on a particular side, i.e. A or B. When RS0 is high, the control register is addressed, when low the data register or the data direction register. For a particular side, the data register and the data direction register have the same address. Which of them is addressed is determined by bit 2 of the control register (see below).

Each of the bits in the A and B control registers is concerned with some features of the operation of the ports. Thus for the A control register we have:

B7	B6	B5	B4	B3	B2	B1	B0
IRQA 1	IRQA 2	CA2 control			DDRA access	CA1 control	

A similar pattern is used for the B control register.

Microprocessor side

Peripheral side

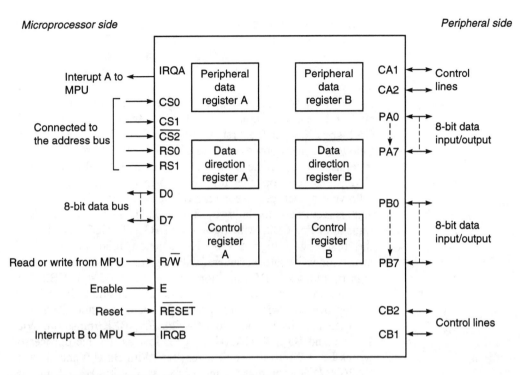

Fig. 16.2 Motorola 6820 PIA

Bits 0 and 1
The first two bits control the way that CA1 or CB1 input hand-shaking lines operates. Bit 0 determines whether the interrupt output is enabled. B0 = 0 disables the IRQA(B) microprocessor interrupt, B0 = 1 enables the interrupt. CA1 and CB1 are not set by the static level of the input but are edge triggered, i.e. set by a changing signal. Bit 1 determines whether bit 7 is set by a high-to-low transition (a trailing edge) or a low-to-high transition (a leading edge). B1 = 0 sets a high-to-low transition, B1 = 1 sets a low-to-high transition.

Bit 2
Bit 2 determines whether data direction registers or peripheral data registers are addressed. With B2 set to 0 data direction registers are addressed, with B2 set to 1 peripheral data registers are selected.

Bits 3, 4 and 5
These bits allow the PIA to perform a variety of functions. Bit 5 determines whether control line 2 is an input or an output. If bit 5 is set to 0 control line 2 is an input, if set to 1 it is an output. In input mode, both CA2 and CB2 operate in the same way. Bits 3 and 4 determine whether the interrupt output is active and which transitions set bit 6.

With B5 = 0, i.e. CA2(CB2) set as an input: B3 = 0 disables

IRQA(B) microprocessor interrupt by CA2(CB2), B3 = 1 enables IRQA(B) microprocessor interrupt by CA2(CB2); B4 = 0 determines that the interrupt flag IRQA(B), bit B6, is set by a high-to-low transition on CA2(CB2), B4 = 1 determines that it is set by a low-to-high transition.

B5 = 1 sets CA2(CB2) as an output. In output mode CA2 and CB2 behave differently. For CA2: with B4 = 0 and B3 = 0, CA2 goes low on the first high-to-low ENABLE (E) transition following a microprocessor read of peripheral data register A and is returned high by the next CA1 transition; B4 = 0 and B3 = 1, CA2 goes low on the first high-to-low ENABLE transition following a microprocessor read of the peripheral data register A and is returned to high by the next high-to-low ENABLE transition. For CB2: with B4 = 0 and B3 = 0, CB2 goes low on the first low-to-high ENABLE transition following a microprocessor write into peripheral data register B and is returned to high by the next CB1 transition; B4 = 0 and B3 = 1, CB2 goes low on the first low-to-high ENABLE transition following a microprocessor write into peripheral data register B and is returned high by the next low-to-high ENABLE transition. With B4 = 1 and B3 = 0, CA2(CB2) goes low as the microprocessor writes B3 = 0 into the control register. With B4 = 0 and B3 = 1, CA2(CB2) goes high as the microprocessor writes B3 = 1 into the control register.

Bit 6
This is the CA2(CB2) interrupt flag, being set by transitions on CA2(CB2). With CA2(CB2) as an input (B5 = 0), it is cleared by a microprocessor read of the data register A(B). With CA2(CB2) as an output (B5 = 1), the flag is 0 and is not affected by CA2(CB2) transitions.

Bit 7
This is the CA1(CB1) interrupt flag, being cleared by a microprocessor read of data register A(B).

The process of selecting which options are to be used is termed *configuring* or *initialising* the PIA. The RESET connection is used to clear all the registers of the PIA. The PIA must then be configured.

16.3.1 Initialising the PIA

Before the PIA can be used, a program has to be written and used so that the conditions are set for the desired peripheral data flow. The PIA program is placed at the beginning of the main program so that, thereafter, the microprocessor can read peripheral data. The initialisation program is thus only run once.

The initialisation program to set which port is to be input and which is to be output can have the following steps:

1 Clear bit 2 of each control register by a reset, so that data direction registers are addressed. Data direction register A is addressed as XXX0 and data direction register B as XXX2.
2 For A to be an input port, load all 0s into direction register A.
3 For B to be an output port, load all 1s into direction register B.
4 Load 1 into bit 2 of both control registers. Data register A is now addressed as XXX0 and data register B as XXX2.

Thus an initialisation program to make side A an input and side B an output could be, following a reset:

INIT	LDAA	#$00	Loads 0s
	STAA	$2000	Make side A input port
	LDAA	#$FF	Load 1s
	STAA	$2000	Make side B output port
	LDAA	#$04	Load 1 into bit 2, all other bits 0
	STAA	$2000	Select port A data register
	STAA	$2002	Select port B data register

Peripheral data can now be read from input port A with the instruction LDAA 2000 and the microprocessor can write peripheral data to output port B with the instruction STAA 2002.

16.3.2 Interfacing examples

To illustrate how a PIA can be used, consider the following simple examples.

There are many situations where the input signals to a PIA will be from mechanical switches or contacts. These may be push-button switches, keyboards, switches used for position sensing, etc. These might just be connected to a PIA in the way shown in figure 16.3. When a switch closes a 0 V signal is applied to the input to the A-side of the PIA.

Another situation which can arise is where the output from a PIA is used to drive a display consisting of light-emitting diodes used singly in seven-segment units (figure 16.4). A single LED is an on–off indicator and thus the display number indicated will depend on which LEDs are on. To display a 0 requires LEDs a, b, c, d, e, f to be on. This requires that the outputs from the PIA should be 1 at PB0, PB1, PB2, PB3, PB4, PB5 and 0 at PB6, i.e. an output of 0111111 or 3F. To display a 1 requires b and c to be

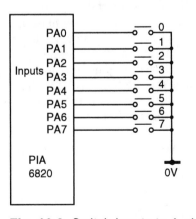

Fig. 16.3 Switch inputs to A-side

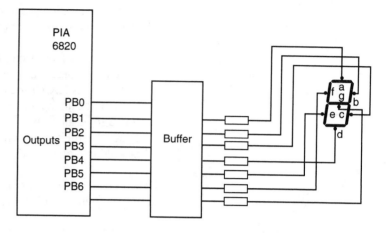

Fig. 16.4 Driving a display

on and so outputs of 1s from PB1 and PB2 with 0s from the other outputs, i.e. 0000110 or 06.

16.4 Polling and interrupts

Consider the situation where all input/output transfers of data are controlled by the program. When peripherals need attention they signal the microprocessor by changing the voltage level of an input line. The microprocessor can then respond by jumping to a program service routine for the device. On completion of the routine, a return to the main program occurs. Program control of inputs/outputs is thus a loop to continuously read inputs and update outputs, with jumps to service routines as required. This process of repeated checking each peripheral device to see if it is ready to send or accept a new byte of data is called *polling*.

An alternative to program control is *interrupt control*. An interrupt involves a peripheral device activating a separate interrupt request line. The reception of an interrupt results in the microprocessor suspending execution of its main program and jumping to the service routine for the peripheral. Figure 16.5 illustrates this by means of a flow diagram. The interrupt must not lead to a loss of data and an interrupt handling routine has to be incorporated in the software so that the state of processor registers and the last address accessed in the main program are stored in dedicated locations in memory. After the interrupt service routine, the contents of the memory are restored and the microprocessor can continue executing the main program from where it was interrupted.

Typically a microprocessor has three different types of interrupt lines:

1 Reset
2 Interrupt request
3 Non-maskable interrupt

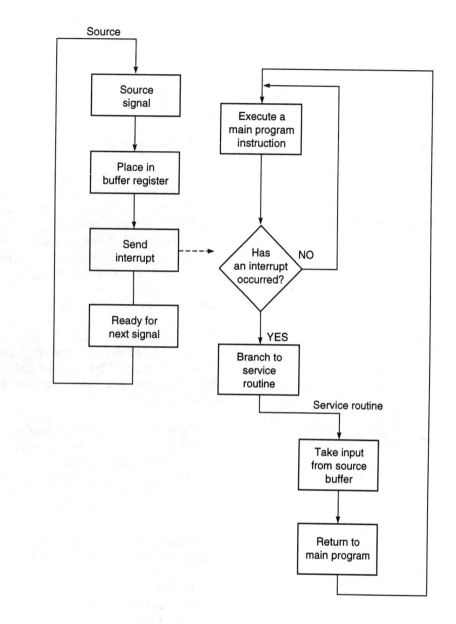

Fig. 16.5 Interrupt

These, on the Motorola 6800 microprocessor, are the pins labelled $\overline{\text{RES}}$, $\overline{\text{IRQ}}$ and $\overline{\text{NMI}}$. When any of these pins is driven to the active LOW state then an interrupt is requested. The reset interrupt is the system reset and when this is activated then all activity in the system stops, the starting address of the main program is loaded and the start-up routine is executed. The interrupt request is the line most commonly used by peripherals to interrupt the microprocessor main program. A LOW signal on this line will initiate the interrupt. However, it will be ignored if the I bit in the condition code register is set to 1, the interrupt then being said to be masked. The non-maskable interrupt cannot be masked and so there is no method of preventing the interrupt service routine being executed when it is connected to this line. An interrupt of this type is usually reserved for emergency routines such as those required when there is a power failure, e.g. switching to a back-up power supply.

16.4.1 Connecting interrupt signals via a PIA

The Motorola 6820 PIA (figure 16.2) has two connections IRQA and IRQB through which interrupt signals can be sent to the microprocessor so that an interrupt request from CA1, CA2 or CB1, CB2 can drive the $\overline{\text{IRQ}}$ pin of the microprocessor to the active LOW state. When, in section 16.2.1, the initialisation program for a PIA was considered, only bit 2 of the control register was set as 1, the other bits being 0. These 0s disabled interrupt inputs. In order to use interrupts, the initialisation step which stores $04 into the control register must be modified. The form of the modification will depend on the type of change in the input which is required to initiate the interrupt. Suppose, for example, we want CA1 to enable an interrupt when there is a high-to-low transition, CA2 and CB1 not used and CB2 enabled and used for a set/reset output. The control register format to meet this specification is, for CA:

> B0 is 1 to enable interrupt on CA1
> B1 is 0 so that the interrupt flag IRQA1 is set by a high-to-low transition on CA1
> B2 is 1 to give access to the data register
> B3, B4, B5 are 0 because CA2 is disabled.
> B6, B7 are read-only flags and thus a 0 or 1 may be used.

Hence the format for CA1 can be 00000101 which is 05 in hexadecimal notation. The control register format for CB2 is:

> B0 is 0 to disable CB1
> B1 may be 0 or 1 since CB1 is disabled
> B2 is 1 to give access to the data register
> B3 is 0, B4 is 1 and B5 is 1, to select the set/reset
> B6, B7 are read-only flags and thus a 0 or 1 may be used.

Hence the format for CA1 can be 00110100 which is 34 in hexadecimal notation. The initialisation program might then read:

```
INIT    LDAA    #$00     Loads 0s
        STAA    $2000    Make side A input port
        LDAA    #$FF     Load 1s
```

STAA	$2000	Make side B output port
LDAA	#$05	Load the required control register format
STAA	$2000	Select port A data register
LDAA	#$34	Load the required control register format
STAA	$2002	Select port B data register

16.5 Speed control of motors

As an illustration of the use of microprocessors with PIAs, consider their use to control the speed of a permanent magnet motor. With closed-loop control the output from the microprocessor can be used to supply a reference signal to the PIA which is then converted to an analogue voltage by a digital-to-analogue converter. This then becomes the set value input to the closed-loop control system (figure 16.6(a)). There it is compared by an operational amplifier with the feedback signal from a tacho-generator mounted on the motor shaft and the error signal used to control the speed of the motor. Alternatively, the microrprocessor can be in the control loop, as shown in figure 16.6(b). The feedback signal from the tachogenerator is fed, after being converted to a digital signal by a digital-to-analogue converter, as an input to the PIA and hence the microprocessor. The microprocessor compares this input with the required speed located in its memory

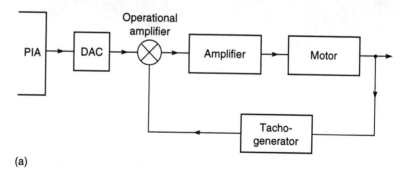

(a)

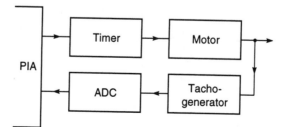

Fig. 16.6 Speed control   (b)

and then supplies an output, which, after passing through the PIA, is used to control a timer. This varies the time for which a voltage/current is applied to the motor and hence the average voltage/current.

Problems

1 Describe the functions that can be required of an interface.
2 Explain the difference between a parallel and a serial interface.
3 Explain what is meant by a memory-mapped system for inputs/outputs.
4 What is the function of a peripheral interface adapter?
5 Describe the architecture of the Motorola 6800 PIA.
6 Explain the function of an initialisation program for a PIA.
7 What are the advantages of using external interrupts rather than software polling as a means of communication with peripherals?
8 For a 6820 PIA, what value should be stored in the control register if CA1 is to be disabled, CB1 is to be an enabled interrupt input and set by a low-to-high transition, CA2 is to be enabled and used as a set/reset output, and CB2 is to be enabled and go low on first low-to-high E transition following a microprocessor? Write into peripheral data register B and return high by the next low-to-high E transition.
9 Write, in assembly language, a program to initialise the 6820 PIA to achieve the specification given in problem 8.
10 Write, in assembly language, a program to initialise the 6820 PIA to read eight bits of data in from port A.

17 Programmable logic controllers

17.1 Programmable logic controllers

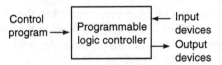

Fig. 17.1 Programmable logic controller

A *programmable logic controller* (PLC) can be defined as a digital electronic device that uses a programmable memory to store instructions and to implement functions such as logic, sequencing, timing, counting and arithmetic in order to control machines and processes (figure 17.1). The term *logic* is used because programming is primarily concerned with implementing logic and switching operations. Inputs devices (e.g. switches) and output devices in the system being controlled (e.g. motors) are connected to the PLC. The operator then enters a sequence of instructions, i.e. a program, into the memory of the PLC. The controller then monitors the inputs and outputs according to this program and so controls the machine or process. Originally they were designed as a replacement for hard-wired relay and timer logic control systems. PLCs have the great advantage that it is possible to modify a control system without having to rewire the input and output devices, the only requirement being that an operator has to key in a different set of instructions. The result is a flexible system which can be used to control systems which vary quite widely in their nature and complexity.

PLCs are similar to computers but have certain features which are specific to their use as controllers. These are:

1. They are rugged and designed to withstand vibrations, temperature, humidity and noise.
2. The interfacing for inputs and outputs is inside the controller.
3. They are easily programmed and have an easily understood programming language. Programming is primarily concerned with logic and switching operations.

PLCs were first conceived in 1968. They are now widely used and extend from small self-contained units for use with perhaps 20 digital input/outputs to modular systems which can be used for large numbers of inputs/outputs, handle digital or analogue inputs/outputs, and also carry out PID control modes.

This chapter is a discussion of the basic structure of PLCs and how they can be used to control machines or processes. For a more detailed discussion the reader is referred to specialist texts such as *Programmable Controllers, Operation and Application* by I.G. Warnock (Prentice Hall 1988).

17.2 Basic structure

Figure 17.2 shows the basic internal structure of a PLC. It consists essentially of a central processing unit (CPU), memory, and input/output circuitry. The CPU controls and processes all the operations within the PLC. It is supplied with a clock with a frequency of typically between 1 and 8 MHz. This frequency determines the operating speed of the PLC and provides the timing and synchronisation for all elements in the system. A bus system carries information and data to and from the CPU, memory and input/output units.

There are several memory elements: a system ROM to give permanent storage for the operating system and fixed data, RAM for the user's program, and temporary buffer stores for the input/output channels. The programs in RAM can be changed by the user. However, to prevent the loss of these programs when the power supply is switched off, a battery is likely to be used in the PLC to maintain the RAM contents for a period of time. After a program has been developed in RAM it may be loaded into an EPROM memory chip and so made permanent. Specifications for small PLCs often specify the program memory size in terms of the number of program steps that can be stored. A program step is an instruction for some event to occur. A program task might consist of a number of steps and could be, for example: examine the state of switch A, examine the state of switch B, if A and B are closed then energise solenoid P which then might result in the operation of some actuator. When this happens another task might then be started. The term *ladder programming* is used since each of the above tasks can be considered as a rung on a ladder. Typically the number of steps that can be handled by a small PLC is of the order

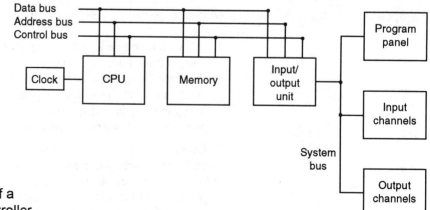

Fig. 17.2 Architecture of a programmable logic controller

of 300 to 1000, which is generally more than adequate for most control situations.

The input/output unit provides the interface between the system and the outside world. Programs are entered into the input/output unit from a panel which can vary from small keyboards with liquid crystal displays to those using a visual display unit (VDU) with keyboard and screen display. The screens can show the program in ladder program format (see section 17.5) as the programmer builds it up using symbolic keys on the keyboard. A VDU can considerably help in the readability of a program by also providing screen graphics and text comments. Alternatively the programs can be entered into the system by means of a link to a personal computer (PC) which is loaded with an appropriate software package.

The input/output channels provide signal conditioning and isolation functions so that sensors and actuators can be generally directly connected to them without the need for other circuitry. With larger units a choice of input/output voltages/currents is likely to be provided. Common input voltages are 5 V and 24 V. Common output voltages are 24 V and 240 V. Outputs are often specified as being of relay type, transistor type or triac type. With the relay type, the signal from the PLC output is used to operate a relay and so is able to switch currents of the order of a few amperes in an external circuit. The relay isolates the PLC from the external circuit. Relays are, however, relatively slow to operate. The transistor type of output uses a transistor to switch current through the external circuit. This gives a faster switching action. Optoisolators (see section 3.3) are used with transistor switches to provide isolation between the external circuit and the PLC. Triac outputs can be used to control external loads which are connected to the a.c. power supply. Optoisolators are again used to provide isolation.

17.3 Input/output processing

The sequence followed by a PLC when carrying out a program can be summarised as:

1 Scan the inputs associated with one rung of the ladder program.
2 Solve the logic operation involving those inputs.
3 Set/reset the outputs for that rung.
4 Move on to the next rung and repeat operations 1, 2, 3.
5 Move on to the next rung and repeat operations 1, 2, 3.
6 Move on to the next rung and repeat operations 1, 2, 3.
⋮

And so on, until the end of the program, i.e. the end rung on the ladder, is reached.

Each rung of the ladder program is thus scanned in turn.

There are two methods that can be used for input/output processing:

1 *Continuous updating* This involves the CPU scanning the input channels as they occur in the program instructions. Each input point is examined individually and its effect on the program determined. There will be a built-in delay, typically about 3 ms, when each input is examined in order to ensure that only valid input signals are read by the microprocessor. This delay enables the microprocessor to avoid counting an input signal twice, or more frequently, if there is contact bounce at a switch. A number of inputs may have to be scanned, each with a 3 ms delay, before the program has the instruction for a logic operation to be executed and an output to occur. The outputs are latched so that they retain their status until the next updating.

2 *Mass input/output copying* Because, with continuous updating, there has to be a 3 ms delay on each input, the time taken to examine several hundred input/output points can become comparatively long. To allow a more rapid execution of a program, a specific area of RAM is used as a buffer store between the control logic and the input/output unit. Each input/output has an address in this memory. At the start of each program cycle the CPU scans all the inputs and copies their status into the input/output addresses in RAM. As the program is executed the stored input data is read, as required, from RAM and the logic operations carried out. The resulting output signals are stored in the reserved input/output section of RAM. At the end of each program cycle all the outputs are transferred from RAM to the appropriate output channels. The outputs are latched so that they retain their status until the next updating.

17.4 Example of a PLC

The following are some of the features of a typical small PLC, a Mitsubishi F2-20MR-ES.

Power supply	110–120 V/220–240 V a.c., single phase 50/60 Hz
Program language	Ladder logic
Programming capacity	1000 steps
Execution speed	Average 7 µs/step
Program memory	CMOS-RAM built in, EPROM cassette can be added
Battery back-up	Lithium battery, approx. 5 years life

Timers	0.1 s timer: 24 points, on-delay timers (0.1 to 999 s) 0.01 s timer: 8 points, on-delay timers (0.01 to 99.9 s)
Counters (retentive)	32 points, down counter (0 to 999)
Number of inputs	12 points, all optoisolated
Input voltage	Built-in 24 V d.c., external 24 V d.c.
Number of outputs	8 points
Choice of output	Relay output: relay isolated Transistor output: optoisolated Triac output: optoisolated

17.5 Programming

PLC programming is based on the use of *ladder diagrams*. Using these, writing a program is equivalent to drawing a switching circuit. The language can thus be considered to be a high-level language (see section 15.2) in that the instruction are very close to the functions required. The ladder diagram consists of two vertical

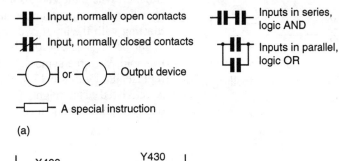

(a)

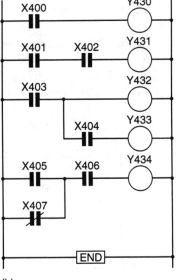

Fig. 17.3 Programmable logic controller: (a) standard symbols, (b) example of ladder diagram

(b)

lines representing the power rails. Circuits are connected as horizontal lines, i.e. the rungs of the ladder, between these two verticals. Figure 17.3 shows standard symbols that are used and an example of a ladder diagram. In drawing the circuit line for a rung, inputs must always precede outputs and there must be at least one output on each line. Each rung must start with an input or a series of inputs and end with an output. The inputs and outputs are numbered, the notation used depending on the PLC manufacturer. The Mitsubishi F series of PLCs precedes input elements by an X and output elements by a Y and uses the following numbers:

Inputs	X400–407, 410–413
	X500–507, 510–513
	(24 possible inputs)
Outputs	Y430–437
	Y530–537
	(16 possible outputs)

To illustrate the drawing of a ladder diagram, consider a situation where the energising of a solenoid output device depends on a normally open start switch being activated by being closed. Starting with the input, we have the normally open symbol ‖. This might be labelled X400. The line terminates with the output, the solenoid, with the symbol O. This might be labelled Y430. This is represented by figure 17.4. When the switch is closed the solenoid is activiated. This might, for example, be a solenoid valve which opens to allow water to enter a vessel.

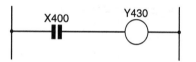

Fig. 17.4 Switch controlling a solenoid

17.5.1 Logic functions

Figure 17.5(a) shows a situation where a coil is not energised unless two, normally open, switches are both closed. Switch A and switch B have both to be closed, which thus gives an AND logic situation. The truth table is (section 1.6.1):

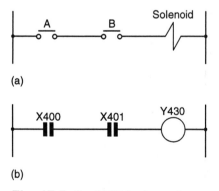

(a)

(b)

Fig. 17.5 An AND logic system

Inputs		Outputs
A	B	
0	0	0
0	1	0
1	0	0
1	1	1

The ladder diagram starts with ‖, labelled X400, to represent switch A and in series with it ‖, labelled X401, to represent switch

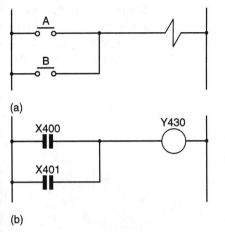

(a)

(b)

Fig. 17.6 An OR logic system

B. The line then terminates with O to represent the output. Figure 17.5(b) shows the line.

Figure 17.6(a) shows a situation where a coil is not energised until either, normally open, switch A or B is closed. The situation is an OR logic gate. The truth table is:

Inputs		Outputs
A	B	
0	0	0
0	1	1
1	0	1
1	1	1

The ladder diagram starts with ‖, labelled X400, to represent switch A and in parallel with it ‖, labelled X401, to represent switch B. The line then terminates with O to represent the output. Figure 17.6(b) shows the line.

Consider a situation where a normally open switch A must be activated and either of two other, normally open, switches B and C must be activated for a coil to be energised. We can represent this arrangement of switches as switch A in series with two parallel switches B and C (figure 17.7(a)). For the coil to be energised we require A to be closed and either B or C to be closed. Switch A when considered with the parallel switches gives an AND logic situation. The two parallel switches give an OR logic situation. We thus have a combination of two gates. The truth table is:

	Inputs		Output
A	B	C	
0	0	0	0
0	0	1	0
0	1	0	0
0	1	1	0
1	0	0	0
1	0	1	1
1	1	0	1
1	1	1	1

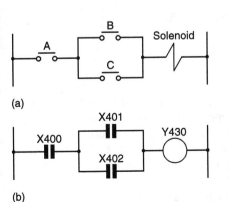

(a)

(b)

Fig. 17.7 Switches controlling a solenoid

For the ladder diagram, we start the ladder diagram with ‖, labelled X400, to represent switch A. This is in series with two ‖ in parallel, labelled X401 and X402, for switches B and C. The line then terminates with O to represent the output, the coil. Figure 17.7(b) shows the line.

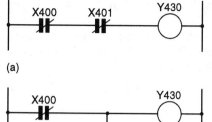

(a)

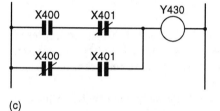

(b)

(c)

Fig. 17.8 Logic gates: (a) NOR, (b) NAND, (c) EXCLUSIVE-OR

Figure 17.5 has an AND gate and figure 17.6 an OR gate. Figure 17.8 shows how we can represent a NOR gate, a NAND gate and an EXCLUSIVE-OR (XOR) gate. These gates include switches which are not normally open. The NOR gate has the truth table:

Inputs		Outputs
X400	X401	Y430
0	0	1
0	1	0
1	0	0
1	1	0

The NAND gate has the truth table:

Inputs		Outputs
X400	X401	Y430
0	0	1
0	1	1
1	0	1
1	1	0

The XOR gate has the truth table:

Inputs		Outputs
X400	X401	Y430
0	0	0
0	1	1
1	0	1
1	1	0

17.5.2 Latching

There are often situations where it is necessary to hold a coil energised, even when the input which energised it ceases. The term *latch circuit* is used for the circuit used to carry out such an operation. It is a self-maintaining circuit in that, after being energised, it maintains that state until another input is received. It remembers its last state.

An example of a latch circuit is shown in figure 17.9. When X400 is energised and closes, there is an output Y430. However, when there is an output, another set of contacts associated with

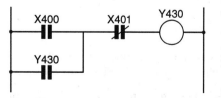

Fig. 17.9 A latch circuit

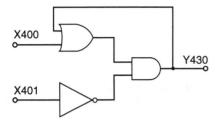

Fig. 17.10 Logic circuit

Y430 is energised and closes. These contacts OR the X400 contacts. Thus, even if the input X400 opens, the circuit will still maintain the output energised. The only way to release Y430 is by operating the normally closed contact X401. Figure 17.10 shows the logic circuit corresponding to figure 17.9.

17.6 Entering the program

Each horizontal line, i.e. rung on the ladder, represents a line in the program and the entire ladder can be translated into a program. While this can be done by hand, the programmer usually enters the program using a keyboard with the graphic symbols for the ladder elements and the program panel then translates these symbols into machine language that is then stored in the PLC memory.

For a small PLC, to key a ladder program into the PLC mnemonic code keys are used, each code corresponding to a ladder element. For the Mitsubishi F series PLCs, the mnemonics are:

LD	Start a rung with an open contact.
OUT	An output.
AND	A series element and so an AND logic instruction.
OR	Parallel elements and so an OR logic instruction.
I	A NOT logic instruction.
... I	Used in conjunction with other instructions to indicate the inverse.
ORI	An OR NOT logic function.
ANI	An AND NOT logic function.
LDI	Start a rung with a closed contact.
ANB	AND used with two subcircuits.
ORB	OR used with two subcircuits.
RST	Reset shift register/counter.
SHF	Shift.
K	Insert a constant.
END	End ladder

In the examples discussed in the rest of this chapter, the above mnemonics will be used. However, those used by other manufacturers do not differ widely from these. For example, for OMRON PLCs the codes are:

LD	Start a rung with an open contact.	
OUT	An output.	
TIM	A timer output.	
CNT	A counter output.	
AND	A series element and so an AND logic instruction.	
OR	Parallel elements and so an OR logic instruction.	
NOT	A NOT logic instruction.	
... NOT	Used in conjunction with other instructions to indicate the inverse.	
OR NOT	An OR NOT logic function.	
AND NOT	An AND NOT logic function.	
LD NOT	Start a rung with a closed contact.	
AND LD	AND used with two subcircuits.	
OR LD	OR used with two subcircuits.	
#	Insert a constant.	
END	End ladder	

The following shows how individual rungs on a ladder are entered. Using the Mitsubishi mnemonics, the AND gate shown in figure 17.5(b) would be entered as:

Step	Instruction	
0	LD	X400
1	AND	X401
2	OUT	Y430

The OR gate shown in figure 17.6(b) would be entered as:

Step	Instruction	
0	LD	X400
1	OR	X401
2	OUT	Y430

The NOR gate shown in figure 17.8(a) would be entered as:

Step	Instruction	
0	LDI	X400
1	ANI	X401
2	OUT	Y430

The NAND gate shown in figure 17.8(b) would be entered as:

Step	Instruction	
0	LDI	X400
1	ORI	X401
2	OUT	Y430

The XOR gate shown in figure 17.8(c) would be entered as:

Step	Instruction	
0	LD	X400
1	ANI	X401
2	LDI	X400
3	AND	X401
4	ORB	
5	OUT	Y430

After reading the first two instructions, the third instruction starts a new line. But the first line has not been ended by an output. The CPU thus recognises a parallel line is involved for the second line and reads together the listed elements until the ORB instruction is reached. The mnemonic ORB (OR branches/blocks together) indicates to the CPU that it should OR the results of steps 0 and 1 with that of steps 2 and 3.

17.7 Timers, markers and counters

The previous sections in this chapter have been concerned with tasks requiring the series and parallel connections of input contacts. However, there are tasks which can involve time delays and event counting. These requirements can be met by the timers, counters and markers which are supplied as a feature of PLCs. They can be controlled by logic instructions and represented on ladder diagrams. The numbering system used for the functions varies from one PLC manufacturer to another. For the Mitsubishi F series, numbers used are:

Timers	450–457, 8 points, delay-on period 0.1–999 s
T	550–557, 8 points, delay-on period 0.1–999 s
Markers	100–107, 170–177, 200–207, 270–277,
M	128 points
	300–307, 370–377, battery backed, 64 points
Counters	460–467, 8 points, 1 to 999
C	560–567, 8 points, 1 to 999

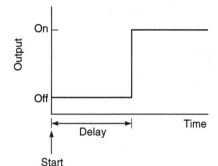

Fig. 17.11 Delay-on timer

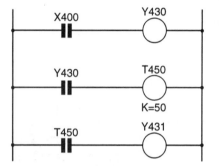

Fig. 17.12 Timer circuit

The term *point* is used for a data point and so is a timing, marker or counter element. Thus, the 16 points for timers means that there are 16 timer circuits. The term *delay-on* is used to indicate that this type of timer waits for a fixed delay period before turning on (figure 17.11). In the data given above, this is a period which can be set between 0.1 and 999 s in steps of 0.1 s. Other time delay ranges and steps are possible. In this chapter the timer examples are based on the above.

17.7.1 Timers

A timer circuit is specified by stating the interval to be timed and the conditions or events that are to start and/or stop the timer. Consider an example of a timer that is required to switch on some output 5 s after receiving a start signal. Figure 17.12 shows a ladder circuit, consisting of three rungs, for such a timer. It uses a delay-on timer T450. The timer is represented as an output that has its delay period specified by a constant K. This specifies the number of units of the smallest time interval that constitute the delay period. Thus, when the smallest time interval is 0.1 s and we have K = 50, there is a time delay of 5 s.

Line 1: When contacts X400 are closed then output Y430 is energised.
Line 2: The closing of the associated Y430 contact starts the timer T450 which will close its normally open contact after a delay of 5 s.
Line 3: The opening of the timer contact turns on the output Y431.

The instruction sequence would thus be:

Step	Instruction	
0	LD	X400
1	OUT	Y430
2	LD	Y430
3	OUT	T450
4	K	50
5	LD	T450
6	OUT	Y431

Step 4 gives the time delay as K = 50, this representing 50 times the smallest delay interval of 0.1 s, i.e. a delay of 5 s.

Timers can be linked together, the term is *cascaded*, to give larger delay times than is possible with just one timer. Figure 17.13 shows such an arrangement. Timer T450 has a delay of 999 s, i.e.

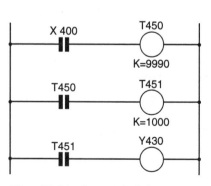

Fig. 17.13 Cascaded timers

K = 9990 when the smallest time interval is 0.1 s, and is started by the closure of contact X400. When timer Y430 contact closes it starts timer T451. This timer has a delay of 100 s, i.e. K = 1000. Output Y430 is energised when the T451 contact closes. The result is thus a total delay of the sum of the delays of the two timers, i.e. 10990 s. The instruction sequence is:

Step	Instruction	
0	LD	X400
1	OUT	T450
2	K	9990
3	LD	T450
4	OUT	T451
5	K	1000
6	LD	T451
7	OUT	Y430

Figure 17.14 shows a circuit that can be used to cause an output to go on for 0.5 s, then off for 0.5 s, then on for 0.5 s, then off for 0.5 s, and so on. When X400 contact closes, timer T450 is started and comes on after 0.5 s. After this time the T450 contact closes and starts timer T451. It comes on after 0.5 s and opens its contact, which results in timer T450 being switched off. This results in its contact opening and switching off timer T451. This then closes its contact and so starts the entire cycle again. The result is that timer contact T450 is switched on for 0.5 s, then off for 0.5 s, on for 0.5 s, and so on. Thus the output Y430 is switched on for 0.5 s, then off for 0.5 s, on for 0.5 s, and so on. The instruction sequence is:

Step	Instruction	
0	LD	X400
1	ANI	T451
2	OUT	T450
3	K	5
4	LD	T450
5	OUT	T451
6	K	5
7	LD	T450
8	OUT	Y430

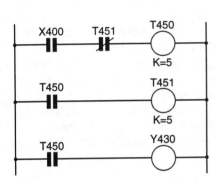

Fig. 17.14 On/off cycle timer

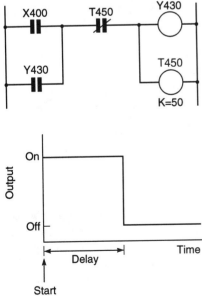

Fig. 17.15 Delay-on timer

PLCs are generally provided with only delay-on timers, i.e. a timer which comes on after a time delay. Figure 17.15 shows how a delay-off timer, i.e. a timer which goes off after a time delay, can be devised. When contact X400 is momentarily closed the output Y430 is energised and the timer T450 started. The Y430 contact latches and keeps Y430 on. After 5 s the timer comes on, breaks the latch circuit and output Y430 is switched off.

17.7.2 Markers

The term *marker* M or *auxiliary relay* is used for what can be considered as an internal relay in the PLC. These markers behave like relays with their associated contacts, but in reality are not actual relays but simulations by the software of the PLC. Some of the markers have battery back-up so that they can be used in circuits to ensure a safe shut-down of plant in the event of a power failure. Markers/internal relays can be very useful aids in the implementation of switching sequences.

Consider the situation where the excitation of an output depends on two different input arrangements. Figure 17.16 shows how we can draw a ladder diagram using internal relays. The first rung shows one input arrangement being used to control the coil of internal relay M100. The second rung shows the other input arrangement controlling the coil of internal relay M101. The contacts of the two relays are then put in an OR situation to control the output Y430. The instruction sequence for this ladder diagram is:

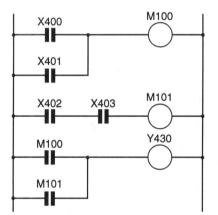

Fig. 17.16 An output controlled by two input arrangements

Step	Instruction	
0	LD	X400
1	OR	X401
2	OUT	M100
3	LD	X402
4	AND	X403
5	OUT	M101
6	LD	M100
7	OR	M101
8	OUT	Y430

Another example of the use of markers/internal relays is resetting a latch. Figure 17.17 shows the ladder diagram. When contact X400 is momentarily pressed then output Y430 is energised. The Y430 contact is then closed and so latches the output, i.e. keeps it on even when X400 is no longer closed. Y430 can be unlatched by M100 contact opening. This will occur if X401 is closed and energises the coil of M100.

An example of the use of a battery-backed marker/internal

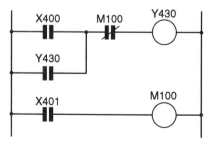

Fig. 17.17 Resetting a latch

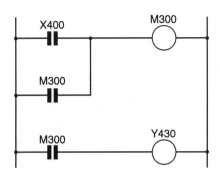

Fig. 17.18 Coping with a power failure

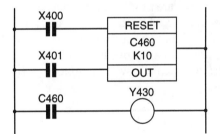

Fig. 17.19 Counter

relay is shown in figure 17.18. When contact X400 closes, the coil of the internal relay M300 (a battery-backed internal relay address) is energised. This closes the M300 contact and so even if X400 opens as a result of power failure the M300 contact remains closed. This means that the output controlled by M300 remains energised, even when there is a power failure.

17.7.3 Counters

Counters are used when there is a need to count a specified number of contact operations, e.g. where items pass along a conveyor into boxes, and when the specified number of items has passed into a box the next item is diverted into another box. Counter circuits are supplied as an internal feature of PLCs. In most cases the counter operates as a *down-counter*. This means that the counter counts down from the present value to zero, i.e. events are subtracted from the set value. When zero is reached the counter's contact changes state. An *up-counter* would count up to the preset value, i.e. events are added until the number reaches the set value. When the set value is reached the counter's contact changes state.

Figure 17.19 shows a basic counting circuit. In the ladder diagram the counter is represented by a rectangle spanning two lines. On one line is the reset which is used to reset the counter. The other line is the out line and the K10 indicates that the counter contact will change state on the 10th pulse. When contact X400 momentarily closes the counter is reset to the set value. The count will then count the number of pulses from the X401 contact. When this reaches the set value, in this case 10, the counter contact closes. Thus output Y430 is switched on after 10 pulses have been received by X401. If the X400 contact is momentarily closed any time during the count, the counter will reset to 10. The instruction sequence is:

Step	Instruction	
0	LD	X400
1	RST	C460
2	LD	X401
3	OUT	C460
4	K	10
5	LD	C460
6	OUT	Y430

Consider the problem of the control for a machine which is required to direct 6 items along one path for packaging in a box, and then 12 items along another path for packaging in another

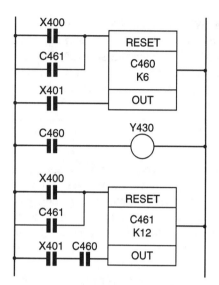

Fig. 17.20 Counter

box. Figure 17.20 shows the circuit that could be used. Contact X400 when closed starts the counting cycle. Contact X401 could be activated by microswitch which is activated every time an item passes up to the junction in the paths. C460 counts 6 items and then closes its contact. This activates Y430, which might be a solenoid used to activate a flap which closes one path and opens another. It also has a contact which closes and enables C461 to start counting. When C461 has counted 12 items it resets both the counters and opens the C460 contact which then results in Y430 becoming deactivated.

17.8 Shift registers

A number of auxiliary relays can be grouped together to form a register which can provide a storage area for a series sequence of individual bits. Thus a 4-bit register would be formed by using four auxiliary registers, an 8-bit using eight. The term *shift register* is used because the bits can be shifted along by one bit when there is a suitable input to the register. For example, with an 8-bit register we might initially have

| 1 | 0 | 1 | 1 | 0 | 1 | 0 | 1 |

Then there is an input of a 0 shift pulse.

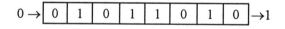

$$0 \rightarrow \boxed{0 \mid 1 \mid 0 \mid 1 \mid 1 \mid 0 \mid 1 \mid 0} \rightarrow 1$$

All the bits shift along one place and the last bit overflows.

The grouping together of a number of auxiliary registers to form a shift register is done automatically by a PLC when the shift register function is selected at the control panel. With the Mitsubishi PLC, this is done by using the programming function SFT (shift) against the auxiliary relay number that is to be the first in the register array. This then causes the block of relays, starting from that initial number, to be reserved for the shift register. Thus, if we select M140 to be the first relay then the shift register will consist of M140, M141, M142, M143, M144, M145, M146 and M147. Shift registers have three inputs: one to load data into the first element of the register (OUT), one as the shift command

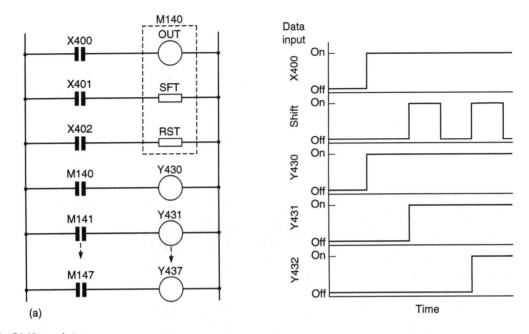

(a)

Fig. 17.21 Shift register

(SFT) and one for resetting (RST). With OUT, a logic level 0 or 1 is loaded into the first element of the shift register. With SFT, a pulse moves the contents of the register along one bit at a time, the final bit overflowing and being lost. With RST, a pulse of a closure of a contact resets the register contents to all 0s.

Figure 17.21(a) gives a ladder diagram involving a shift register. M140 has been designated as the first relay of the register. When X400 is switched on, a logic 1 is loaded into the first element of the shift register, i.e. M140. We thus have the register with 10000000. The circuit shows that each element of the shift register has been connected as a contact in the circuit. Thus M140 contact closes and Y430 is switched on (figure 17.21(b)). When contact X401 is closed, then the bits in the register are shifted along the register by one place to give 11000000, a 1 being shifted into the register because X400 is still on. Contact M141 thus closes and Y430 is switched on. As each bit is shifted along, the outputs are energised in turn. The instruction sequence for this ladder is:

Step	Instruction	
0	LD	X400
1	OUT	M140
2	LD	X401
3	SFT	M140
4	LD	X402
5	RST	M140
6	LD	M140
7	OUT	Y430
8	LD	M141
9	OUT	Y431
10	LD	M142
11	OUT	Y432
etc.		
20	LD	M147
21	OUT	Y437

As illustrated above, a shift register can be used to sequence events.

17.9 Data handling

With the exception of the shift register, the previous parts of this chapter have been concerned with the handling of individual bits of information, e.g. a switch being closed or not. There are, however, some control tasks where it is useful to deal with related groups of bits, e.g. a block of 8 inputs, and so operate on them as a data word. Such a situation can arise when a sensor supplies an analogue signal which is converted to, say, an 8-bit word before becoming an input to a PLC. The operations that may be carried out with a PLC on data words normally include:

1 Moving data.
2 Comparison of magnitudes of data, i.e. greater than, equal to, or less than.
3 Addition and subtraction.
4 Conversions between binary-coded-decimal (BCD), binary and octal.

Table 17.1 shows, for the Mitsubishi PLCs, the code symbols used on the keyboard, the form of the operation and the ladder symbols for each of the above operations.

As discussed earlier, individual bits have been stored in memory locations specified by unique addresses. For example, for the Mitsubishi PLC, input memory addresses have been preceded

Table 17.1 Data handling instructions

Code	Operation	Ladder symbol
MOV	Move data $S \rightarrow D$	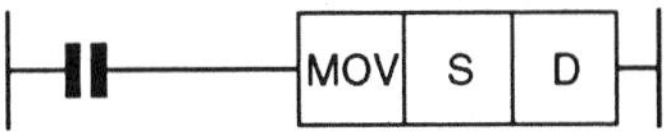
>	Compare, more than $S > D$	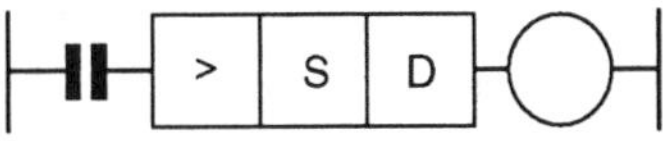
<	Compare, less than $S < D$	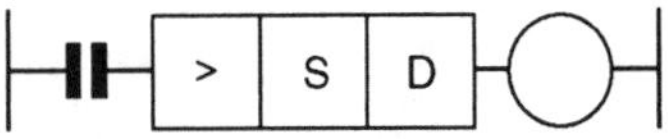
=	Compare, equal $S = D$	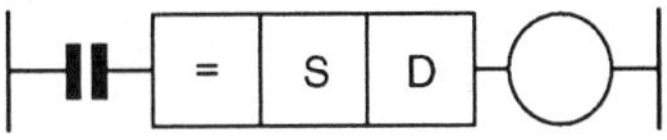
+	Add $S + D \rightarrow D$	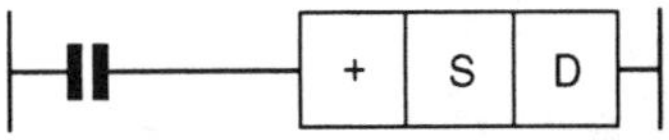
–	Subtract $S - D \rightarrow D$	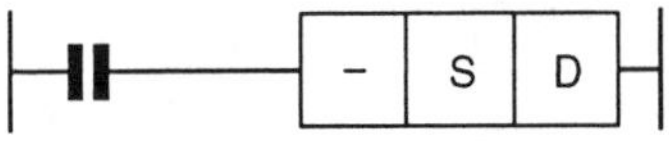
BCD	Converts $BIN \rightarrow BCD$ $S \rightarrow BCD \rightarrow D$	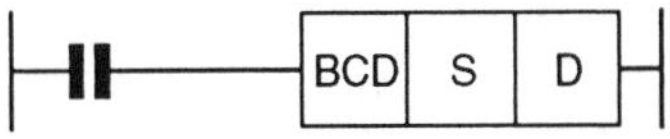
BIN	Converts $BCD \rightarrow BIN$ $S \rightarrow BIN \rightarrow D$	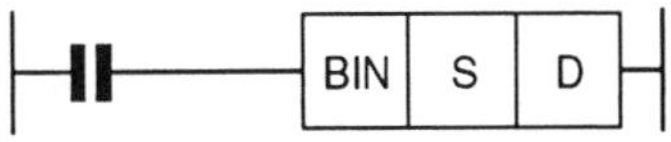

Note: >, < and = cannot handle negative values. S is the source, D the destination.

by an A, outputs by a Y, timers by a T, auxiliary relays by an M, etc. Data instructions also require memory addresses and the locations in the PLC memory allocated for data is termed a *data register*. Each data register can store a binary word of, usually, 8 or 16 bits and is given an address such as D0, D1, D2, etc. An 8-bit word means that a quantity is specified to a precision of 1 in 256, a 16-bit a precision of 1 in 65 536.

Each instruction has to specify the form of the operation, the source of the data used in terms of its data register and the destination data register of the data. Thus the instructions for the ladder diagram shown for the MOV item in table 17.1 could be, when data is to be moved from data register D1 to data register D2:

Step	Instruction	
0	LD	X300
1	MOV	
2		D1
3		D2

Other types of moves might be a constant into a data register, a time or counter to a data register, a data register to a timer or counter, a data register to an output, an input to a data register, a data register to a marker, a marker to a data register. Likewise, to compare data in data register D1 to see if it is greater than data in data register D2, the instructions would be:

Step	Instruction	
0	LD	X300
1	>	
2		D1
3		D2

Such a comparison might be used when the signals from two sensors are to be compared by the PLC before action is taken. To add some constant K to the data in register D1 the instruction would be:

Step	Instruction	
0	LD	X300
1	+	
2		K
3		D1

Addition, or subtraction, might be used to add or subtract an offset to data before it is acted on by other instructions. This might be to alter the preset values of counters or times.

All the internal operations in the CPU of a PLC are carried out using binary numbers. Thus, where the input is a signal which is decimal, conversion to binary-coded decimal (BCD) is used. Likewise, where a decimal output is required, conversion to decimal is required. The *binary-coded decimal* system codes decimal numbers into 4-bit binary numbers. The decimal digits 0 to 9 are represented by the binary numbers 0000 to 1001. Thus the decimal number 92 has the 2 represented by 0010 and the 9 by 1001 and so the BCD number is 1001 0010.

17.9.1 Analogue input/outputs

Many sensors generate analogue signals and many actuators require analogue signals. Thus, some PLCs may have analogue-to-digital converters fitted to input channels and digital-to-analogue converters fitted to output channels. An example of where such an item might be used is for the control of the speed of a motor so that its speed moves up to its steady value at a steady rate. The input is an on/off switch to start the operation. The output can be a word which is then transformed by a digital-to-analogue converter to an analogue signal to drive the motor. By building up the bits in the word at a controlled rate the analogue signal used to drive the motor builds up at a controlled rate.

A PLC equipped with analogue input channels can be used to carry out a continuous control function, i.e. PID control (see chapter 13). Thus, for example, to carry out proportional control on an analogue input the following set of operations can be used:

1   Convert the sensor output to a digital signal.
2   Compare the converted actual sensor output with the required sensor value, i.e. the set point, and obtain the difference. This difference is the error.
3   Multiply the error by the proportional constant K_p.
4   Move this result to the digital-to-analogue converter output and use the result as the correction signal to the actuator.

An example of where such a control action might be used is with a temperature controller. The input could be from a thermocouple, which after amplification is fed through an analogue-to-digital converter into the PLC. The PLC is programmed to give an output proportional to the error between the input from the sensor and the required temperature. The output word is then fed through a digital-to-analogue converter to the actuator, a heater, in order to reduce the error.

Some PLCs have add-on modules which more easily enable PLC control to be used, without the need to write lists of instructions in the way outlined above.

17.10 Selection of a PLC

In considering the size and type of PLC required for a particular task or tasks, criteria that need to be considered are:

1   What input/output capacity is required, i.e. the number of inputs/outputs, capability of expansion for future needs?
2   What types of inputs/outputs are required, i.e. isolation, on-board power supply for inputs/outputs, signal conditioning?

3   What size of memory is required? This is linked to the number of inputs/outputs and the complexity of program used.
4   What speed and power is required of the CPU? This is linked to the number of types of instruction that can be handled by a PLC. As the number of types increases so a faster CPU is required. Likewise, the greater the number of inputs/outputs to be handled the faster the CPU required.

Problems

1   What are the logic functions used for switches (a) in series, (b) in parallel?
2   Draw the ladder rungs to represent:
(a) Two switches are normally open and both have to be closed for a motor to operate.
(b) Either of two, normally open, switches have to be closed for a coil to be energised and operate an actuator.
(c) A motor is switched on by pressing a spring return push-button start switch, and the motor remains on until another spring-return push-button stop switch is pressed.
3   Write the program instructions corresponding to the latch circuit in figure 17.9.
4   Write the program instructions corresponding to the delay-off timer circuit in figure 17.15.
5   Write the program instruction corresponding to the reset latch circuit in figure 17.17.
6   Devise a timing circuit that will switch an output on for 1 s then off for 20 s, then on for 1 s, then off for 20 s, and so on.
7   Devise a timing circuit that will switch an output on for 10 s then switch it off.
8   Devise a circuit that can be used to start a motor and then after a delay of 100 s start a pump. When the motor is switched off there should be a delay of 10 s before the pump is switched off.
9   Devise a circuit that could be used with a domestic washing machine to switch on a pump to pump water for 100 s into the machine, then switch off and swich on a heater for 50 s to heat the water. The heater is then switched off and another pump is to empty the water from the machine for 100 s.
10 Devise a circuit that could be used with a conveyor belt which is used to move an item to a work station. The presence of the item at the work station is detected by means of breaking a contact activated by a beam of light to a photosensor. There it stops for 100 s for an operation to be carried out before moving on and off the conveyor. The motor for the belt is started by a normally open start switch and stopped by a normally closed switch.

11 How would the timing pattern for the shift register in figure 17.21(a) change if the data input X400 was of the form shown in figure 17.22?

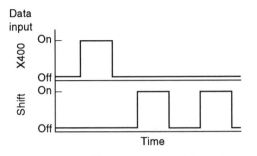

Fig. 17.22 Problem 11

12 Explain how a PLC can be used to handle an analogue input.
13 Devise a system, using a PLC, which can be used to control the movement of a piston in a cylinder so that when a switch is momentarily pressed, the piston moves in one direction and when a second switch is momentarily pressed the piston moves in the other direction. Hint: you might consider using a 4/2 solenoid controlled valve.
14 Devise a system, using a PLC, which can be used to control the movement of a piston in a cylinder using a 4/2 solenoid-operated pilot valve. The piston is to move in one direction when a proximity sensor at one end of the stroke closes contacts and in the other direction when a proximity sensor at the other end of the stroke indicates its arrival there.

18 Communication systems

18.1 Digital communications

This chapter is about data communications involving computers. In automated plant there is not only a need for data to pass between programmable logic controllers, displays, sensors, and actuators and allow for data and programs to be inputted by operator, but there can also be data communications with other computers. There may, for example, be a need to link a PLC into a control system involving a number of PLCs and computers. With small PLCs, the necessary communications hardware and software is incorporated in the PLC; with larger PLCs it is usually a bolt-on module which can be selected from a range of modules according to the communication requirements.

This chapter is a consideration of how such data communications between computers can take place, whether it is just a simple machine to machine or a large network involving large numbers of machines linked together. Computer integrated manufacturing (CIM) is an example of a large network which can involve large numbers of machines linked together.

For more details concerning digital data communications, the reader is referred to specialist texts such as *Computer Communications and Networks* by J. Freer (Pitman 1988).

18.2 Centralised, hierarchial and distributed control

Centralised computer control involves the use of one central computer to control an entire plant. This has the problem that failure of the computer results in the loss of control of the entire plant. This can be avoided by the use of dual computer systems. If one computer fails the other one takes over. Such centralised systems were common in the 1960s and 1970s. The development of the microprocessor and the ever-reducing costs of computers has led to multi-computer systems becoming more common and the development of hierarchical and distributed systems.

With the *hierarchical system*, there is a hierarchical system of computers according to the tasks they carry out. The computers handling the more routine tasks are supervised by computers which have a greater decision-making role. For example, the computers which are used for direct digital control of systems are subservient to a computer which performs supervisory control of the entire system. The work is divided between the computers according to the function involved. There is specialisation of computers with some computers only receiving some information and others different information.

With the *distributed system*, each computer system carries out essentially similar tasks to all the other computer systems. In the event of a failure of one, or overloading of a particular computer, work can be transferred to other computers. The work is spread across all the computers and not allocated to specific computers according to the function involved. There is no specialisation of computers. Each computer thus needs access to all the information in the system.

In most modern systems there is generally a mixture of distributed and hierarchical systems. For example, the work of measurement and actuation may be distributed among a number of microprocessors/computers which are linked together and provide the data base for the plant. These may these be overseen by a computer used for direct digital control or sequencing and this in turn may be supervised by one used for supervisory control of the plant as a whole. The following are typical levels in such a scheme:

Level 1 Measurement and actuators
Level 2 Direct digital and sequence control
Level 3 Supervisory control
Level 4 Management control and design

Distributed/hierarchical systems have the advantage of allowing the task of measurement scanning and signal conditioning in control systems to be carried out by sharing it between a number of microprocessors. This can involve a large number of signals with a high frequency of scanning. If extra measurement loops are required, it is a simple matter to increase the capacity of the system by adding microprocessors. The units can be quite widely dispersed, being located near the source of the measurements. Failure of one unit does not result in failure of the entire system.

18.2.1 Data transmission

Distributed and distributed/hierarchical systems require communications between computers. Such communication can be via parallel or serial transmission links.

Within computers, data transmission is usually by *parallel data paths*. Parallel data buses transmit 8, 16 or 32 bits simultaneously, a separate bus wire for each data bit and the control signals. Thus, if there are 8 data bits to be transmitted, e.g. 11000111, then 8 data wires are needed. The entire 8 data bits are transmitted in the same time as it takes to transmit 1 data bit because each bit is on a parallel wire. Parallel data transmission permits high data transfer rates but is expensive because of the cabling and interface circuitry required. It is thus normally only used over short distances or where high transfer rates are essential.

Serial data transmission involves the transmission of data which, together with control signals, is sent bit by bit in sequence along a single line. Data transmission can be by:

1   *Simplex mode* in which the transmission is only possible in one direction, from device A to device B, where device B is not capable of transmitting back to device A. This is usually only used for transmission to devices such as printers which never transmit information.
2   *Half-duplex mode* in which data is transmitted in one direction at a time but the direction can be changed. Terminals at each end of the link can be switched from transmit to receive. Thus device A can transmit to device B and device B to device A but not at the same time.
3   *Full duplex mode* in which data may be transmitted simultaneously in both directions between devices A and B.

Consider the problem of sending a sequence of characters along a serial link. The receiver needs to know where one character starts and stops. Serial data transmission can be either asynchronous or synchronous transmission. *Asynchronous transmission* implies that both the transmitter and receiver computers are not synchronised, each having its own independent clock signals. The time between transmitted characters is arbitrary. Each character transmitted along the link is thus preceded by a start bit to indicate to the receiver the start of a character, and followed by a stop bit to indicate its completion. This method has the disadvantage of requiring extra bits to be transmitted along with each character and thus reduces the efficiency of the line for data transmission. With *synchronous transmission* there is no need for start and stop bits since the transmitter and receiver have a common clock signal and thus characters automatically start and stop always at the same time in each cycle.

18.3 Networks

The term *network* is used for a system which allows two or more computers/microprocessors to be linked for the interchange of data. The logical form of the links is known as the network *topology*. The term *node* is used for a point in a network where one or more communication lines terminate or a unit is connected to the communication lines. Commonly used forms are:

1 *Data bus* This has a linear bus (figure 18.1(a)) into which all the stations are plugged. This system is often used for multipoint terminal clusters. It is generally the preferred method for distances between nodes of more than 100 m.

2 *Star* This has dedicated channels between each station and a central switching hub (figure 18.1(b)) through which all communications must pass. This is the type of network used in the telephone systems (private branch exchanges PBXs) used in many companies, all the lines passing through a central exchange. This system is also often used to connect remote and local terminals to a central mainframe computer. There is a major problem with this system in that if the central hob fails then the entire system fails.

3 *Hierarchy* or *tree* This consists of a series of branches converging indirectly to a point at the head of the tree (figure 18.1(c)). With this system there is only one transmission path between any two stations. This arrangement may be formed from a number of linked data bus systems. Like the bus method, it is often used for distances between nodes of more than 100 m.

4 *Ring* This is a very popular method for local area networks, involving each station being connected to a ring (figure 18.1(d)). The distances between nodes is generally less than 100 m. Data put into the ring system continues to circulate round the ring until some system removes it. The data is available to all the stations.

5 *Mesh* This method (figure 18.1(e)) has no formal pattern to the connections between stations and there will be multiple data paths between them.

The term *local area network* (LAN) is used for a network over a local geographic area such as a building or a group of buildings on one site. The topology is commonly bus/tree or ring. A *wide area network* is one that interconnects computers, terminals and local area networks over a national or internation level. This chapter is primarily concerned with local area networks.

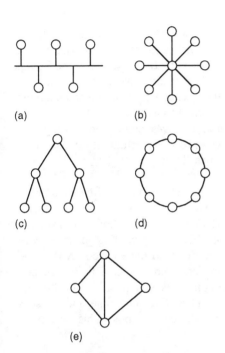

Fig. 18.1 Network topologies: (a) data bus, (b) star, (c) hierarchy, (d) ring, (e) mesh

18.3.1 Network access control

Access control methods are necessary with a network to ensure that only one user of the network is able to transmit at any one time. With ring-based local area networks, two commonly used methods are:

1 *Token passing* With this method a token, a special bit pattern, is circulated. When a station wishes to transmit it waits until it receives the token then transmits the data with the token attached to its end. Another station wishing to transmit, removes the token from the package of data and transmits its own data with the token attached to its end.
2 *Slot passing* This method involves empty slots being circulated. When a station wises to transmit data it deposits it in the first empty slot that comes along.

With bus or tree networks a method that is often used is the *carrier sense multiple access and collision detection (CSMA/CD) access method*. With the CSMA/CD access method, stations have to listen for other transmissions before transmitting. If no activity is detected then transmission can occur. If there is activity then the system has to wait until it can detect no further activity. Despite this listening before transmission, it is still possible for two or more systems to start to transmit at the same time and cause a collision of their transmitted data on the bus. This results in corruption of the data.

18.4 Protocols

Transmitted data will contain two types of information. One is the data which one computer wishes to send to another, the other is information termed *protocol data* and is used by the interface between a computer and the network to control the transfer of the data into the network or from the network into the computer. A protocol is a formal set of rules governing data format, timing, sequencing, access control and error control. The three elements of a protocol are:

1 *Syntax*, which defines data format, coding and signal levels.
2 *Semantics*, which deals with synchronisation, control and error handling.
3 *Timing*, which deals with the sequencing of data and the choice of data rate.

Communication protocols have to exist on a number of levels. The International Standards Organisation (ISO) has defined a seven-layer standard protocol system known as the *Open*

Systems Interconnection (OSI) model. The model is a framework for developing a coordinated system of standards. The layers are:

1 *Physical layer* This layer describes the physical circuits which provide the means of transmitting information between two users. It is concerned with such items as the type of cable used, input/output ports, line voltage levels and pin connections.
2 *Data link layer* This layer defines the protocols for sending and receiving messages, error detection and correction and the proper sequencing of transmitted data. It is concerned with packaging data into packets and placing them on the cable and then taking them off the cable at the receiving end.
3 *Network layer* This layer deals with communication paths and the addressing, routeing and control of messages on the network and thus making certain that the messages get to the right destinations.
4 *Transport layer* This layer provides for reliable end-to-end message transport. It is concerned with establishing and maintaining the connection between transmitter and receiver.
5 *Session layer* This layer is concerned with the establishment of dialogues between application processes which are connected together by the network. It is responsible for determining when to turn a communication between two stations on or off.
6 *Presentation layer* This layer is concerned with allowing the encoded data transmitted to be presented in a suitable form for user manipulation.
7 *Application layer* This layer provides the actual user information processing function and application specific services. It provides such functions as file transfer or electronic mail which a station can use to communicate with other systems on the network.

18.4.1 Network standards

The term *broadband transmission* is used for a network in which information is modulated onto a radio frequency carrier which passes through the transmission medium such as a coaxial cable. Typically the topology of broadband local area networks is a bus with branches. Broadband transmission allows a number of modulated radio frequency carriers to be simultaneously transmitted and so offers a multichannel capability. The term *baseband transmission* is used when digital information is passed directly through the transmission medium. Baseband transmission networks can only support one information signal at a time.

The IEEE, in the United States, has developed a number of standards for local area networks and these are frequently used to define the lower levels of network standards developed for particular purposes.

IEEE 802.1 (A)	Local and metropolitan area network standard. Overview and architecture.
IEEE 802.1 (B)	Addressing, internetworking and network management.
IEEE 802.2	Local area network standard for logical link control. This performs the function of the OSI data link layer.
IEEE 802.3	Carrier sense multiple access and collision detection (CSMA/CD) access method and physical layer specification.
IEEE 802.4	Token passing bus access method and physical layer specifications.
IEEE 802.5	Token passing ring access method and physical layer specifications.
IEEE 802.6	Metropolitan area network standards.
IEEE 802.7	Broadband local area network standards.
IEEE 802.8	Fibre optic standards.

For further details of the above standards, the reader is referred to the standards themselves or specialist texts such as *Computer Communications and Networks* by J. Freer (Pitman 1988).

There are a number of network standards that are based on the OSI layer model and commonly used. In the United States, General Motors realised that the automation of their manufacturing activities posed a problem of equipment being supplied with a variety of non-standard protocols. They thus developed a standard communication system for factory automation applications. The standard is referred to as the *manufacturing automation protocol* (MAP). This standard is now widely accepted in the manufacturing industry. The choice of protocols at the different layers reflects the requirement for the system to fit the manufacturing environment. Layers 1 and 2 are implemented in hardware electronics and levels 3 to 7 using software. For the physical layer broadband transmission is used, the standard followed being IEEE 802.4. The broadband method allows the system to be used for services in addition to those required for MAP communications. For the data link layer the token system with a bus is used, the standard adopted being IEEE 802.4, with logical link control (LLC) to implement such functions as error checking, etc., the standard used by MAP being IEEE 802.2 class 1. For the other layers ISO standards are used. At layer 7, MAP includes manufacturing message services (MMS), an application

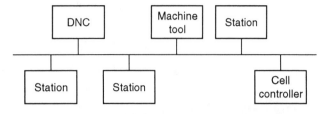

Fig. 18.2 MAP

relevant to factory floor communications which defines inter-actions between programmable logic controllers and numerically controlled machines or robots. Figure 18.2 illustrates the form part of a MAP network could take.

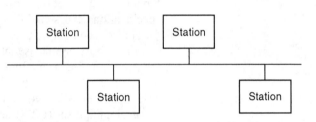

Fig. 18.3 TOP

Technical and office protocol (TOP) is a standard that was developed by Boeing Computer Services. It has much in common with MAP but can be implemented at a lower cost because it is a baseband system. Figure 18.3 shows the form that part of a TOP network could take. It differs from MAP in layers 1 and 2, using either the token with a ring IEEE 802.5 or the carrier sense multiple access and collision detection (CSMA/CD) access method (IEEE 802.3) with a bus network. Also, at layer 7, it specifies application protocols that concern office requirements, rather than factory floor requirements. With the CSMA/CD access method, stations have to listen for other transmissions before transmitting. TOP and MAP networks are compatible and a gate- way device can be used to connect TOP and MAP networks. This device carries out the appropriate address conversions and protocol changes.

Systems network architecture (SNA) is a system developed by IBM as a design standard for IBM products. SNA is divided into seven layers. These layers do, however, differ from the OSI layers in two areas.

OSI layers		SNA layers	
7	Application	7	Transaction services/application
6	Presentation	6	Presentation services
5	Session	5	Data flow control
		4	Transmission control
			OSI 5 split into two layers

4	Transport	3	Path control
3	Network		OSI 3 and 4 combined
2	Data link	2	Data link control
1	Physical	1	Physical control

18.4.2 Interface standards

The physical interface layer in the Open Service Interconnection (OSI) is concerned with the mechanical specification for the cable and connectors, electrical specifications for the connection, functional descriptions of the interface including signal-pin specifications and signal definitions, and procedural specifications for control and data transfer.

The most popular serial interface is RS-232, first defined by the American Electronic Industries Association (EIA) in 1962, and the CCITT (Comité Consultatif International Télégraphique et Téléphonique) V24 equivalent. The standard relates to data terminal equipment (DTE) and data circuit-terminating equipment (DCE). Data terminal equipment can send or receive data via the interface and is, for example, a microcomputer. Data circuit-terminating equipment are devices which facilitate communication and a typical example is a modem. This forms an essential link between a microcomputer and a conventional analogue telephone line.

The connector to a RS-232 serial port is via a 25-pin D-type connector (figure 18.4). Usually, though not always, a male plug is used on cables and a female socket on the DCE or DTE. Figure 18.5 shows how the connectors might be used for a simple set up

Pins 1 ──────► 13

Pins 14 ──────► 25

Fig. 18.4 RS-232 connector (pin view)

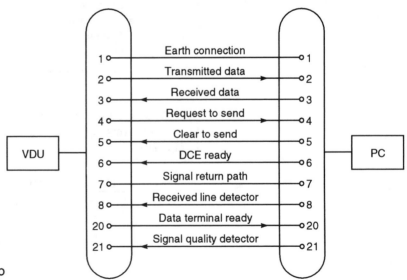

Fig. 18.5 A minimum set up

involving a personal computer (PC) being linked with a visual display unit (VDU). A positive signal between 3 and 25 V is interpreted as 'on' for control circuits or as binary 0 for data circuits. A negative voltage between −3 and −25 V is interpreted as 'off' for control circuits or as binary 1 for data circuits.

Problems

1 Explain the difference between centralised and distributed communication systems.
2 Explain the forms of bus/tree and ring networks.
3 A LAN is required with a distance between nodes of more than 100 m. Should the choice be bus or ring topology?
4 A multichannel LAN is required. Should the choice be broadband or baseband transmission?
5 What are MAP and TOP?
6 Explain what is meant by communication protocol.

19 Fault finding

19.1 Fault detection techniques

This chapter is a brief consideration of the problems of fault detection with measurement, control and communication systems. For details of the fault-finding checks required for specific systems or components, the manufacturer's manuals should be used.

There are a number of techniques that can be used to detect faults in measurement, control and communication systems:

1 *Replication checks* This involves duplicating or replicating an activity and comparing the results. In the absence of faults it is assumed that the results should be the same. It could mean, with transient errors, just repeating an operation twice and comparing the results or it could involve having duplicate systems and comparing the results given by the two. This can be an expensive option.

2 *Expected value checks* Software errors are commonly detected by checking whether an expected value is obtained when a specific numerical input is used. If the expected value is not obtained then there is a fault.

3 *Timing checks* This involves the use of timing checks that some function has been carried out within a specified time. These checks are commonly referred to as *watchdog timers*. For example, with a PLC, when an operation starts a timer is also started and if the operation is not completed within the specified time a fault is assumed to have occurred. The watchdog timer trips, sets off an alarm and closes down part or the entire plant.

4 *Reversal checks* Where there is a direct relationship between input and output values, the value of the output can be taken and the input which should have caused it computed. This can then be compared with the actual input.

5 *Parity and error coding checks* This form of checking is commonly used for detecting memory and data transmission errors. Communication channels are frequently subject to

interference which can affect data being transmitted. To detect whether data has been corrupted a parity bit is a bit added to the transmitted data word. The parity bit is chosen to make the resulting number of 1s in the group either odd (odd parity) or even (even parity). If odd parity then the word can be checked after transmission to see if it is still odd. Other forms of checking involve codes added to transmitted data in order to detect corrupt bits.

6   *Diagnostic checks*   Diagnostic checks are used to test the behaviour of components in a system. Inputs are applied to a component and the outputs compared with that which should occur.

19.2 Common faults

The following are some of the commonly encountered faults that can occur with specific types of components and systems.

19.2.1 Sensors

If there are faults in a measurement system then the sensor might be at fault. A simple test is to substitute the sensor with a new one and see what effect this has on the results given by the system. If the results change then it is likely that the original sensor was faulty; if the results do not change then the fault is elsewhere in the system. It is also possible to check that the voltage/current sources are supplying the correct voltages/currents, whether there is electrical continuity in connecting wires, that the sensor is correctly mounted and used under the conditions specified by the manufacturer's data sheet, etc.

19.2.2 Switches and relays

Dirt and particles of waste material between switch contacts is a common source of incorrect functioning of mechanical switches. A voltmeter used across a switch should indicate the applied voltage when the contacts are open and very nearly zero when they are closed. Mechanical switches used to detect the position of some item, e.g. the presence of a work piece on a conveyor, can fail to give the correct responses if the alignment is incorrect or if the actuating lever is bent.

Inspection of a relay can disclose evidence of arcing or contact welding. The relay should then be replaced. If a relay fails to operate then a check can be made for the voltage across the coil. If the correct voltage is present then coil continuity can be checked with an ohmmeter. If there is no voltage across the coil then the fault is likely to be the switching transistor used with the relay.

19.2.3 Motors

Maintenance of both d.c. and a.c. motors involves correct lubrication. With d.c. motors the brushes wear and can require changing. Setting of new brushes needs to be in accordance with the manufacturer's specification. A single-phase capacitor start a.c. motor that is sluggish in starting probably needs a new starting capacitor. The three-phase induction motor has no brushes, commutator, slip rings or starting capacitor and short of a severe overload, the only regular maintenance that is required is periodic lubrication.

19.2.4 Hydraulic and pneumatic systems

A common cause of faults with hydraulic and pneumatic systems is dirt. Small particles of dirt can damage seals, block orifices, cause valve spools to jam, etc. Thus filters should be regularly checked and cleaned, components should only be dismantled in clean conditions, and oil should be regularly checked and changed. With an electrical circuit a common method of testing a circuit is to measure the voltages at a number of test points. Likewise, with a hydraulic and pneumatic system there needs to be points at which pressures can be measured. Damage to seals can result in hydraulic and pneumatic cylinders leaking, beyond that which is normal, and result in a drop in system pressure when the cylinder is actuated. This can be remedied by replacing the seals in the cylinders. The vanes in vane-type motors are subject to wear and can then fail to make a good seal with the motor housing with the result of a loss of motor power. The vanes can be replaced. Leaks in hoses, pipes and fittings are common faults.

19.3 PLC systems

Programmable logic controllers have a high reliability. The PLC is electrically isolated by optoisolator or relays from potentially damaging voltages and currents at input/output ports; battery-backed RAM protects the application software from power failures or corruption; and the construction is so designed that the PLC can operate reliably in industrial conditions for long periods of time. PLCs generally have several built-in fault procedures. Critical faults cause the CPU to stop, while other less critical faults cause the CPU to continue running but display a fault code on a display. The PLC manual will indicate the remedial action required when a fault code is displayed.

The PLC can also be used as a monitor of the system being controlled. It can be used to sound an alarm or light up a red light if inputs move outside prescribed limits, using the greater than, equal to, or less than functions, or its operations take longer than a prescribed time. It can also keep a log of time and indicate when maintenance of items is due.

Problems

1   Explain what is meant by (a) replication checks, (b) expected value checks, (c) reversal checks, (d) parity checks.

2   Explain how a watchdog timer can be used with a PLC controlled plant in order to indicate the presence of faults.

3   The F2 series Mitsubishi PLC is specified as having:

Diagnosis: Programmable check (sum, syntax, circuit check), watchdog timer, battery voltage, power supply voltage

Explain the significance of the terms.

20 Design and mechatronics

20.1 Designing

This chapter is a brief review of the design process and brings together many of the topics discussed in this book in the consideration of both traditional and mechatronics solutions to design problems.

The design process can be considered to have a number of stages:

1 *The need* The design process begins with a need from, perhaps, a customer or client. This may be identified by market research being used to establish the needs of potential customers.

2 *Analysis of the problem* The first stage in developing a design is to find out the true nature of the problem, i.e. analysing it. This is an important stage in that not defining the problem accurately can lead to wasted time on designs that will not fulfil the need.

3 *Preparation of a specification* Following the analysis a specification of the requirements can be prepared. This will state the problem, any constraints placed on the solution, and the criteria which may be used to judge the quality of the design. In stating the problem, all the functions required of the design, together with any desirable features, should be specified. Thus there might be a statement of mass, dimensions, types and range of motion required, accuracy, input and output requirements of elements, inter- faces, power requirements, operating environment, relevant standards and codes of practice, etc.

4 *Generation of possible solutions* This is often termed the *conceptual stage*. Outline solutions are prepared which are worked out in sufficient detail to indicate the means of obtaining each of the required functions. Thus there could be the approximate sizes, shapes, materials and costs.

5   *Selections of a suitable solution*   The various solutions are evaluated and the most suitable one selected.

6   *Produce a detailed design*   The detail of the selected design has now to be worked out. This might require the production of prototypes or mock-ups in order to determine the optimum details of a design.

7   *Produce working drawings*   The selected design is then translated into working drawings, circuit diagrams, etc. so that the item can be made.

It should not be considered that each stage of the design process just flows on stage by stage. There will often be the need to return to an earlier stage and give it further consideration. Thus when at the stage of generating possible solutions there might be a need to go back and reconsider the analysis of the problem.

20.1.1 Traditional and mechatronics designs

Engineering design is a complex process involving interactions between many skills and disciplines. The basis of the mechatronics approach is considered to lie in the inclusion of the disciplines of electronics, computer technology and control engineering. For example, the design of bathroom scales might be considered only in terms of the compression of springs and a mechanism used to convert the motion into rotation of a shaft and hence movement of a pointer across a scale. A problem that has to be taken into account in the design is that the weight indicated should not depend on the person's position on the scales. With mechatronics, other possibilities can be considered. For example, the springs might be replaced by load cells with strain gauges and the output from them used with a microprocessor to provide a digital readout of the weight on an LED display. The resulting scales might be mechanically simpler, involving fewer components and moving parts. The complexity has, however, been transferred to the electronics.

The traditional design of the temperature control for a domestic central heating system has been the bimetallic thermostat in a closed-loop control system. The bending of the bimetallic strip changes as the temperature changes and is used to operate an on/off switch for the heating system. A mechatronics solution to the problem might be to use a microprocessor-controlled system employing perhaps a thermocouple as the sensor. Such a system has many advantages over the bimetallic thermostat system. The bimetallic thermostat is comparatively crude and the temperature is not accurately controlled, also devising a method for having different temperatures at different times of the day is complex and not easily achieved. The microprocessor-controlled system can, however, cope easily with giving precision and programmed control. The system is much more flexible. This improvement in flexibility is a common characteristic of mechatronics systems when compared with traditional systems.

Microprocessors are increasingly finding a use in home appliances, cars, factory machines, etc., as an embedded controller. Virtually every mechanical design can consider a microprocessor subsystem as a possible solution.

20.2 Mechanisms

Mechanisms have a role in both traditional and mechatronics systems. For example, the mechatronics system in use in an automatic camera for adjusting the aperture for correct exposures involves a mechanism for adjusting the size of the diaphragm. While electronics might now be used often for many functions that previously were fulfilled by mechanisms, mechanisms might still be used to provide such functions as:

1   Force amplification, e.g. that given by levers.
2   Change of speed, e.g. that given by gears.
3   Action at a distance, e.g. that given by hydraulics or belts.
4   Particular types of motion, e.g. that given by a quick-release mechanism.

In considering mechanisms, consideration has to be given to such things as loading and degrees of freedom.

20.2.1 Loading

Mechanisms are structures and as such transmit and support loads. Analysis is thus necessary to determine the loads to be carried by individual elements. Then consideration can be given to the dimensions of the element so that it might, for example, have sufficient strength and perhaps stiffness under such loading.

20.2.2 Freedom and constraints

An important aspect in the design of mechanical elements is the orientation and arrangement of the elements and parts. A body that is free in space can move in three, independent, mutually perpendicular directions and rotate in three ways about those directions (figure 20.1). It is said to have six *degrees of freedom*. If a point is constrained to lie on a line then its translational degrees of freedom are reduced to one; if it is constrained to lie on a plane then they are reduced to two. The problem in design is often to reduce the number of degrees of freedom and this then requires an

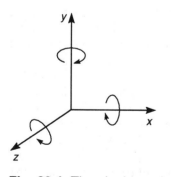

Fig. 20.1 The six degrees of freedom

appropriate number and orientation of constraints. For constraints on a single rigid body we have the basic rule

6 – number of constraints = number of degrees of freedom
– number of redundancies

Thus if a body is required to be fixed, i.e. have zero degrees of freedom, then if no redundant constraints are introduced the number of constraints required is 6.

A concept that is used in design is that of the *principle of least constraint*. This states that in fixing a body or guiding it to a particular type of motion, the minimum number of constraints should be used, i.e. there should be no redundancies. This is often referred to as *kinematic design*.

For example, to have a shaft which only rotates about one axis with no translational motions, we have to reduce the number of degrees of freedom to 1. The minimum number of constraints to do this is 5. Any more constraints than this will give redundancies. The mounting that might be used to mount the shaft has a ball bearing at one end and a roller bearing at the other (figure 20.2). The pair of bearings together prevent translational motion at right angles to the shaft, the y-axis, and rotations about the z-axis and the y-axis. The ball bearing prevents translational motion along the x-axis and along the z-axis. Thus there is a total of five constraints. This leaves just one degree of freedom, the required rotation about the x-axis. If there had been a roller bearing at each end of the shaft then both the ball bearings could have prevented translational motion along the x-axis and the z-axis and thus there would have been redundancy. Such redundancy might cause damage. If ball bearings are used at both ends of the shaft, then in order to prevent redundancy one of the bearings would have its outer race not fixed in its housing so that it could slide to some extent in an axial direction.

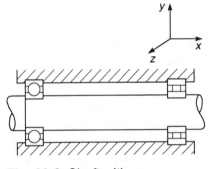

Fig. 20.2 Shaft with no redundancies

20.3 Examples of designs

Consider the design of bathroom scales. The main requirements are that a person can stand on a platform and the weight of that person will be displayed on some form of readout. The weight should be given with reasonable speed and accuracy and be independent of where on the platform the person stands. The following are possible solutions.

20.3.1 A mechanical solution

A possible solution is to use the weight of the person on the platform to deflect an arrangement of two parallel leaf springs (figure 20.3(a)). With such an arrangement the deflection is virtually independent of where on the platform the person stands.

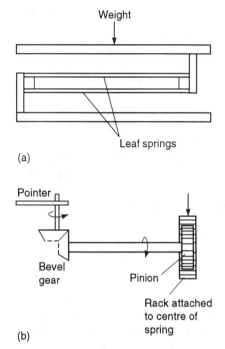

Weight

Leaf springs

(a)

Pointer

Bevel
gear

Pinion

Rack attached
to centre of
spring

(b)

Fig. 20.3 A solution for bathroom scales

The deflection can be transformed into movement of a pointer across a scale by using the arrangement shown in figure 20.3(b). A rack-and-pinion is used to transform the linear motion into a circular motion about a horizontal axis. This is then transformed into a rotation about a vertical axis, and hence movement of a pointer across a scale, by means of a bevel gear.

20.3.2 A mechatronics solution

Another possible solution involves the use of a microprocessor. The platform can be mounted on load cells employing electrical resistance strain gauges. When the person stands on the platform the gauges suffer strain and change resistance. If the gauges are mounted in a four-active arm Wheatstone bridge then the out-of-balance voltage output from the bridge is a measure of the weight of the person. This can be amplified by a differential operational amplifier. The resulting analogue signal can then be fed through an analogue-to-digital converter for inputting to the PIA and hence the microprocessor, e.g. the Motorola 6820 with the Motorola 6802. There will also be a need to provide a non-erasable memory and this can be provided by an EPROM chip. However, if a microcontroller is used then memory is present within the single microprocessor chip, and by a suitable choice of microcontroller we can obtain analogue-to-digital conversion for the inputs. For example, the 8-bit micro-processor Motorola 6800, and its later version of 6802, have microcontroller equivalent of the Motorola 6805. A number of versions of this exist with different amounts and types of memory, different numbers of input/output ports and some with analogue-to-digital conversion built in. A possibility is the Motorola 68705R3. This has 112 bytes of RAM, 3776 bytes of EPROM, 32 input/output ports and analogue-to-digital conversion. The system then becomes: strain gauges feeding through an operational amplifier a voltage to the microcontroller, with the output passing through suitable drives to a LED display (figure 20.4). With suitable programming, the display might be adjustable to read either pounds and stones or kilograms, display the message 'WAIT' when a person steps on the scale while it averages out the results from a number of load cells under the platform, then display the weight.

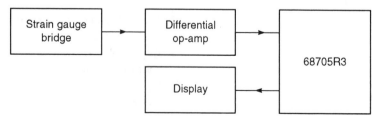

Strain gauge bridge

Differential op-amp

Display

68705R3

Fig. 20.4 A solution for bathroom scales

Problems

1   Explain what is meant by the principle of least constrain and its relevance to the design of mechanisms.

2   Present outline solutions of designs for the following:
    (a) A temperature controller for an oven.
    (b) A mechanism for sorting small, medium and large size objects moving along a conveyor belt so that they each are diverted down different chutes for packaging.
    (c) An x-y plotter (such a machine plots graphs showing how an input to x varies as the input to y changes).

Appendix A
The Laplace transform

This appendix gives more details of the Laplace transform than appears in chapter 10. For a more detailed discussion of the Laplace transform, and examples of their use, the reader is referred to *Laplace and z-Transforms* by W. Bolton (Mathematics for Engineers Series, Longman 1994).

Consider a quantity which is a function of time. We can talk of this quantity being in the *time domain* and represent such a function as $f(t)$. In many problems we are only concerned with values of time greater than or equal to 0, i.e. $t \geq 0$. To obtain the Laplace transform of this function we multiply it by e^{-st} and then integrate with respect to time from zero to infinity. Here s is a constant with the unit of 1/time. The result is what we now call the *Laplace transform* and the equation is then said to be in the *s-domain*. Thus the Laplace transform of the function of time $f(t)$, which is written as $\mathcal{L}\{f(t)\}$, is given by

$$\mathcal{L}\{f(t)\} = \int_0^\infty e^{-st} f(t)\, dt$$

The transform is *one-sided* in that values are only considered between 0 and $+\infty$, and not over the full range of time from $-\infty$ to $+\infty$.

We can carry out algebraic manipulations on a quantity in the *s*-domain, i.e. adding, subtracting, dividing and multiplying in the normal way we do any algebraic quantities. We could not have done this on the original function, assuming it to be in the form of a differential equation, when in the time domain. By this means we can obtain a considerably simplified expression in the *s*-domain. If we want to see how the quantity varies with time in the time domain then we have to carry out the inverse transformation. This involves finding the time domain function that could have given the simplified *s*-domain expression.

When in the *s*-domain a function is usually written, since it is a function of *s*, as $F(s)$. It is usual to use a capital letter F for the

Laplace transform and a lower-case letter f for the time-varying function $f(t)$. Thus

$$\mathcal{L}\{f(t)\} = F(s)$$

For the inverse operation, when the function of time is obtained from the Laplace transform, we can write

$$f(t) = \mathcal{L}^{-1}\{F(s)\}$$

This equation thus reads as: $f(t)$ is the inverse transform of the Laplace transform $F(s)$.

A.1.1 The Laplace transform from first principles

To illustrate the transformation of a quantity from the time domain into the s-domain, consider a function that has the constant value of 1 for all values of time greater than 0, i.e. $f(t) = 1$ for $t \geq 0$. This describes a *unit step* function and is shown in figure Ap.A.1. The Laplace transform is then

$$\mathcal{L}\{f(t)\} = F(s) = \int_0^\infty 1\, e^{-st}\, dt \ = -\frac{1}{s}[e^{-st}]_0^\infty$$

Since with $t = \infty$ the value of e is 0 and with $t = 0$ the value of e^{-0} is -1, then

$$F(s) = \frac{1}{s}$$

As another example, the following shows the determination, from first principles, of the Laplace transform of the function e^{at}, where a is a constant. The Laplace transform of $f(t) = e^{at}$ is thus

$$F(s) = \int_0^\infty e^{at} e^{-st}\, dt$$

$$= \int_0^\infty e^{-(s-a)t}\, dt = -\frac{1}{s-a}[e^{-(s-a)t}]_0^\infty$$

When $t = \infty$ the term in the brackets becomes 0 and when $t = 0$ it becomes -1. Thus

$$F(s) = \frac{1}{s-a}$$

A.2 Unit steps and impulses

Common input functions to systems are the unit step and the impulse. The following indicates how their Laplace transforms are obtained.

A.2.1 The unit-step function

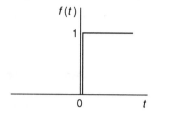

Fig. Ap.A.1 Unit step function

Figure Ap.A1 shows a graph of a unit-step function. Such a function, when the step occurs at $t = 0$, has the equation

$f(t) = 1$ for all values of t greater than 0
$f(t) = 0$ for all values of t less than 0

The step function describes an abrupt change in some quantity from zero to a steady value, e.g. the change in the voltage applied to a circuit when it is suddenly switched on.

The unit-step function thus cannot be described by $f(t) = 1$ since this would imply a function that has the constant value of 1 at all values of t, both positive and negative. The unit-step function that switches from 0 to +1 at $t = 0$ is conventionally described by the symbol $u(t)$ or $H(t)$, the H being after the originator O. Heaviside. It is thus sometimes referred to as the *Heaviside function*.

The Laplace transform of this step function is, as derived in the previous section,

$$F(s) = \frac{1}{s}$$

The Laplace transform of a step function of height a is

$$F(s) = \frac{a}{s}$$

A.2.2 Impulse function

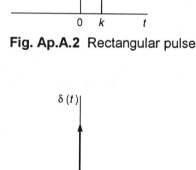

Fig. Ap.A.2 Rectangular pulse

Consider a rectangular pulse of size $1/k$ that occurs at time $t = 0$ and which has a pulse width of k, i.e. the area of the pulse is 1. Figure Ap.A.2 shows such a pulse. The pulse can be described as

$f(t) = \frac{1}{k}$ for $0 \leq t < k$
$f(t) = 0$ for $t > k$

Fig. Ap.A.3 Impulse

If we maintain this constant pulse area of 1 and then decrease the width of the pulse (i.e. reduce k), the height increases. Thus, in the limit as $k \rightarrow 0$ we end up with just a vertical line at $t = 0$ with the height of the graph going off to infinity. The result is a graph that is zero except at a single point where there is an infinite spike (figure Ap.A.3). Such a graph can be used to represent an impulse.

The impulse is said to be a unit impulse because the area enclosed by it is 1. This function is represented by $\delta(t)$, the *unit impulse function* or the *Dirac-delta function*.

The Laplace transform for the unit area rectangular pulse in figure Ap.A3 is given by

$$F(s) = \int_0^\infty f(t) e^{-st}\, dt$$

$$= \int_0^k \frac{1}{k} e^{-st}\, dt + \int_k^\infty 0\, e^{-st}\, dt$$

$$= \left[-\frac{1}{sk} e^{-st} \right]_0^k$$

$$= -\frac{1}{sk}(e^{-sk} - 1)$$

To obtain the Laplace transform for the unit impulse we need to find the value of the above in the limit as $k \to 0$. We can do this by expanding the exponential term as a series. Thus

$$e^{-sk} = 1 - sk + \frac{(-sk)^2}{2!} + \frac{(-sk)^3}{3!} + \dots$$

and so we can write

$$F(s) = 1 - \frac{sk}{2!} + \frac{(sk)^2}{3!} + \dots$$

Thus in the limit as $k \to 0$ the Laplace transform tends to the value 1.

$$\mathcal{L}\{\delta(t)\} = 1$$

Since the area of the above impulse is 1 we can define the size of such an impulse as being 1. Thus the above equation gives the Laplace transform for a *unit impulse*. An impulse of size a is represented by $a\,\delta(t)$ and the Laplace transform is

$$\mathcal{L}\{a\,\delta(t)\} = a$$

A.3 Standard Laplace transforms

In determining the Laplace transforms of functions it is not usually necessary to evaluate integrals since tables are available that give the Laplace transforms of commonly occurring functions. These, when combined with a knowledge of the properties of such transforms (see the next section), enable most commonly encountered problems to be tackled. Table Ap.A.1 lists some of the more common time functions and their Laplace transforms.

Table Ap.A.1 Laplace transforms

Time function $f(t)$	Laplace transform $F(s)$
1 $\delta(t)$, unit impulse	1
2 $\delta(t-T)$, delayed unit impulse	e^{-sT}
3 $u(t)$, a unit step	$\dfrac{1}{s}$
4 $u(t-T)$, a delayed unit step	$\dfrac{e^{-sT}}{s}$
5 t, a unit ramp	$\dfrac{1}{s^2}$
6 t^n, nth-order ramp	$\dfrac{n!}{s^{n+1}}$
7 e^{-at}, exponential decay	$\dfrac{1}{s+a}$
8 $1-e^{-at}$, exponential growth	$\dfrac{a}{s(s+a)}$
9 te^{-at}	$\dfrac{1}{(s+a)^2}$
10 $t^n e^{-at}$	$\dfrac{n!}{(s+a)^{n+1}}$
11 $t-\dfrac{1-e^{-at}}{a}$	$\dfrac{a}{s^2(s+a)}$
12 $e^{-at}-e^{-bt}$	$\dfrac{b-a}{(s+a)(s+b)}$
13 $(1-at)e^{-at}$	$\dfrac{s}{(s+a)^2}$
14 $1-\dfrac{b}{b-a}e^{-at}+\dfrac{a}{b-a}e^{-bt}$	$\dfrac{ab}{s(s+a)(s+b)}$
15 $\dfrac{e^{-at}}{(b-a)(c-a)}+\dfrac{e^{-bt}}{(c-a)(a-b)}+\dfrac{e^{-ct}}{(a-c)(b-c)}$	$\dfrac{1}{(s+a)(s+b)(s+c)}$
16 $\sin\omega t$, a sine wave	$\dfrac{\omega}{s^2+\omega^2}$
17 $\cos\omega t$, a cosine wave	$\dfrac{s}{s^2+\omega^2}$
18 $e^{-at}\sin\omega t$, a damped sine wave	$\dfrac{\omega}{(s+a)^2+\omega^2}$
19 $e^{-at}\cos\omega t$, a damped cosine wave	$\dfrac{s+a}{(s+a)^2+\omega^2}$

Time function $f(t)$	Laplace transform $F(s)$
20 $1 - \cos\omega t$	$\dfrac{\omega^2}{s(s^2 + \omega^2)}$
21 $t\cos\omega t$	$\dfrac{s^2 - \omega^2}{(s^2 + \omega^2)^2}$
22 $t\sin\omega t$	$\dfrac{2\omega s}{(s^2 + \omega^2)^2}$
23 $\sin(\omega t + \theta)$	$\dfrac{\omega\cos\theta + s\sin\theta}{s^2 + \omega^2}$
24 $\cos(\omega t + \theta)$	$\dfrac{s\cos\theta - \omega\sin\theta}{s^2 + \omega^2}$
25 $e^{-at}\sin(\omega t + \theta)$	$\dfrac{(s+a)\sin\theta + \omega\cos\theta}{(s+a)^2 + \omega^2}$
26 $e^{-at}\cos(\omega t + \theta)$	$\dfrac{(s+a)\cos\theta - \omega\sin\theta}{(s+a)^2 + \omega^2}$
27 $\dfrac{\omega}{\sqrt{1-\zeta^2}} e^{-\zeta\omega t}\sin\omega\sqrt{1-\zeta^2}\, t$	$\dfrac{\omega^2}{s^2 + 2\zeta\omega s + \omega^2}$
28 $1 - \dfrac{1}{\sqrt{1-\zeta^2}} e^{-\zeta\omega t}\sin\left(\omega\sqrt{1-\zeta^2}\, t + \phi\right)$, $\cos\phi = \zeta$	$\dfrac{\omega^2}{s(s^2 + 2\zeta\omega s + \omega^2)}$
29 $\sinh\omega t$	$\dfrac{\omega}{s^2 - \omega^2}$
30 $\cosh\omega t$	$\dfrac{s}{s^2 - \omega^2}$
31 $e^{-at}\sinh\omega t$	$\dfrac{\omega}{(s+a)^2 - \omega^2}$
32 $e^{-at}\cosh\omega t$	$\dfrac{s+a}{(s+a)^2 - \omega^2}$
33 Half-wave rectified sine, period $T = 2\pi/\omega$	$\dfrac{\omega}{(s^2 + \omega^2)(1 - e^{-\pi s/\omega})}$
34 Full-wave rectified sine, period $T = 2\pi/\omega$	$\dfrac{\omega}{(s^2 + \omega^2)}\dfrac{(1 + e^{-\pi s/\omega})}{(1 - e^{-\pi s/\omega})}$
35 Rectangular pulses, period T, amplitude +1 to 0	$\dfrac{1}{s(1 + e^{-sT/2})}$

Note: $f(t) = 0$ for all negative values of t. The $u(t)$ terms have been omitted from most of the time functions and have to be assumed.

A.3.1 Properties of Laplace transforms

In this section the basic properties of the Laplace transform are outlined. These properties enable the table of standard Laplace transforms to be used in a wide range of situations.

Linearity property

If two separate time functions, e.g. $f(t)$ and $g(t)$, have Laplace transforms then the transform of the sum of the time functions is the sum of the two separate Laplace transforms.

$$\mathcal{L}\{af(t) + bg(t)\} = a\mathcal{L}f(t) + b\mathcal{L}g(t)$$

a and b are constants.

Thus, for example, the Laplace transform of $1 + 2t + 4t^2$ is given by the sum of the transforms of the individual terms in the expression. Thus, using items 1, 5 and 6 in table Ap.A.1,

$$F(s) = \frac{1}{s} + \frac{2}{s^2} + \frac{8}{s^3}$$

s-Domain shifting property

This property is used to determine the Laplace transform of functions that have an exponential factor and is sometimes referred to as the *first shifting property*. If $F(s) = \mathcal{L}\{f(t)\}$ then

$$\mathcal{L}\{e^{at}f(t)\} = F(s - a)$$

For example, the Laplace transform of $e^{at}t^n$ is, since the Laplace transform of t^n is given by item 6 in table Ap.A.1 as $n!/s^{n+1}$, given by

$$\mathcal{L}\{e^{at}t^n\} = \frac{n!}{(s - a)^{n+1}}$$

Time domain shifting property

If a signal is delayed by a time T then its Laplace transform is multiplied by e^{-sT}. If $F(s)$ is the Laplace transform of $f(t)$ then

$$\mathcal{L}\{f(t - T)u(t - T)\} = e^{-sT}F(s)$$

This delaying of a signal by a time T is referred to as the *second shift theorem*.

The time domain shifting property can be applied to all Laplace transforms. Thus for an impulse $\delta(t)$ which is delayed by a time T to give the function $\delta(t - T)$, the Laplace transform of $\delta(t)$, namely 1, is multiplied by e^{-sT} to give $1e^{-sT}$ as the transform for the delayed function.

Periodic functions

For a function $f(t)$ which is a periodic function of period T, the Laplace transform of that function is

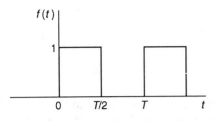

Fig. Ap.A.4 Rectangular pulses

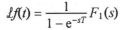

$$\mathcal{L}f(t) = \frac{1}{1 - e^{-sT}} F_1(s)$$

where $F_1(s)$ is the Laplace transform of the function for the first period.

Thus, for example, consider the Laplace transform of a sequence of periodic rectangular pulses of period T, as shown in figure Ap.A.4. The Laplace transform of a single pulse is given by $(1/s)(1 - e^{-sT/2})$. Hence, using the above equation, then the Laplace transform is

$$\frac{1}{1 - e^{-sT}} \times \frac{1}{s}(1 - e^{-sT/2}) = \frac{1}{s(1 + e^{-sT/2})}$$

Initial and final value theorems
The initial value theorem can be stated as: if a function of time $f(t)$ has a Laplace transform $F(s)$ then in the limit as the time tends to zero the value of the function is given by

$$\lim_{t \to 0} f(t) = \lim_{s \to \infty} sF(s)$$

The final value theorem can be stated as: if a function of time $f(t)$ has a Laplace transform $F(s)$ then in the limit as the time tends to infinity the value of the function is given by

$$\lim_{t \to \infty} f(t) = \lim_{s \to 0} sF(s)$$

Derivatives
The Laplace transform of a derivative of a function $f(t)$ is given by

$$\mathcal{L}\left\{\frac{d}{dt}f(t)\right\} = sF(s) - f(0)$$

where $f(0)$ is the value of the function when $t = 0$. For a second derivative

$$\mathcal{L}\left\{\frac{d^2}{dt^2}f(t)\right\} = s^2 F(s) - sf(0) - \frac{d}{dt}f(0)$$

where $df(0)/dt$ is the value of the first derivative at $t = 0$. Examples of the Laplace transforms of derivatives are given in chapter 10.

Integrals
The Laplace transform of the integral of a function $f(t)$ which has a Laplace transform $F(s)$ is given by

$$\mathcal{L}\left\{\int_0^t f(t)\, dt\right\} = \frac{1}{s}F(s)$$

For example, the Laplace transform of the integral of the function e^{-t} between the limits 0 and t is given by

$$\mathcal{L}\left\{\int_0^t e^{-t}\,dt\right\} = \frac{1}{s}\,\mathcal{L}\left\{e^{-t}\right\} = \frac{1}{s(s+1)}$$

A.4 The inverse transform

The inverse Laplace transformation is the conversion of a Laplace transform $F(s)$ into a function of time $f(t)$. This operation can be written as

$$\mathcal{L}^{-1}\{F(s)\} = f(t)$$

The inverse operation can generally be carried out by using table Ap.A.1. The linearity property of Laplace transforms means that if we have a transform as the sum of two separate terms then we can take the inverse of each separately and the sum of the two inverse transforms is the required inverse transform.

$$\mathcal{L}^{-1}\{aF(s) + bG(s)\} = a\mathcal{L}^{-1}F(s) + b\mathcal{L}^{-1}G(s)$$

Thus, to illustrate how rearrangement of a function can often put it into the standard form shown in the table, the inverse transform of $3/(2s + 1)$ can be obtained by rearranging it as

$$\frac{3(1/2)}{s+(1/2)}$$

The table (item 7) contains the transform $1/(s + a)$ with the inverse of e^{-at}. Thus the inverse transformation is just this multiplied by the constant $(3/2)$ with $a = (1/2)$, i.e. $(3/2)e^{-t/2}$.

As another example, consider the inverse Laplace transform of $(2s + 2)/(s^2 + 1)$. This expression can be rearranged as

$$2\left[\frac{s}{s^2+1} + \frac{1}{s^2+1}\right]$$

The first term in the brackets has the inverse transform of $\cos t$ (item 17 in table Ap.A.1) and the second term $\sin t$ (item 16 in table Ap.A.1). Thus the inverse transform of the expression is

$$2\cos t + 2\sin t$$

A.4.1 Partial fractions

Often $F(s)$ is a ratio of two polynomials and cannot be readily identified with a standard transform in table Ap.A.1. It has to be converted into simple fraction terms before the standard transforms can be used. The process of converting an expression into simple fraction terms is called decomposing into *partial fractions*. This technique can be used provided the degree of the numerator is less than the degree of the denominator, i.e. in the above equation n is less than m.   The degree of a polynomial is the highest power of s in the expression. When the degree of the numerator is equal to or higher than that of the denominator, the denominator must be divided into the numerator until the result is the sum of terms with the remainder fractional term having a numerator of lower degree than the denominator.

We can consider there to be basically three types of partial fractions:

1 The denominator contains factors which are only of the form $(s + a)$, $(s + b)$, $(s + c)$, etc. The expression is of the form

$$\frac{f(s)}{(s+a)(s+b)(s+c)}$$

and has the partial fractions of

$$\frac{A}{(s+a)} + \frac{B}{(s+b)} + \frac{C}{(s+c)}$$

2 There are repeated $(s + a)$ factors in the denominator, i.e. the denominator contains powers of such a factor, and the expression is of the form

$$\frac{f(s)}{(s+a)^n}$$

This then has partial fractions of

$$\frac{A}{(s+a)^1} + \frac{B}{(s+a)^2} + \frac{C}{(s+a)^3} + \cdots + \frac{N}{(s+a)^n}$$

3 The denominator contains quadratic factors and the quadratic does not factorise without imaginary terms. For an expression of the form

$$\frac{f(s)}{(as^2 + bs + c)(s + d)}$$

the partial fractions are

$$\frac{As+B}{as^2 + bs + c} + \frac{C}{s+d}$$

The values of the constants A, B, C, etc. can be found by either making use of the fact that the equality between the expression and the partial fractions must be true for all values of s or that the coefficients of s^n in the expression must equal those of s^n in the partial fraction expansion. The use of the first method is illustrated by the following example where the partial fractions of

$$\frac{3s+4}{(s+1)(s+2)} \text{ are } \frac{A}{s+1} + \frac{B}{s+2}$$

Then, for the expressions to be equal, we must have

$$\frac{3s+4}{(s+1)(s+2)} = \frac{A(s+2)+B(s+1)}{(s+1)(s+2)}$$

and consequently

$$3s+4 = A(s+2)+B(s+1)$$

This must be true for all values of s. The procedure is then to pick values of s that will enable some of the terms involving constants to become zero and so enable other constants to be determined. Thus if we let $s = -2$ then we have

$$3(-2)+4 = A(-2+2)+B(-2+1)$$

and so $B = 2$. If we now let $s = -1$ then

$$3(-1)+4 = A(-1+2)+B(-1+1)$$

and so $A = 1$. Thus

$$\frac{3s+4}{(s+1)(s+2)} = \frac{1}{s+1} + \frac{2}{s+2}$$

Appendix B
6800/6802 instruction set

B.1 Motorola 6800/6802 architecture

The Motorola 6802 has the same basic architecture and instructions as the 6800 but includes a built-in clock generator and a small amount of chip RAM. Figure Ap.B.1 shows the basic

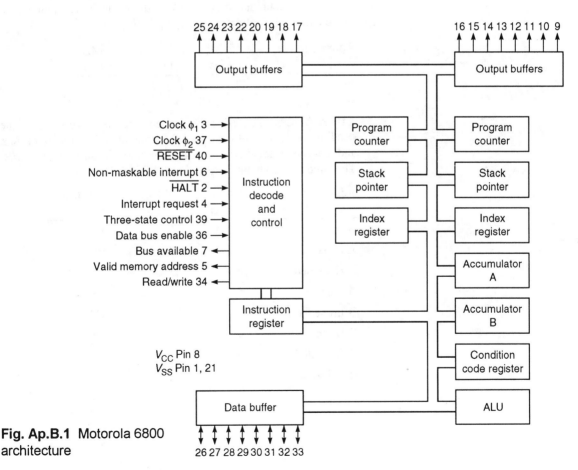

Fig. Ap.B.1 Motorola 6800 architecture

architecture. The register set is:

1 Two 8-bit accumulators ACCA and ACCB
2 A 16-bit index register X
3 A 16-bit stack pointer
4 A status register with the flags:
C = carry, bit 0
V = overflow, bit 1
Z = zero indicator, bit 2
N = negative, bit 3
I = interrupt mask, bit 4
H = half carry, bit 5
5 A 16-bit program counter

B.2 Instruction set

The following, in alphabetical sequence, are the instruction set for the Motorola 6800/6802 microprocessor.

ABA This instruction adds the contents of accumulator B to the contents of accumulator A and place the result in accumulator A. The condition bits affected are H, N, Z, V and C.

Address mode	No. of cycles	No. of bytes	Machine code
Inherent	2	1	1B

ADCA/ADCB This adds the contents of the carry bit C in the condition register to the sum of the contents of accumulator A/B and a memory location and place the result in accumulator A/B. The condition code bits affected are H, N, Z, V and C.

Address mode	No. of cycles	No. of bytes	Machine code
A Immediate	2	2	89
A Direct	3	2	99
A Extended	4	3	B9
A Indexed	5	2	A9
B Immediate	2	2	C9
B Direct	3	2	D9
B Extended	4	3	F9
B Indexed	5	2	E9

ADDA/ADDB This adds the contents of accumulator A/B and the contents of a memory location and place the result in accumulator A/B. The condition code bits affected are H, N, Z, V and C.

Address mode	No. of cycles	No. of bytes	Machine code
A Immediate	2	2	8B
A Direct	3	2	9B
A Extended	4	3	BB
A Indexed	5	2	AB
B Immediate	2	2	CB
B Direct	3	2	DB
B Extended	4	3	FB
B Indexed	5	2	EB

AND This instruction carries out the logical AND between the contents of accumulator A/B and the contents of the memory and places the result in accumulator A/B. The condtion code bits affected are N and Z, the V bit being cleared.

Address mode	No. of cycles	No. of bytes	Machine code
A Immediate	2	2	84
A Direct	3	2	94
A Extended	4	3	B4
A Indexed	5	2	A4
B Immediate	2	2	C4
B Direct	3	2	D4
B Extended	4	3	F4
B Indexed	5	2	E4

ASL(A/B) This shifts all the bits of accumulator A/B or memory one place to the left. The C bit is loaded from the most significant bit. The condition code bits affected are N and Z. The V bit is set, if, after completion of the shift, either N is set and C is cleared or N is cleared and C is set, otherwise it is cleared.

Address mode	No. of cycles	No. of bytes	Machine code
ASLA	2	1	48
ASLB	2	1	58
ASL Direct	6	3	78
ASL Indexed	7	2	68

ASR(A/B) This shifts all the bits of accumulator A/B or memory one place to the right. The least significant bit is loaded into the C bit. The condition bits affected are N and Z. The V bit is set, if, after completion of the shift, either N is set and C is cleared or N is cleared and C is set, otherwise it is cleared.

Address mode	No. of cycles	No. of bytes	Machine code
ASRA	2	1	47
ASRB	2	1	57
ASR Direct	6	3	77
ASR Indexed	7	2	67

BCC This results in a branch if the C bit is clear. The condition code bits are not affected.

Address mode	No. of cycles	No. of bytes	Machine code
Relative	4	2	24

BCS This results in a branch if the C bit is set. The condition code bits are not affected.

Address mode	No. of cycles	No. of bytes	Machine code
Relative	4	2	25

BEQ This results in a branch if the Z bit is set. The branch is if equal. The condition code bits are not affected.

Address mode	No. of cycles	No. of bytes	Machine code
Relative	4	2	27

BGE This results in a branch if N is set and V is set or N is clear and V is clear. The branch is if greater than or equal to zero. The condition code bits are not affected.

Address mode	No. of cycles	No. of bytes	Machine code
Relative	4	2	2C

BGT This results in a branch if Z is clear and N is set and V is set or N is clear and V is clear. The branch is if greater than zero. The condition code bits are not affected.

Address mode	No. of cycles	No. of bytes	Machine code
Relative	4	2	2E

BHI This results in a branch if executed immediately after execution of any of the instructions CBA, CMP, SUB and the unsigned number represented by the minuend was greater than the unsigned number represented by the subtrahend. The branch is if higher. The condition code bits are not affected.

Address mode	No. of cycles	No. of bytes	Machine code
Relative	4	2	22

BITA/B This performs the logical AND comparison of the contents of ACCA/B and the contents of the memory and modifies flags accordingly. It is is termed the bit test. The condition code bit N is set if the most significant bit of the result of the AND would be set, otherwise it is cleared. The Z bit is set if all the bits of the AND would be cleared, otherwise it is cleared. The V bit is cleared.

Address mode	No. of cycles	No. of bytes	Machine code
A Immediate	2	2	85
A Direct	3	2	95
A Extended	4	3	B5
A Indexed	5	2	A5
B Immediate	2	2	C5
B Direct	3	2	D5
B Extended	4	3	F5
B Indexed	5	2	E5

BLE This causes a branch if Z is set or N is set and V is clear or N is clear and V is set. This is branch if less than or equal to zero. The condition code bits are not affected.

Address mode	No. of cycles	No. of bytes	Machine code
Relative	4	2	2F

BLS This causes a branch if C is set or Z is set. This is branch if lower or the same. The condition code bits are not affected.

Address mode	No. of cycles	No. of bytes	Machine code
Relative	4	2	23

BLT This causes a branch if N is set and V is clear or N is clear and V is set. It is branch if less than zero. The condition code bits are not affected.

Address mode	No. of cycles	No. of bytes	Machine code
Relative	4	2	2D

BMI This causes a branch if N is set. It is branch if minus. The condition code bits are not affected.

Address mode	No. of cycles	No. of bytes	Machine code
Relative	4	2	2B

BNE This causes a branch if the Z bit is clear. It is branch if not equal. The condition code bits are not affected.

Address mode	No. of cycles	No. of bytes	Machine code
Relative	4	2	26

BPL This causes a branch if N is clear. It is branch if plus. The condition code bits are not affected.

Address mode	No. of cycles	No. of bytes	Machine code
Relative	4	2	2A

BRA This is an unconditional branch. It is branch always. The condition code bits are not affected.

Address mode	No. of cycles	No. of bytes	Machine code
Relative	4	2	20

BSR This is branch to a subroutine. The condition code bits are not affected.

Address mode	No. of cycles	No. of bytes	Machine code
Relative	8	2	8D

BVC This causes a branch if V is clear. It is branch if overflow clear. The condition code bits are not affected.

Address mode	No. of cycles	No. of bytes	Machine code
Relative	4	2	28

BVS This causes a branch if V is set. It is branch if overflow set. The condition code bits are not affected.

Address mode	No. of cycles	No. of bytes	Machine code
Relative	4	2	29

CBA   This compares the contents of accumulator A and accumulator B and sets the flags. The condition code bits affected are N, Z and V. The C bit is clear if the subtraction requires a borrow in the most significant bit of the result, otherwise it is cleared.

Address mode	No. of cycles	No. of bytes	Machine code
Inherent	2	1	11

CLC This clears the carry bit. The other condition code bits are not affected.

Address mode	No. of cycles	No. of bytes	Machine code
Inherent	2	1	0C

CLI This clears the interrupt mask flag. The other condition code bits are not affected.

Address mode	No. of cycles	No. of bytes	Machine code
Inherent	2	1	0E

CLR(A/B) This causes the contents of accumulator A/B or memory to be replaced by zeros. This clears the condition code bits N, V and C and sets the Z bit.

Address mode	No. of cycles	No. of bytes	Machine code
CLRA	2	1	4F
CLRB	2	1	5F
CLR Extended	6	3	7F
CLR Indexed	7	2	6F

CLV This clears the twos complement overflow bit in the condition codes register. This clears the V condition code bit, the other bits not being affected.

Address mode	No. of cycles	No. of bytes	Machine code
Inherent	2	1	0A

CMPA/B This compares the contents of accumulator A or B and the contents of the memory and determines the condition codes. The condition code bit N is set if the most significant bit of the result of the subtraction would be set, otherwise it is cleared. The Z bit is set if all the bits of the result would be cleared, otherwise it is cleared. The V bit is set if the result would cause twos complement overflow, otherwise it is cleared. The C bit is set if the absolute value of the contents of a specified memory location is larger than the absolute value of the selected accumulator, otherwise it is cleared.

Address mode	No. of cycles	No. of bytes	Machine code
A Immediate	2	2	81
A Direct	3	2	91
A Extended	4	3	B1
A Indexed	5	2	A1
B Immediate	2	2	C1
B Direct	3	2	D1
B Extended	4	3	F1
B Indexed	5	2	E1

COM(A/B) This replaces the contents of accumulator A or B or memory with its ones complement, i.e. each bit is replaced with the complement of that bit. This clears the V condition code bit and sets the C bit.

Address mode	No. of cycles	No. of bytes	Machine code
COMA Implied	2	1	43
COMB Implied	2	1	53
COM Extended	6	3	73
COM Indexed	7	2	63

CPX This compares the more significant byte of the contents of the index register with the contents of the byte of memory at the address specified by the instruction. The less significant byte of the contents of the index register is compared with the contents of the next byte of memory, at one plus the address specified by the instruction. CPX stands for compare index register. The N condition code bit is set if the most significant bit of the result of the subtraction would be set, otherwise it is cleared. The Z bit is set if all the bits of the results of both subtractions would be cleared, otherwise it is cleared. The V bit is set if the subtraction from the most significant byte of the index register would cause twos complement overflow, otherwise it is cleared.

Address mode	No. of cycles	No. of bytes	Machine code
Immediate	3	3	8C
Direct	4	2	9C
Extended	5	3	BC
Indexed	6	2	AC

DAA This instruction results in the hexadecimal numbers 00, 06, 60, or 66 being added to accumulator A and may also set the C bit. DAA stands for decimal adjust ACCA. Condition code bits affected are N, Z and V. The C bit is set according to the following table:

State of C bit before DAA	Upper half-byte of A (bits 4-7)	Initial half-carry H bit	Lower half-byte of A (bits 0-3)	Number added after DAA	State of C bit after DAA
0	0-9	0	0-9	00	0
0	0-8	0	A-F	06	0
0	0-9	1	0-3	06	0
0	A-F	0	0-9	60	1
0	9-F	0	A-F	66	1
0	A-F	1	0-3	66	1
1	0-2	0	0-9	60	1
1	0-2	0	A-F	66	1
1	0-3	1	0-3	66	1

Address mode	No. of cycles	No. of bytes	Machine code
Inherent	2	1	19

DEC(A/B) This subtracts one from the contents of accumulator A, acumulator B or memory, i.e. decrement. The condition code bits affected are N and Z. The V bit is set if there was a twos complement overflow as a result of the operation, otherwise it is cleared.

Address mode	No. of cycles	No. of bytes	Machine code
DECA	2	1	4A
DECB	2	1	5A
DEC extended	6	3	7A
DEC indexed	7	2	6A

DES This subtracts one from the stack pointer, i.e. decrement stack pointer. The condition code bits are not affected.

Address mode	No. of cycles	No. of bytes	Machine code
Inherent	4	1	34

DEX This subtracts one from the index register, i.e. decrement index register. The Z condition code bit is affected.

Address mode	No. of cycles	No. of bytes	Machine code
Inherent	4	1	09

EORA/B This carries out logical EXCLUSIVE-OR between the contents of accumulator A, or accumulator B, and the contents of memory and place the result in accumulator A, or accumulator B. Each bit of the accumulator A/B after the operation will be the EXCLUSIVE-OR of the corresponding bit of the memory and accumulator A/B. The condition code bits affected are N and Z. The V bit is cleared.

Address mode	No. of cycles	No. of bytes	Machine code
EORA Immediate	2	2	88
EORA Direct	3	2	98
EORA Extended	4	3	B8
EORA Indexed	5	2	A8
EORB Immediate	2	2	C8
EORB Direct	3	2	D8
EORB Extended	4	3	F8
EORB Indexed	5	2	E8

INC(A/B) This instruction adds one to the contents of accumulator A, accumulator B or memory. INC stands for increment. The condition code bits affected are N, Z and V.

Address mode	No. of cycles	No. of bytes	Machine code
INCA	2	1	4C
INCB	2	1	5C
INC Extended	6	3	7C
INC Indexed	7	2	6C

INS This adds one to the stack pointer. The condition code bits are not affected.

Address mode	No. of cycles	No. of bytes	Machine code
Inherent	4	1	31

INX This adds one to the index register. The Z condition code bit is affected.

Address mode	No. of cycles	No. of bytes	Machine code
Inherent	4	1	08

JMP This causes a jump to occur to the instruction stored at the numerical address. The condition code bits are not affected.

Address mode	No. of cycles	No. of bytes	Machine code
Extended	3	3	7E
Indexed	4	2	6E

JSR This is a jump to a subroutine instruction stored at the indicated numerical address. The condition code bits are not affected.

Address mode	No. of cycles	No. of bytes	Machine code
Extended	9	3	BD
Indexed	8	2	AD

LDAA/B This is load accumulator A or B with the contents of the memory address indicated. The condition code bits affected are N and Z. The V bit is cleared.

Address mode	No. of cycles	No. of bytes	Machine code
LDAA Immediate	2	2	86
LDAA Direct	3	2	96
LDAA Extended	4	3	B6
LDAA Indexed	5	2	A6
LDAB Immediate	2	2	C6
LDAB Direct	3	2	D6
LDAB Extended	4	3	F6
LDAB Indexed	5	2	E6

LDS This is the instruction to load the stack pointer from the byte of memory at the address specified by the program. The N condition code bit is set if the most significant bit of the stack pointer is set by the operation, otherwise it is cleared. The Z bit is set if all the bits of the stack pointer are cleared by the operation, otherwise it is cleared. The V bit is cleared.

Address mode	No. of cycles	No. of bytes	Machine code
Immediate	3	3	8E
Direct	4	2	9E
Extended	5	3	BE
Indexed	6	2	AE

LDX This is the instruction to load the index register from the byte of memory at the address specified by the program. The N condition code bit is set if the most significant bit of the index register is set by the operation, otherwise it is cleared. The Z bit is set if all the bits of the index register are cleared. The V bit is cleared.

Address mode	No. of cycles	No. of bytes	Machine code
Immediate	3	3	CE
Direct	4	2	DE
Extended	5	3	FE
Indexed	6	2	EE

LSR(A/B) This shifts all bits of accumulator A/B or memory one place to the right. Bit 7 is loaded with a zero and the C bit is loaded from the least significant bit of accumulator A/B or memory. The condition code bit N is cleared. The Z bit is set if all the bits of the result are cleared, otherwise it is cleared. The V bit is set if, after completion of the shift, either N is set and C cleared or N is cleared and C set, otherwise it is cleared.

Address mode	No. of cycles	No. of bytes	Machine code
LSRA	2	1	44
LSRB	2	1	54
LSR Extended	6	3	74
LSR Indexed	7	2	64

NEG(A/B) This instruction replaces the contents of accumulator A/B or memory with its twos complement. The N and Z condition code bits are affected. The V condition code bit is set if there is a twos complement overflow as a result of the implied subtraction from zero. The C bit is set if there would be a borrow in the implied subtraction. The C bit is set in all cases except when the contents of the selected accumulator or memory is zero.

Address mode	No. of cycles	No. of bytes	Machine code
NEGA	2	1	40
NEGB	2	1	50
NEG Extended	6	3	70
NEG Indexed	7	2	60

NOP This instruction causes only the program counter to be incremented. No other registers are affected. The condition code bits are not affected.

Address mode	No. of cycles	No. of bytes	Machine code
Inherent	2	1	01

ORAA/B This carries out the logical OR between the contents of accumulator A/B and the contents of the memory and places the result in the same accumulator. Each bit of the accumulator after the operation will be the logical OR of the corresponding bits of the memory and the accumulator before the operation. The N and Z condition code bits are affected. The V bit is cleared.

Address mode	No. of cycles	No. of bytes	Machine code
ORAA Immediate	2	2	8A
ORAA Direct	3	2	9A
ORAA Extended	4	3	BA
ORAA Indexed	5	2	AA
ORAB Immediate	2	2	CA
ORAB Direct	3	2	DA
ORAB Extended	4	3	FA
ORAB Indexed	5	2	EA

PSHA/B This stores the contents of accumulator A/B in the stack at the address contained in the stack pointer. The stack pointer is then decremented. The condition code bits are not affected.

Address mode	No. of cycles	No. of bytes	Machine code
PSHA	4	1	36
PSHB	4	1	37

PULA/B This causes the stack pointer to be incremented. Accumulator A/B is then loaded from the stack, from the address contained in the stack pointer. The condition code bits are not affected.

Address mode	No. of cycles	No. of bytes	Machine code
PULA	4	1	32
PULB	4	1	33

ROL(A/B) This is the rotate left instruction. All the bits of accumulator A/B or memory are shifted one place to the left. Bit 0 is loaded from the C bit. The C bit is loaded from the most significant bit of accumulator A/B or memory. The N and Z condition code bits are affected. The V bit is set if, after completion of the shift, either N is set and C is cleared or N is cleared and C is set, otherwise it is cleared.

Address mode	No. of cycles	No. of bytes	Machine code
ROLA	2	1	49
ROLB	2	1	59
ROL Extended	6	3	79
ROL Indexed	7	2	69

ROR(A/B) This is the rotate right instruction. All the bits of accumulator A/B or memory are shifted one place to the right. Bit 7 is loaded from the C bit. The C bit is loaded from the least significant bit of accumulator A/B or memory. The N and Z condition code bits are affected. The V bit is set if, after completion of the shift, either N is set and C is cleared or N is cleared and C is set, otherwise it is cleared.

Address mode	No. of cycles	No. of bytes	Machine code
RORA	2	1	46
RORB	2	1	56
ROR Extended	6	3	76
ROR Indexed	7	2	66

RTI This is the instruction for return from interrupt. The condition codes, accumulators A and B, the index register and the program counter will all be restored to a state pulled from the stack. The condition code bits are restored to the states pulled from the stack.

Address mode	No. of cycles	No. of bytes	Machine code
Inherent	10	1	3B

RTS This is the instruction for return from subroutine. The stack pointer is incremented by 1 and then the contents of the byte of memory at the resulting stack pointer address is loaded into the 8 bits of highest significance in the program counter. The stack pointer is then incremented again by 1 and the contents of the byte of memory at the resulting stack pointer address is loaded into the 8 bits of lowest significance in the program counter. The condition code bits are not affected.

Address mode	No. of cycles	No. of bytes	Machine code
Inherent	5	1	39

SBA This causes the contents of accumulator B to be subtracted from the contents of accumulator A. The result is placed in accumulator A. The contents of accumulator B remain unaffected. The condition code bits affected are N, Z, V and C.

Address mode	No. of cycles	No. of bytes	Machine code
Inherent	2	1	10

SBCA/B This causes the contents of the memory and the carry to be subtracted from the contents of accumulator A/B and the result placed in accumulator A/B. The condition code bits affected are N, Z and V. The C bit is set if the absolute value of the contents of the memory plus the previous carry is larger than the absolute value of the contents of accumulator A/B, otherwise it is cleared.

Address mode	No. of cycles	No. of bytes	Machine code
SBCA Immediate	2	2	82
SBCA Direct	3	2	92
SBCA Extended	4	3	B2
SBCA Indexed	5	2	A2
SBCB Immediate	2	2	C2
SBCB Direct	3	2	D2
SBCB Extended	4	3	F2
SBCB Indexed	5	2	E2

SEC This instruction sets the carry bit in the conditions code register. The C bit is set with all other condition code bits unaffected.

Address mode	No. of cycles	No. of bytes	Machine code
Inherent	2	1	0D

SEI This instruction sets the interrupt mask in the condition codes register. The microprocessor will continue carrying out the instructions of the program until the interrupt mask has been cleared. The I bit is set with all other condition code bits un-affected.

Address mode	No. of cycles	No. of bytes	Machine code
Inherent	2	1	0F

SEV This instruction sets the twos complement bit in the condition codes register. The V bit is set with all other condition code bits unaffected.

Address mode	No. of cycles	No. of bytes	Machine code
Inherent	2	1	0B

STAA/B This stores the contents of accumulator A/B in memory, the contents of the accumulator remaining unchanged. The N and Z condition code bits are affected. The V bit is cleared.

Address mode	No. of cycles	No. of bytes	Machine code
STAA Direct	4	2	97
STAA Extended	5	3	B7
STAA Indexed	6	2	A7
STBB Direct	4	2	D7
STBB Extended	5	3	F7
STAB Indexed	6	2	E7

STS This instruction stores the more significant byte of the stack pointer in memory at the address specified by the program and stores the less significant byte of the stack pointer at the next location in memory at one plus the address specified. The N condition code bit is set if the most significant bit of the stack pointer is set, otherwise it is cleared. The Z bit is set if all bits of the stack pointer are cleared, otherwise it is cleared. The V bit is cleared.

Address mode	No. of cycles	No. of bytes	Machine code
STS Direct	5	2	9F
STS Extended	6	3	BF
STS Indexed	7	2	AF

STX This stores the more significant byte of the index register in memory at the address specified by the program and stores the less significant byte of the index register at the next location in memory at one plus the address specified. The N condition code bit is set if the most significant bit of the index register is set, otherwise it is cleared. The Z bit is set if all bits of the index register are cleared, otherwise it is cleared. The V bit is cleared.

Address mode	No. of cycles	No. of bytes	Machine code
STX Direct	5	2	DF
STX Extended	6	3	FF
STX Indexed	7	2	EF

SUBA/B This subtracts the contents of memory from the contents of accumulator A or B and places the result in accumulator A/B. The condition code bits affected are N, Z and V. The C bit is set if the absolute value of the contents of the memory is larger than the absolute value of the contents of accumulator A/B, otherwise it is cleared.

Address mode	No. of cycles	No. of bytes	Machine code
SUBA Immediate	2	2	80
SUBA Direct	3	2	90
SUBA Extended	4	3	B0
SUBA Indexed	5	2	A0
SUBB Immediate	2	2	C0
SUBB Direct	3	2	D0
SUBB Extended	4	3	F0
SUBB Indexed	5	2	E0

SWI This is the software interrupt instruction. The program counter is incremented by 1. The program counter, index register, and accummulators A and B, are pushed into the stack. The condition codes register is then pushed into the stack. The stack pointer is decremented by 1 after each byte of data is stored in the stack. The interrupt mask bit is then set and the program counter loaded with the address stored in the software interrupt pointer. The I condition code bit is set.

Address mode	No. of cycles	No. of bytes	Machine code
Inherent	12	1	3F

TAB This instruction transfers the contents of accumulator A into accumulator B. The contents of accumulator A are not affected but the contents of B are lost. The N and Z condition code bits are affected. The V bit is cleared.

Address mode	No. of cycles	No. of bytes	Machine code
Inherent	2	1	16

TAP This instruction transfers the contents of bit positions 0 to 5 of accumulator A to the corresponding bit positions of the condition codes register. Bit 0 to C, bit 1 to V, bit 2 to Z, bit 3 to N, bit 4 to I and bit 5 to H. The contents of accumulator A remain unchanged.

Address mode	No. of cycles	No. of bytes	Machine code
Inherent	2	1	06

TBA This instruction transfers the contents of accumulator B to accumulator A. The contents of accumulator B are not affected but those of A are lost. The N and Z condition code bits are affected. The V bit is cleared.

Address mode	No. of cycles	No. of bytes	Machine code
Inherent	2	1	17

TPA This transfers the contents of the condition codes register to corresponding bit positions 0 to 5 in accumulator A. C goes to bit 0, V to bit 1, Z to bit 2, N to bit 3, I to bit 4 and H to bit 5. Bits 6 and 7 are set as 1s. The condition code bits are not affected.

Address mode	No. of cycles	No. of bytes	Machine code
Inherent	2	1	07

TST(A/B) This instruction, test, sets the condition codes N and Z according to the contents of accumulator A/B or memory. The N and Z condition code bits are affected. The V and C bits are cleared.

Address mode	No. of cycles	No. of bytes	Machine code
TSTA	2	1	4D
TSTB	2	1	5D
TST Extended	6	3	7D
TST Indexed	7	2	6D

TSX This instruction loads the index register with one plus the contents of the stack pointer, the contents of the stack pointer remaining unchanged. The condition code bits are not affected.

Address mode	No. of cycles	No. of bytes	Machine code
Inherent	4	1	30

TXS This instruction loads the stack pointer with the contents of the index register minus one. The contents of the index register remain unchanged. The condition code bits are not affected.

Address mode	No. of cycles	No. of bytes	Machine code
Inherent	4	1	35

WAI This is the instruction for wait for interrupt. The program counter is increment by 1. The program counter, index register, accumulators A and B, are pushed into the stack. The condition codes register is then pushed into the stack with the condition codes going into bits 0 to 5 and bits 6 and 7 set to 1. The stack pointer is decremented by 1 after each byte of data is stored in the stack. Execution of the program is then suspended until an interrupt from a peripheral is signalled by the interrupt request control input going to a low state.

When an interrupt is signalled, and provided the I bit is clear, the interrupt mask bit is set. The program is then loaded with the address stored in the internal interrupt pointer.

Address mode	No. of cycles	No. of bytes	Machine code
Inherent	9	1	3E

Appendix C
Number systems

C.1 Number systems

The *decimal system* is based on the use of ten symbols or digits: 0, 1, 2, 3, 4, 5, 6, 7, 8, 9. When a number is represented by this system, the digit position in the number indicates that the weight attached to each digit increases by a factor of 10 as we proceed from right to left.

...	10^3	10^2	10^1	10^0
	thousands	hundreds	tens	units

The *binary system* is based on just two symbols or states: 0 and 1. These are termed *binary digits* or *bits*. When a number is represented by this system, the digit position in the number indicates that the weight attached to each digit increases by a factor of 2 as we proceed from right to left.

...	2^3	2^2	2^1	2^0
	bit 3	bit 2	bit 1	bit 0

For example, the decimal number 15 in the binary system is 1111. In a binary number the bit 0 is termed the *least significant bit* (LSB) and the highest bit the *most significant bit* (MSB).

The *octal system* is based on eight digits: 0, 1, 2, 3, 4, 5, 6, 7. When a number is represented by this system, the digit position in the number indicates that the weight attached to each digit increases by a factor of 8 as we proceed from right to left.

...	8^3	8^2	8^1	8^0

For example, the decimal number 15 in the octal system is 17.

The *hexadecimal system* is based on 16 digits/symbols: 0, 1, 2, 3, 4, 5, 6, 7, 8, 9, A, B, C, D, E, F. When a number is represented by this system, the digit position in the number indicates that the weight attached to each digit increases by a factor of 16 as we proceed from right to left.

$$... \qquad 16^3 \qquad 16^2 \qquad 16^1 \qquad 16^0$$

For example, the decimal number 15 is F in the hexadecimal system.

The *binary coded decimal system* (BCD) is a widely used system with computers. Each decimal digit is coded separately in binary. For example, the decimal number 15 in BCD is 0001 0101. However, before a computer will operate on such a number, it will convert it to binary. It does its arithmetic in binary. It can then transform the outcome into BCD to give an output in this form.

Table Ap.C.1 gives examples of numbers in the above systems.

Table Ap.C.1 Number systems

Decimal	Binary	BCD	Octal	Hexadecimal
0	0000	0000 0000	0	0
1	0001	0000 0001	1	1
2	0010	0000 0010	2	2
3	0011	0000 0011	3	3
4	0100	0000 0100	4	4
5	0101	0000 0101	5	5
6	0110	0000 0110	6	6
7	0111	0000 0111	7	7
8	1000	0000 1000	10	8
9	1001	0000 1001	11	9
10	1010	0001 0000	12	A
11	1011	0001 0001	13	B
12	1100	0001 0010	14	C
13	1101	0001 0011	15	D
14	1110	0001 0100	16	E
15	1111	0001 0101	17	F

C.2 Binary mathematics

Addition of binary numbers follows the following rules:

$$0 + 0 = 0$$

$$0 + 1 = 1 + 0 = 1$$

$$1 + 1 = 10 \qquad \text{i.e. } 0 + \text{carry } 1$$

$$1 + 1 + 1 = 11 \qquad \text{i.e. } 1 + \text{carry } 1$$

With an 8-bit microprocessor we are limited to a binary number of 11111111, i.e. decimal number of 255. Thus, if in adding two binary numbers we end with a number which is greater than this, then we have a carry which is placed in a separate register. The following example illustrates this. In decimal numbers the addition of 14 and 19 gives 33. In binary numbers this addition becomes

Augend	01110
Addend	10111
Sum	100001

For bit 0, $0 + 1 = 1$. For bit 1, $1 + 1 = 10$ and so we have 0 with 1 carried to the next column. For bit 3, $1 + 0 +$ carried $1 = 10$. For bit 4, $1 + 0 +$ carried $1 = 10$. We continue this through the various bits and end up with the sum plus a carry 1. The final number is thus 100001. When adding binary numbers A and B to give C, i.e. $A + B = C$, then A is termed the *augend*, B the *addend* and C the *sum*.

Subtraction of binary numbers follows the following rules:

$$0 - 0 = 0$$

$$1 - 0 = 1$$

$$1 - 1 = 0$$

$$0 - 1 = 10 - 1 + \text{borrow} = 1 + \text{borrow}$$

When evaluating $0 - 1$, a 1 is borrowed from the next column on the left containing a 1. The following example illustrates this. In decimal numbers the subtraction of 14 from 27 gives 13.

Minuend	11011
Subtrahend	01110
Difference	01101

For bit 0 we have $1 - 0 = 1$. For bit 1 we have $1 - 1 = 0$. For bit 2 we have $0 - 1$. We thus borrow 1 from the next column and so have $10 - 1 = 1$. For bit 3 we have $0 - 1$, remember we borrowed the 1. Again borrowing 1 from the next column, we then have $10 - 1 = 1$. For bit 4 we have $0 - 0 = 0$, remember we borrowed the 1. When subtracting binary numbers A and B to give C, i.e. we have $A - B = C$, then A is termed the *minuend*, B the *subtrahend* and C the *difference*.

The subtraction of binary numbers is more easily carried out electronically when an alternative method of subtraction is used. The subtraction example above can be considered to be the addition of a positive number and a negative number. The following techniques indicates how we can specify negative numbers and so turn subtraction into addition. It also enables us to deal with negative numbers in any circumstances.

The numbers used so far are referred to as *unsigned*. This is because the number itself contains no indication whether it is negative or positive, the sign symbol is separate from the number. When a number is said to be *signed* then the most significant bit is used to indicate the sign of the number, a 0 being used if the number is positive and a 1 if it is negative. When we have a positive number then we write it in the normal way with a 0 preceding it. Thus a positive binary number of 10010 would be written as 010010. A negative number of 10010 would be written as 110010. However this is not the most useful way of representing negative numbers for ease of manipulation by computers.

A more useful way of representing negative numbers is to use the twos complement method. A binary number has two complements, known as the *ones complement* and the *twos complement*. The ones complement of a binary number is obtained by changing all the 1s in the signed number into 0s and the 0s into 1s. The twos complement is then obtained by adding 1 to the ones complement. When we have a negative number then we obtain the twos complement and then sign it with a 1. Consider the representation of the decimal number −6 as a signed twos complement number. We first write the binary number for +6, i.e. 000 0110, then obtain the ones complement of 111 1001, then add 1 to give 111 1010, and finally sign it with a 1 to indicate it is negative. The result is thus 1111 1010.

Unsigned binary number when sign ignored	000 0110
Ones complement	111 1001
Add 1	1
Unsigned twos complement	111 1010
Signed twos complement	1111 1010

Table Ap.C.2 Signed numbers

Decimal number	Signed number	Comment
+127	0111 1111	Positive numbers are
⋮	⋮	just the same as the
+6	0000 0110	binary number with a 0
+5	0000 0101	as the sign at the left
+4	0000 0100	end of the number.
+3	0000 0011	
+2	0000 0010	
+1	0000 0001	
+0	0000 0000	
−1	1111 1111	Negative numbers are
−2	1111 1110	the twos complement
−3	1111 1101	form with a 1 as the
−4	1111 1100	sign at the left end of
−5	1111 1011	the number.
−6	1111 1010	
⋮	⋮	
−128	1000 0000	

When we have a positive number then we write it in the normal way with a 0 preceding it. Thus a positive binary number of 100 1001 would be written as 01001001. Table Ap.C.2 shows some examples of numbers on this system.

Subtraction of a positive number from a positive number involves obtaining the signed twos complement of the subtrahend and then adding it to the signed minuend. Hence, for the subtraction of the decimal number 6 from the decimal number 4 we have

Signed minuend	0000 0100
Subtrahend, signed twos complement	1111 1010
Sum	1111 1110

The most significant bit of the outcome is 1 and so the result is negative. This is the signed twos complement for −2.

Consider another example, the subtraction of 43 from 57. The signed positive number of 57 is 0011 1001. The signed twos complement for −43 is given by:

Unsigned binary number for 43 when negative sign ignored	010 1011
Ones complement	101 0100
Add 1	1
Unsigned twos complement	101 0101
Signed twos complement	1101 0101

Thus we obtain by the addition of the signed positive number and the signed twos complement number:

Signed minuend	0011 1001
Subtrahend, signed twos complement	1101 0101
Sum	0000 1110 + carry 1

The carry 1 is ignored. The result is thus 0000 1110 and since the most significant bit is 0 the result is positive. The result is the decimal number 14.

If we wanted to add two negative numbers then we would obtain the signed twos complement for each number and then add them. Whenever a number is negative we use the signed twos complement, when positive just the signed number.

Answers

The following are answers to numerical problems and brief clues as to possible answers with descriptive problems.

Chapter 1

1 (a) Sensor – mercury, signal conditoner – fine bore stem, display – marks on the stem, (b) sensor – curved tube, signal conditioner – gears, display – pointer moving across a scale.

2 See text

3 (a) Comparison/controller – thermostat, correction – perhaps a relay, process – heat, variable – temperature, measurement – a temperature-sensitive device, perhaps a bimetallic strip.

4 See Figure A.1

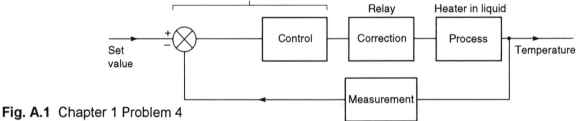

Fig. A.1 Chapter 1 Problem 4

5 See text

6 See text

7 For example: water in, rinse, water out, water in, heat water, rinse, water out, water in, rinse, water out.

8 For example: (a) ticket selected AND correct money in, correct money decided by OR gates analysis among possibilities.
(b) AND with safety guards, lubricant, coolant, workpiece, power, etc. all operating or in place.

9 Traditional: bulky, limited functions, requires rewinding. Mechatronics: compact, many functions, no rewinding, cheaper.

10 Bimetallic element: slow, limited accuracy, simple functions, cheap. Mechatronics: fast, accurate, many functions, getting cheaper.

Chapter 2

1 See the text for explanation of the terms
2 −3.9%
3 67.5 s
4 0.73%
5 0.105 Ω
6 Incremental − angle from some datum, not absolute; absolute − unique identification of an angle
7 162
8 (a) ±1.2°, (b) 3.3 mV
9 See text
10 2.8 kPa
11 19.6 kPa
12 −0.89%
13 +1.54°C
14 Yes
15 −9.81 N, −19.62 N, e.g. strain gauges
16 For example: differential pressure cell
17 for example: LVDT displacement sensor

Chapter 3

1 As figure 3.2 with $R_2/R_1 = 50$, e.g. $R_1 = 1$ kΩ, $R_2 = 50$ kΩ
2 Figure 3.4 with two inputs. E.g. $V_A = 1$ V, $V_B = 0$ to 100 mV, $R_A = R_2 = 40$ kΩ, $R_B = 1$ kΩ
3 Figure 3.8 with $R_1 = 1$ kΩ and $R_2 = 2.32$ kΩ
4 $V = K\sqrt{I}$
5 Figure A.2. Fuse to safeguard against high current, limiting resistor to reduce currents, diode to rectify a.c., Zener diode circuit for voltage and polarity protection, low-pass filter to remove noise and interference, optoisolator to isolate the high voltages from the microprocessor.

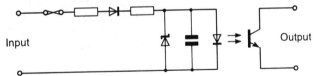

Input
Output

Fig. A.2 Chapter 3 Problem 5

6 0.059 V
7 5.25×10^{-5} V
8 24.4 mV
9 9
10 See text
11 Buffer, digital-to-analogue converter, protection

Chapter 4

1 See text
2 See section 4.1

3 For example: (a) a galvanometric or potentiometric recorder, (b) a moving coil meter, (c) a magnetic tape recorder, (d) a storage oscilloscope

4 Could be four active arm bridge, differential operational amplifier, display of a voltmeter. The values of components will depend on the thickness chosen for the steel and the diameter of a load cell. You might choose to mount the tank on three cells.

5 Could be as in figure 3.7 with cold junction compensation by a bridge (see section 3.5.2). Linearity might be achieved by suitable choice of thermocouple materials.

6 Could be thermistors with a sample and hold element followed by an analogue-to-digital converter for each sensor. This would give a digital signal for transmission, so reducing the effects of possible interference. Optoisolators could be used to isolate high voltages/currents, followed by a multiplexer feeding digital meters.

7 This is based on Archimedes' principle, the upthrust on the float equals the weight of fluid displaced.

8 Could use a LVDT or strain gauges with a Wheatstone bridge.

9 For example: (a) Bourdon gauge, (b) thermistors, galvanometric chart recorder, (c) strain-gauged load cells, Wheatstone bridge, differential amplifier, digital voltmeter, (d) tachogenerator, signal conditioning to shape pulses, counter.

Chapter 5

1 See section 5.4
2 See section 5.4.1
3 See section 5.4.3
4 See (a) figure 5.14, (b) figure 5.10, (c) figure 5.8(a), (d) figure 5.13
5 See text and figure 5.4
6 0.0057 m^2
7 124 mm
8 1.27 MPa, 3.9×10^{-5} m^3/s
9 (a) 0.05 m^3/s, (b) 0.10 m^3/s
10 (a) 0.42 m^3/s, (b) 0.89 m^3/s
11 960 mm
12 See section 5.5 and figure 5.12

Chapter 6

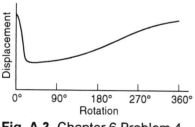

Fig. A.3 Chapter 6 Problem 4

1 (a) A system of elements arranged to transmit motion from one form to another form. (b) A sequence of joints and links to provide a controlled output in response to a supplied input motion.

2 See section 6.2.1
3 Quick-return
4 See figure A.3
5 60 mm
6 Heart-shaped with distance from axis of rotation to top of heart 40 mm and to base 100 mm. See figure 6.6(b).

7 For example: (a) cams on a shaft, (b) quick-return mechanism, (c) eccentric cam, (d) rack and pinion, (e) belt drive, (f) bevel gears

8 41.1 rev/min

9 1/24

Chapter 7

1 See text and figure 7.7

2 (a) Series wound, (b) shunt wound

3 (a) D.C. shunt wound, (b) induction or synchonous motor with an inverter, (c) d.c., (d) a.c.

4 See section 7.5

5 See section 7.4.3

Chapter 8

1 (a) $m\dfrac{d^2x}{dt^2} + c\dfrac{dx}{dt} = F$, (b) $m\dfrac{d^2x}{dt^2} + c\dfrac{dx}{dt} + (k_1 + k_2)x = F$

2 Two torsional springs in series with a moment of inertia block,
$$T = I\frac{d^2\theta}{dt^2} + k_1(\theta_1 - \theta_2) = m\frac{d^2\theta}{dt^2} + \frac{k_1 k_2}{k_1 + k_2}\theta_1$$

3 $v = v_R + \dfrac{1}{RC}\int v_R\,dt$

4 $v = \dfrac{L}{R}\dfrac{dv_R}{dt} + \dfrac{1}{CR}\int v_R\,dt + v_R$

5 $v = R_1 C\dfrac{dv_C}{dt} + \left(\dfrac{R_1}{R_2} + 1\right)v_C$

6 $RA_2\dfrac{dh_2}{dt} + h_2\rho g = h_1$

7 $RC\dfrac{dT}{dt} + T = T_r$, charged capacitor discharging through a resistor

8 $RC\dfrac{dT_1}{dt} = Rq - 2T_1 + T_2 + T_3$, $RC\dfrac{dT_2}{dt} = T_1 - 2T_2 + T_3$

9 $pA = m\dfrac{d^2x}{dt^2} + R\dfrac{dx}{dt} + \dfrac{1}{c}x$, where R = resistance to stem movement, c = capacitance of spring

10 $T = \left(\dfrac{I_1}{n} + n\right)\dfrac{d^2\theta}{dt^2} + \left(\dfrac{c_1}{n} + nc_2\right)\dfrac{d\theta}{dt} + \left(\dfrac{k_1}{n} + nk_2\right)\theta$

Chapter 9

1 $\Delta F = (2kx_o)\,\Delta x$

2 $\Delta E = (a + 2bT_o)\,\Delta T$

3 $\Delta T = (mgL)\,\Delta\theta$

4 $\dfrac{IR}{k_1 k_2}\dfrac{d\omega}{dt} + \omega = \dfrac{1}{k_2}v$

5 $(L_a + L_L)\dfrac{di_a}{dt} + (R_a + R_L)i_a - k_1\dfrac{d\theta}{dt} = 0$,

$I\dfrac{d^2\theta}{dt^2} + B\dfrac{d\theta}{dt} + k_2 i_a = T$

6 Same as armature controlled motor

7 Inductance in series with resistance and current source, rod with mass, damping and moving against a spring.

Chapter 10

1 $4\dfrac{dx}{dt} + x = 6y$

2 (a) 59.9°C, (b) 71.9°C

3 (a) $i = \dfrac{V}{R}(1 - e^{-Rt/L})$, (b) L/R, (c) V/R

4 (a) Continuous oscillations, (b) under-damped, (c) critically damped, (d) over-damped

5 (a) 4 Hz, (b) 1.25, (c) $i = I\left(\frac{1}{3}e^{-8t} - \frac{4}{3}e^{-2t} + 1\right)$

6 (a) 5 Hz, (b) 1.0, (c) $x = (-32 + 6t)\,e^{-5t} + 6$

7 (a) 9.5%, (b) 0.020 s

8 (a) 4 Hz, (b) 0.625, (c) 1.45 Hz, (d) 0.5 s, (e) 8.1%

9 (a) 0.59, (b) 0.87

Chapter 11

1 (a) $\dfrac{1}{As + \rho g/R}$, (b) $\dfrac{1}{ms^2 + cs + k}$, (c) $\dfrac{1}{LCs^2 + RCs + 1}$

2 (a) 3 s, (b) 0.67 s

3 (a) $1 + e^{-2t}$, (b) $2 + 2\,e^{-5t}$

4 (a) Over-damped, (b) under-damped, (c) critically damped, (d) under-damped

5 $t\,e^{-3t}$

6 $2\,e^{-4t} - 2\,e^{-3t}$

7 $\sqrt{10}$

8 $\dfrac{5}{s + 53}$

9 $\dfrac{5s}{s^2 + s + 10}$

10 $\dfrac{2}{3s + 1}$

Chapter 12

1 (a) $\dfrac{5}{\sqrt{\omega^2 + 4}}$, $\dfrac{\omega}{2}$, (b) $\dfrac{2}{\sqrt{\omega^4 + \omega^2}}$, $\dfrac{1}{\omega}$,

(c) $\dfrac{1}{\sqrt{4\omega^6 - 3\omega^4 + 3\omega^2 + 1}}$, $\dfrac{\omega(3 - 2\omega^2)}{1 - 3\omega^2}$

2 $0.56 \sin (5t - 38°)$

3 $1.18 \sin (2t + 25°)$

4 (a) (i) ∞, 90°, (ii) 0.44, 450°, (iii) 0.12, 26.6°, (iv) 0, 0°,
(b) (i) 1, 0°, (ii) 0.32, −71.6°, (iii) 0.16, −80.5°, (iv) 0, −90°

5 See figure A.4

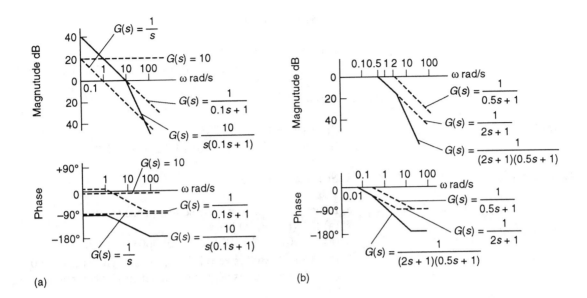

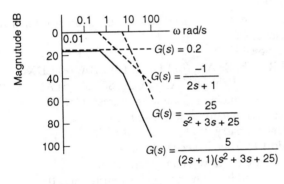

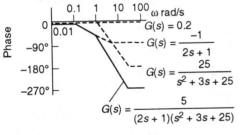

(c)

Fig. A.4 Chapter 11 Problem 5

6 17 dB, 77°
7 (a) 0.83, (b) 0.66

Chapter 13

1 See section 13.2
2 (a) 8 minutes, (b) 20 minutes
3 (a) 12 s, (b) 24 s
4 20%, 5
5 62.5%, 0.63%

6 (a) 51.0%, (b) 51.0%, (c) 9, (d) 49.5%
7 (a) 54%, (b) 66%
8 (a) 60%, (b) 73%, (c) 76.1%, (d) 62%
9 See the text. In particular P - offset, PI and PID no offset.
10 3, 0.0025 s^{-1}, 100 s
11 3, 0.01 s^{-1}, 25 s

Chapter 14

1 See section 14.1.
2 See sections (a) 14.2.1, (b) 14.2.2, (c) 14.2.3

Chapter 15

1 See section 15.1.2
2 256
3 Easier to comprehend and so less prone to errors.
4 (a) D7, (b) CF, (c) B9
5 (a) 10100101, (b) 10010001, (c) 000110111001
6 (a) When a result is zero Z is set to 1, when not zero to 0
(b) When a result is negative N is set to 1, when not negative to 0
7 (a) 89, (b) 99
8 No address has to be specified since the address is implied by the mnemonic.
9 See section 15.1.2
10 (a) CLRA, (b) STAA, (c) LDAA, (d) CBA, (e) LDX
11 (a) LDAA $20, (b) DECA, (c) CLR $0020, (d) ADDA $0020
12 (a) Store accumulator B value at address 0036,
(b) load accumulator A will data F2,
(c) Clear the carry flag,
(d) Add 1 to value in accumulator A,
(e) Compare C5 to value in accumulator A,
(f) Jump to address given by index register plus 05.
13 (a)

DATA1	EQU	$0050	
DATA2	EQU	$0060	
DIFF	EQU	$0070	
	ORG	$0010	
	LDAA	DATA1	Get minuend
	SUBA	DATA2	Subtract subtrahend
	STAA	DIFF	Store difference
	SWI		Program end

(b)

MULT1	EQU	$0020
MULT2	EQU	$0021
PROD	EQU	$0022
	ORG	$0010

	CLR	PROD	Clear product address
	LDAB	MULT1	Get first number
SUM	LDAA	MULT2	Get multiplicand
	ADDA	PROD	Add multiplicand
	STAA	PROD	Store result
	DECB		Decrement acc. B
	BNE	SUM	Branch if adding not complete
	WAI		Program end

(c)

	FIRST	EQU	$0020	
	ORG	$0000	
	CLRA		Clear accumulator
	LDX	#0	
MORE	STAA	$20,X	
	INX		Increment index reg.
	INCA		Increment accumulator
	CMPA	#$10	Compare with number 10
	BNE	MORE	Branch if not zero
	WAI		Program end

(d)

	ORG	$0100	
	LDX	#$2000	Set pointer
LOOP	LDA A	$00,X	Load data
	STA A	$50,X	Store data
	INX		Increment index register
	CPX	$3000	Compare
	BNE	LOOP	Branch
	SWI		Program end

14 YY	EQU	$??	Value chosen to give required time delay
SAVEX	EQU	$0100	
	ORG	$0010	
	STA	SAVEX	Save accumulator A
	LDAA	YY	Load accumulator A
LOOP	DECA		Decrement acc. A
	BNE	LOOP	Branch if not zero

	LDA	SAVEX	Restore accumulator
	RTS		Return to calling program
15	LDA	$2000	Read input data
	AND A	#$01	Mask off all bits but bit 0
	BEQ	$03	If switch low, branch over JMP which is 3 program lines
	JMP	$3000	If switch high no branch and so execute JMP
	Continue		

Chapter 16

1 See section 16.1.1
2 See section 16.1.1. A parallel interface has the same number of input/output lines as the microprocessor. A serial interface has just a single input/output line.
3 See section 16.1.2
4 See section 16.2
5 See section 16.2 and figure 16.2
6 See section 16.2.1
7 See section 16.3. Polling involves the interrogation of all peripherals at frequent intervals, even when some are not activated. It is thus wasteful of time. Interrupts are only initiated when a peripheral requests it and so is more efficient.
8 CRA 00110100, CRB 00101111
9 As the program in 15.3.2 with LDAA #$05 replaced by LDAA #$34 and LDAA #$34 replaced by LDAA #$2F.
10 As the program in 16.2.1 followed by
READ LDAA $2000 Read port A
Perhaps after some delay program there may then be
BRA READ.

Chapter 17

1 (a) AND, (b) OR
2 (a) Figure 17.5(b), (b) figure 17.6(b), (c) a latch circuit, figure 17.9, with X400 the start and X401 the stop switches.
3 0 LD X400, 1 LD Y430, 2 ORB, 3 ANI X401, 4 OUT Y430
4 0 LD X400, 1 OR Y430, 3 OUT Y430, 4 OUT T450, 5 K 50
5 0 LD X400, 1 OR Y430, 2 ANI M100, 3 OUT Y430,
4 LD X401, 5 OUT M100
6 As in figure 17.14 with T450 K = 10 and T451 K = 200
7 Figure A.5

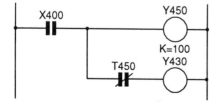

Fig. A.5 Chapter 16 Problem 7

8 Figure A.6

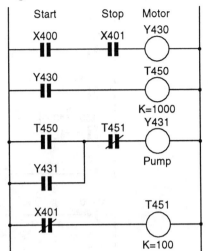

Fig. A.6 Chapter 16 Problem 8

9 Figure A.7

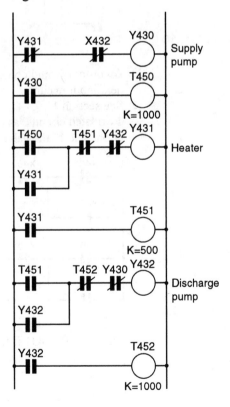

Fig. A.7 Chapter 16 Problem 9

10 Figure A.8

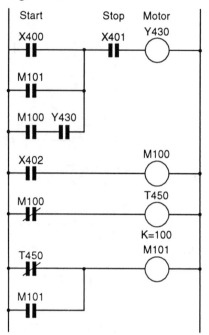

Fig. A.8 Chapter 16 Problem 10

11 An output would come on, as before, but switch off when the next input occurs.
12 See section 17.9.1
13 Two latch circuits, as in figure A.9 with Y431 is the solenoid at one end of the valve, Y432 that at the other.

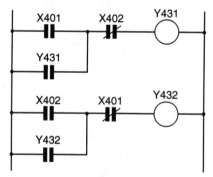

Fig. A.9 Chapter 16 Problem 13

14 Figure A.10, Y431 is the solenoid at one end of the valve, Y432 that at the other.

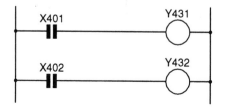

Fig. A.10 Chapter 16 Problem 14

Chapter 18

1 See section 18.2
2 See figure 18.1
3 Bus
4 Broad band
5 See section 18.4.1
6 See section 18.4

Chapter 19

1 See section 19.1
2 See section 19.1
3 See section 19.1

Chapter 20

1 See section 20.2.2
2 Possible solutions might be: (a) thermocouple, cold junction compensation, amplifier, ADC, PIA, microprocessor, DAC, thryristor unit to control the oven heating element, (b) light beam sensors, PLC, solenoid-operated delivery chute deflectors, (c) closed-loop control with, for movement in each direction, a d.c. motor as actuator for movement of pen, microprocessor as comparator and controller, and feedback from an optical encoder.

Index